IEE CONTROL ENGINEERING SERIES 35

Series Editors: Prof. D. P. Atherton
Dr K. Warwick

IMPLEMENTATION
of self-tuning controllers

Other volumes in this series

IMPLEMENTATION
of self-tuning controllers

Edited by
Kevin Warwick

Peter Peregrinus Ltd on behalf of the Institution of Electrical Engineers

Published by: Peter Peregrinus Ltd., London, United Kingdom

British Library Cataloguing in Publication Data

Implementation of self-tuning controllers.
1. Adaptive control systems
I. Warwick, K. II. Series
629.8'36

ISBN 0-86341-127-4

Printed in England by Billing and Sons Ltd.

Contents

List of contributors

Chapter 1

D.W. Clarke, University of Oxford.

Chapter 2

S.L. Shah, University of Alberta and W.R. Cluett, University of Toronto.

Chapter 3

K.J. Hunt and M.J. Grimble, University of Strathclyde.

Chapter 4

C. Mohtadi, University of Oxford.

Chapter 5

K. Warwick, University of Warwick.

Chapter 6

G.C. Goodwin, R.H. Middleton and M. Salgado, University of Newcastle, Australia.

Chapter 7

P.J. Gawthrop, University of Glasgow.

Chapter 8

P.S. Tuffs, Aluminium Company of America.

Chapter 9

B.E. Ydstie, A.H. Kemna and L.K. Liu, University of Massachusetts at Amherst.

Chapter 10

R.M.C. De Keyser, University of Gent, Belgium.

Chapter 11

P.C. Young, M.A. Behzadi and A. Chotai, University of Lancaster.

Chapter 12

J. Byrne and M.R. Katebi, University of Strathclyde.

Chapter 13

D. Peel, University of Oxford and M. Tham, University of Newcastle upon Tyne.

Preface

This book is intended as a useful up-to-date text for those interested in Self-Tuning Control in its widest sense, with a particular emphasis being placed on the basic principles involved with actually getting such a controller to work. The level of the text makes it suitable for practising Control Engineers, final year degree students especially those carrying out a digital control project, or postgraduate students requiring an introductory modern research guide. The book is of use as a Masters course text-book, particularly when the course requires a practical flavour in the study of Self-Tuning Control. The contents of the book have been selected so as to reduce to a minimum, the number of abstract theoretical ideas and to provide, as much as possible, implementation suggestions and application examples.

The topic of Self-Tuning Control has been a major attraction and point of interest for several years now, this being apparent at conferences, workshops and symposia. Indeed several of the chapters, which constitute this book, are based on presentations given at the extremely well-attended tutorial workshop on the 'Implementation of Self-Tuning Controllers', which took place in Athens, December 1986, prior to the 25th IEEE Conference on Decision and Control. Further, there have now been four IEE Workshops on 'Self-Tuning and Adaptive Control' held at the University of Oxford (Easter 1981, 1983, 1985, 1987), and although initially the workshops resulted in the publication of a book with the workshop title, edited by C.J. Harris and S.A. Billings (volume 15 in the Peter Peregrinus series), presentations given of late at the Workshop have concentrated on Self-Tuning Control (rather than other forms of Adaptive Control) and the overall direction has been towards Practical Implementation problems. Those who are familiar with the Harris and Billings text should find this book a most valuable complementary follow-on, in the sense of an up-to-date, applications-oriented work which covers more practical issues.

The first chapter of this book not only provides an overview of the scope and generality provided by Self-Tuning Control schemes, but also indicates where such a scheme might be of benefit and how it complements and augments other control techniques. The opening chapter therefore necessarily discusses how the remaining twelve chapters knit together in terms of their content, to cover the general field of Self-Tuning controller implementation. In general the chapter headings are self-explanatory, being roughly sectioned as follows:

Chapters 2-5. These chapters discuss the basic ideas behind self-tuning control techniques; identification/ parameter estimation (chapter 2), optimal controller design (chapter 3), numerical considerations (chapter 4) and simplified methods (chapter 5).

Chapters 6-9. These chapters discuss advances in the framework employed for self-tuning control; the delta operator (chapter 6), continuous-time control (chapter 7), software aspects (chapter 8) and predictive control (chapter 9).

Chapters 10-13. These chapters are concerned with detailed application examples of; predictive control (chapter 10), PIP control (chapter 11), LQG optimal control (chapter 12) and various SISO/MIMO case studies (chapter 13).

The contributions which make up this book provide an account of the numerous aspects of, and problems involved in, the implementation of self-tuning and related controllers. It is possible for the reader to either make use of the text as a whole, for the purpose of a taught course or to choose appropriate chapters in order to gain an insight into particular implementation problems.

The editor would like to thank all of the authors for their contributions and the prompt attention each gave to the production of their finalised text. In particular the editor wishes to thank Mike Grimble and Sirish Shah for their help with the Athens Workshop, from which the idea for this book originated, with much gratitude also being due to Dave Clarke and Coorous Mohtadi for their suggestions with regard to book content and layout. On behalf of all of the authors, the editor would like to thank those at Peter Peregrinus who were responsible for the production of this book, in particular John St. Aubyn, and Derek Atherton for his help and encouragement as Series Editor.

September, 1987. Kevin Warwick.

Chapter 1

Introduction to self-tuning control

D. W. Clarke

1.1 The need for self-tuning

Most nontrivial control problems involve uncertainty. Signals are corrupted by sensor noise and by quantization in analogue to digital converters. The plant output depends not only on the control input but also on load-disturbances which are mostly unmeasurable and against which the control law has to regulate. Nonlinearities such as saturation, backlash and stiction in the actuator can seriously affect the controlled response yet are difficult to quantify. Moreover, the stability and performance of the closed-loop depends crucially on one major source of uncertainty – the dynamic behaviour of the plant itself. The use of self-tuning to resolve the problem of uncertain dynamics is the theme of this book.

In many cases lack of knowledge of plant dynamics has an *economic* rather than a technical cause. Mathematical modelling of physico-chemical processes is now so well-developed that computational methods, backed up where necessary by experiments (such as wind-tunnel tests for aircraft), can provide accurate simulations of most relevent behaviour. Indeed, safety legislation makes such general models mandatory in the nuclear and aerospace industries. Nevertheless, the cost of this approach is high and not warranted for the vast majority of control loops in industry, where the traditional solution has been to use standard manually-tuned three-term regulators. Good results are obtainable from PID algorithms provided that the plant engineer is familiar with tuning methods, though the procedure can be time-consuming, particularly with 'sluggish' processes, and retuning may be necessary if the dynamics change. Hence manufacturers and users are increasingly adopting controllers with built-in tuning capability. These controllers are then also suitable for non-traditional markets where the black art of manual tuning is lacking. Moreover, the major concern for product quality (as in the rolling of sheet metal) has meant that there is a real financial return on methods which give tighter control than PID on plant with more complex dynamics such as dead-time. These newer methods often involve more than the 3 PID parameters and hence preclude manual approaches to their proper adjustment.

Sometimes the uncertainty in the plant can be quantified, for example by claiming that there are given regions within which the poles of the transfer-function must lie. Alternative design procedures such as the frequency-response approach of

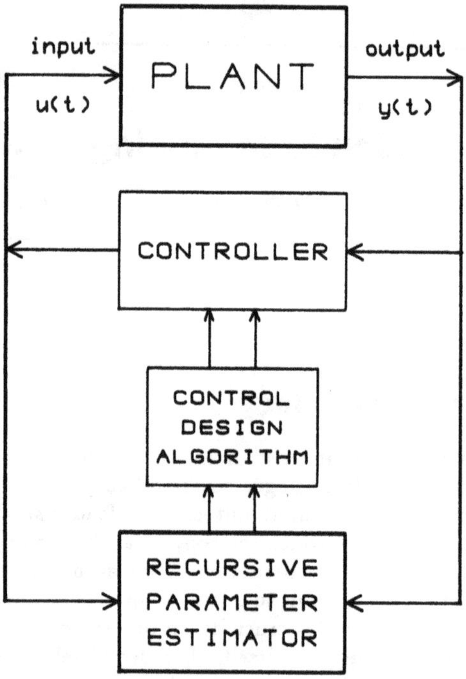

Figure 1.1: Structure of a general self-tuning controller

Horowitz (1963) or H^∞ optimization (see Francis, 1987) can then be used to provide a given *fixed* controller which is 'robustly stable' and yet gives satisfactory performance over all possible plant pole positions. Though such a controller is inevitably a compromise, it must be recalled that one of the historical roots of feedback was to ensure that the gain of each amplifier used in long-distance telephony was made effectively constant despite variations in valve characteristics. Hence fixed-parameter controllers should be considered first for any design problem: indeed one of the goals of self-tuning research is to devise algorithms which provide a robust closed-loop, even when the tuning component is switched off.

The idea behind self-tuning is to adjust the controller settings automatically, based on the measured input/output behaviour of the real plant. This extra level of data-processing inevitably leads to higher complexity, and so the approach had to wait for the advent of microprocessors in the '70s before it became economically feasible. Early designs (Mishkin and Braun, 1961) concentrated on analogue methods with *model-reference* algorithms, though the trend towards digital implementation had started (Westcott, 1962). Indeed Kalman (1958) had built a self-tuner in a hybrid form and it was only the constraints in the existant valve technology which stopped his ideas being exploited sooner. It is interesting that modern VLSI sophistication enables a return to the hybrid approach (Gawthrop, 1980;1982; also Chapter 7), in which the controller and estimator are partly in digital and partly in (pseudo-)analogue form, where there are advantages to be gained in bandwidth and integrity. Chapter 6 unifies the algorithmic development of discrete- and continuous-time methods.

There are two contrasting approaches to the use of this new computational power. The first is to add tuning features to an otherwise standard PID regulator which are as simple to use as possible. For example Bristol in 1977 devised a PID tuner based on *pattern-recognition*: looking for overshoots in responses to step-like disturbances, and this was developed into the successful Foxboro 'Exact' single-loop tuner, as described by Kraus and Myron (1984). The second philosophy is to provide a general-purpose algorithm which is in some sense optimal when used with processes having complex (possibly time-varying) dynamics such as dead-time and which can include tuned feedforward terms to cope with measurable load-disturbances. These self-tuners might involve several design parameters which can be chosen on-site to tailor the loop behaviour according to the plant and the control objectives, though default settings could be adopted. It is said that these controllers have *performance-oriented knobs* as the user is prescribing the character of the closed-loop rather than the K, T_i and T_d of the standard PID law. This operational flexibility carries with it the burden of a significant number of prior parameters, so one objective is to ensure either that their selection is easily understood or that the suboptimal performance resulting from bad choices is still acceptable.

It is useful to distinguish three branches of the subject: *autotuning*, *self-tuning*, and *adaptive* control. Autotuning (Astrom and Hagglund, 1984) describes the approach whereby the parameters of a subsequently fixed PID regulator are determined experimentally. The self-tuner structure of Fig.1 implies that there is a constant but unknown parametric model underlying the plant's control response, an estimator of these parameters based on real plant data, and an analytic control design procedure so that:

> *an algorithm is said to have the* **self-tuning property** *if, as the number of samples approaches infinity, the controller parameters tend to those corresponding to an exactly known plant model.*

Proofs of convergence are available for many algorithms but generally rely on knowledge of the model-order and on time-invariance of the plant. In practice, however, we wish to apply self-tuning to time-varying systems and hence the methods are modified (often heuristically) to enable the tracking of changing plant parameters. This is then adaptive control. As might be expected, proofs of stability for adaptive controllers are much sparser except in the case where changes in the plant are 'slow' in comparison with the current dynamic response. Table 1 outlines some of the design possibilities. The books edited by Narendra and Monopoli (1980); Harris and Billings (1981) describe algorithmic design in more detail, whilst Goodwin and Sin (1984) provide tools for investigating stability and convergence.

1.2 A self-tuned PI regulator for a furnace

As a simple tutorial example of self-tuning design consider an electrically-heated kiln as shown in Fig.2 in which the thermal load (the furnace burden) is unknown

Table 1.1: Possibilities for self-tuning design

Conventional approaches
Manually-tuned PID regulators Pretuning based on physico-chemical models and simulation Scheduled resetting of parameters (e.g. 'Gain-scheduling')
Auto-tuning of PID regulators
Automation of the classical Ziegler-Nichols' (1942) rules Use of operator information ('linguistic' controllers) (Eventual fixed-parameter PID regulator) Recognition of 'patterns' in the system errors (e.g. overshoots) (Slowly adapting PID regulator) On-line model updating ⟶ PID design rules (Fast-adapting PID regulator)
Alternative methods for high-performance
Self-tuning of Smith-predictors; pole-placement; feedforward; ... 'Long-range predictive control' for highly complex systems (Many modes; open-loop unstable; NMP; variable dead-time; ...) Fast-adapting self-tuners (Variable forgetting-factors; 'jacketting' of estimators; ...)

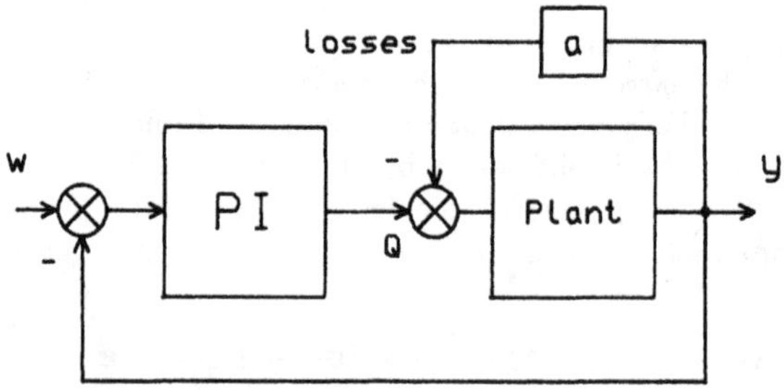

Figure 1.2: Block diagram of the heater controller

and it is desired to regulate the temperature $y(t)^o$ above ambient to some set-point w. If $u(t)$ is the control signal to the heater, $Q(t)$ the heater output and α is the heat-transfer coefficient in Newton's law of cooling, the process equations are:

$$c\frac{dy}{dt} = k_1 u - \alpha y, \text{ or:}$$

$$T\frac{dy}{dt} + y = ku(t) \tag{1.1}$$

where the time-constant T and the gain k of the plant are unknown. Suppose now that a microcomputer monitors the control $u(t)$ and the measured temperature response $y(t)$ at regular sample-intervals h, and that the control between each sample is constant (i.e. a Zero-Order Hold). Then integrating eqn.1 between times a and b gives:

$$T[y(b) - y(a)] + \int_a^b y(t)\,dt = k\int_a^b u(t)\,dt,$$

so if the sample interval h is small compared with T this we have:

$$T[y(b) - y(a)] + h\sum_a^b y(ih) = kh\sum_a^b u(ih); \tag{1.2}$$

and if $c > b$ is a later time another equation can be written:

$$T[y(c) - y(b)] + h\sum_b^c y(ih) = kh\sum_b^c u(ih). \tag{1.3}$$

Note that in equations 2 and 3 all terms except T and k are constructed from data sampled at times $t = ih$. Hence at time $t = c$ the microcomputer has available two equations in the two unknowns which can be solved to obtain *estimated* values $\hat{T}$ and $\hat{k}$. From these estimates of the plant dynamics the control law parameters can be calculated.

The plant has a Laplace transfer-function $k/(1 + sT)$; suppose now that a PI regulator of the form $K(1 + 1/sT_i)$ is proposed. Noting that the PI function can be written as $K(1 + sT_i)/sT_i$ one easy solution is to choose $T_i = \hat{T}$ leading to an open-loop transfer-function (for the case of *exact* estimates):

$$G_o(s) = [K\frac{(1 + sT_i)}{sT_i}].[\frac{\hat{k}}{(1 + s\hat{T})}] = \frac{K\hat{k}}{s\hat{T}},$$

and a closed-loop transfer-function $G_o/(1 + G_o)$, or:

$$G_c(s) = \frac{1}{(1 + s\hat{T}/K\hat{k})}.$$

The time-constant T_c of the closed-loop is therefore $\hat{T}/K\hat{k}$ and hence can be set to any *user-chosen* value by the appropriate value of K. This means that despite having no prior knowledge of the thermal load or the heat-transfer coefficients of the furnace, the closed-loop ends up by having *consistent* dynamics T_c which can be specified beforehand by the operator.

The operation of the self-tuner would be as follows:

1. The user chooses the desired closed-loop time-constant T_c, the sample-interval h, and the three experiment times a, b and c (all these may, of course, have predefined *default* values).
2. The furnace is switched on at a given control setting u_0 and successive measurements $y(ih)$ are then made to construct equations 2 and 3.
3. The values $\hat{T}$ and $\hat{k}$ are computed and from them and T_c the controller parameters T_i and K are deduced.
4. The PI controller is then put into automatic mode to bring the temperature to the set-point with the prescribed dynamic behaviour.

This design is a special case of the self-tuning approach in which the following steps are employed:

1. A candidate input/output structural representation of the plant is proposed together with an analytical design procedure which would produce the 'best' controller coefficients given the correct plant paramters.
2. A parameter estimator using I/O data provides the best-fit model of the plant.
3. The parameter estimates are fed into the design algorithm which accepts them as if they were exact (this is called *certainty equivalence*).
4. The controller coefficients taken from the design process are then used to assert feedback control.

The above describes an *explicit* self-tuner; in an *implicit* self-tuner the estimator directly provides the feedback parameters rather than going through the plant-modelling ⟶ design-algorithm ⟶ controller route.

Many self-tuners can be developed based on the above principles which differ according to the assumptions made at each step. For example the model-fitting can be 'one-shot' as in the heater-controller, or in an adaptive approach the fitting can be refined recursively using new data at each sample as time evolves so that plant variations can be tracked. An example is given in the next section.

1.3 A simple self-tuning controller

Consider the plant which consists of just a gain G as shown in Fig.3. The control problem is to regulate the output $y(t)$ to be as close to the set-point $w(t)$ by a suitable choice of the control $u(t)$ despite the unknown and time-varying gain G and the load-disturbances $d(t)$. Though the example is very simple the solution proposed has many of the characteristics of a more advanced design and shows some of the advantages and disadvantages of the adaptive approach. There are several possible solutions to the problem:

- Take a PI regulator and use manual tuning: this would be satisfactory but slow for the fixed-parameter case.

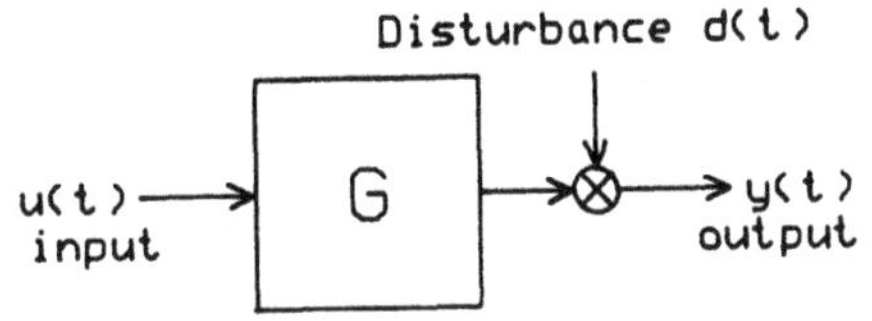

Figure 1.3: A plant with unknown gain

- Measure ('identify') G from I/O data and use an off-line design method: this gives a greater degree of freedom in the control algorithm but again would only be satisfactory for a fixed plant.
- Use self-tuning or adaptive control:
  1. identify G *on-line* $\longrightarrow \hat{G}$;
  2. use $\hat{G}$ in an *on-line* controller design.

As in any self-tuner the first question to ask is: "What is the ***best*** way to control the plant if the value of G were ***known***?". One approach for this example is discussed below.

1.3.1 Known-parameter control

Now the plant model (including the inevitable one-step delay due to the Zero-Order Hold in the loop when a microcomputer is connected) is:

$$\underbrace{y(t+1)}_{\text{next output}} = G \underbrace{u(t)}_{\text{control choice}} + \underbrace{d(t+1)}_{\text{unknown future disturbance}}$$

We want to set $y(t+1) = w(t)$, the set-point, so we define: $e(t) = w(t) - y(t)$ to be the current system error and define an 'optimal control' which minimises:

$$\begin{aligned} J_c = e^2(t+1) &= (y(t+1) - w(t))^2 \\ \text{i.e. } J_c &= (Gu(t) + d(t+1) - w(t))^2 \\ &= \underbrace{(Gu(t) - w)^2}_{\text{can affect}} + \underbrace{d^2(t+1)}_{\text{cannot affect}} - 2\underbrace{(Gu(t) - w)d(t+1)}_{\text{assume 0 on average}} . \end{aligned}$$

If the assumption that $d(t)$ is a zero-mean random signal is justified the minimum value of $E\{J_c\}$ is obtained by the choice $u(t) = w(t)/G$. This is open-loop control which is known to be unsatisfactory. In practice $d(t)$ will be 'step-like' in behaviour and conventional control uses ***integration*** to give offset-free performance. Hence the next step is to provide a ***feedback*** solution.

Here the cost J_c is quadratic in $u(t)$. Suppose $u(t-1)$ is the previous control sent to the plant and the objective is to consider the ***change*** in control (or 'control

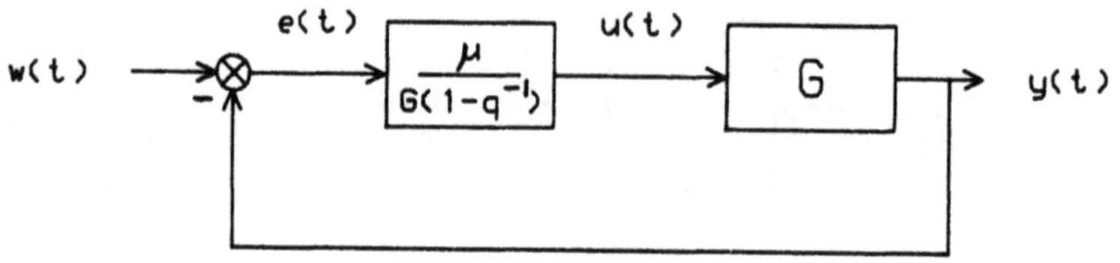

Figure 1.4: Structure of the known-parameter feedback controller

move') $\Delta u(t) \equiv u(t) - u(t-1)$. Now:

$$\begin{aligned} J_c &= (Gu(t) + d(t+1) - w)^2; \\ \text{so:} \quad \frac{\partial J_c}{\partial u} &= 2G(Gu(t) + d - w) \\ &= 2G(y(t+1) - w) = -2G.e(t+1). \end{aligned}$$

One way of minimising J_c is to use a 'steepest-descent' algorithm which *changes* $u(t)$ at a rate depending on the gradient of J_c:

$$\begin{aligned} u(t) &= u(t-1) - k\frac{\partial J_c}{\partial u(t-1)}; \ k \text{ being the descent gain} \\ &= u(t-1) + 2kGe(t) \\ &= u(t-1) + 2kG(w(t) - y(t)). \end{aligned}$$

The 'optimal' value of k of $1/2G^2$ for the deterministic case gives the same value of control as the open-loop form and implies that the controller is:

$$u(t) = u(t-1) + \frac{\mu}{G}e(t),$$

or, if Δ is the difference operator $1 - q^{-1}$: $\Delta u(t) = \mu/G.e(t)$. The structure of this controller is shown in Fig.4. The parameter μ is the 'gain' of the integral controller and has a value of 1 for 'ideal' or single-step control. Note that to achieve this the plant's gain is cancelled by the controller and the forward-path transfer-function between error and output becomes the desired $1/\Delta$. This completes the known-parameter design; next we require an estimator which provides the value $\hat{G}$.

1.3.2 On-line identification

In self-tuning we use a *recursive parameter estimator* or RPE. Fig.5 shows the general block-diagram of the estimator in which a plant model with *current* parameter values is used to generate a predicted plant output, and where the prediction error $\epsilon(t)$ between the predicted and the measured output can be used for refining the estimated coefficients. For an RPE, at each sample t:

1. data $y(t), \ldots$ is acquired;
2. the model generates the predicted output $\hat{y}(t)$, depending on 'old' data and on the unknown parameters $\hat{\theta}$;

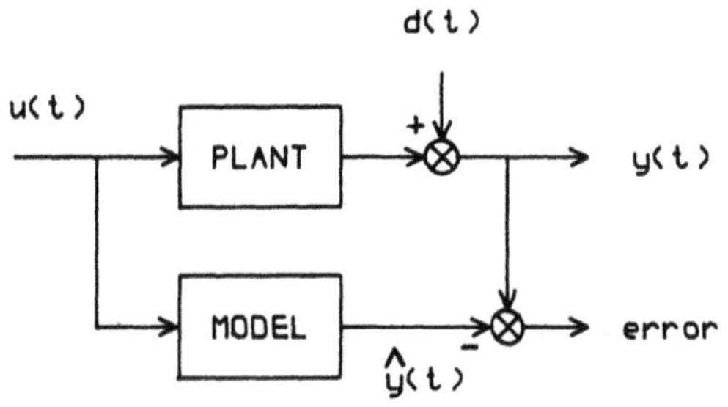

Figure 1.5: The predicted output and prediction error

3. the prediction error $\epsilon(t) = y(t) - \hat{y}(t)$ is evaluated;
4. the parameters are 'updated' using an algorithm of the form:

$$\hat{\theta}(\text{new}) = \hat{\theta}(\text{old}) + f(\text{data}).\epsilon(t).$$

From Fig.3 it is seen that the real plant satisfies:

$$\begin{aligned} y(t) &= Gu(t-1) + d(t), \quad \text{and the model is:} \\ y(t) &= \underbrace{\hat{G}u(t-1)}_{\text{prediction}} + \underbrace{\epsilon(t)}_{\text{prediction error}} \\ \text{Hence: } \epsilon(t) &= (G - \hat{G})u(t-1) + d(t) \\ &= \tilde{G}u(t-1) + d(t), \\ \text{where } \tilde{G} &\equiv G - \hat{G} \quad \text{is the } \textit{error in the model.} \end{aligned}$$

Now $\epsilon(t)$ is seen to depend on *both* modelling errors and the noise, but is the only error signal available to the estimator so that if rapid convergence (high adaptive gains) is required the estimates would become susceptible to the effects of noise. As for the control case we define a 'loss-function':

$$\begin{aligned} J_e = \epsilon^2(t) &= (y(t) - \hat{y}(t))^2 \\ &= (y(t) - \hat{G}(t)u(t-1))^2. \end{aligned}$$

Hence a minimum value of J_e is attained when $\hat{G} = y(t)/u(t-1)$, but as before we prefer an 'integral estimator':

$$\hat{G}(t) = \hat{G}(t-1) + \lambda \frac{\epsilon(t)}{u(t-1)},$$

where the 'adaptive gain' λ is 1 for optimal convergence *in the noise-free case*; the corresponding block-diagram of the estimator is given in Fig.6. Note how closely the RPE and controller structures are related: here the objective is to 'control' the prediction-error $\hat{y}(t)$ so as to follow the 'set-point' $y(t)$.

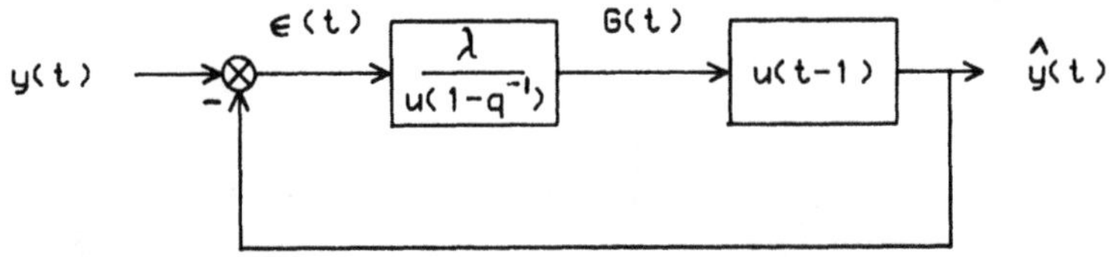

Figure 1.6: The estimator as a feedback loop

Table 1.2: Sequences involved in the simple example

t	$w(t)$	$y(t)$	$e(t)$	$u(t)$	$\epsilon(t)$	$\hat{G}(t)$	Comment
0	10	0	10	0	0	1	initial conditions
1	10	0	10	10	0	1	— switch on —
2	10	20	-10	5	10	2	correct $\hat{G}$ in 1 step
3	10	10	0	5	0	2	to set-point in next step
4	20	10	10	10	0	2	set-point change to 20
5	20	20	0	10	0	2	attained in 1 step
— now there is a load-disturbance d of 1 —							
6	20	21	-1	9.52	1	2.1	estimates $\hat{G}$ affected
7	20	20.05	-0.05	9.5	0.05	2.1	and don't recover, but
8	20	20	0	9.5	0	2.1	controller has no offset

1.3.3 Self-tuning of the simple plant

To self-tune all we need to do is to couple the control and estimation algorithms together as shown in Fig.1, so that the full adaptive set-up is given by the following equations:

$$\begin{aligned}
\text{Plant:} \quad y(t) &= Gu(t) + d(t) \\
\text{System error:} \quad e(t) &= w(t) - y(t) \\
\text{P.E.:} \quad \epsilon(t) &= y(t) - \hat{G}(t-1)u(t-1) \\
\text{RPE:} \quad \hat{G}(t) &= \hat{G}(t-1) + \lambda\frac{\epsilon(t)}{u(t-1)} \\
\text{Control:} \quad u(t) &= u(t-1) + \mu\frac{e(t)}{\hat{G}(t)}.
\end{aligned}$$

Take as an example of the use of these equations a plant with a true gain G of 2 but where the initial assumed value $\hat{G}(0)$ is 1. When the algorithm is started at time $t = 1$ the values of $u(0)$ and $y(0)$ are 0 and the set-point w is 10. The resulting behaviour for the 'optimal' choices of 1 for the gain parameters λ (estimator) and μ (controller) is shown in Table 2. Up to time 6 there is no disturbance $d(t)$ but at time 6 and thereafter a *constant* load-disturbance of 1 is added to the output.

It is seen that with these choices of gains the estimator converges in one step and with the correctly estimated value of plant gain the control signal drives the plant output to w at the next step. The subsequent change in w to 20 at time 4 is again followed exactly by y after one sample. The effect of the load-disturbance is interesting: the integral controller eliminates the offset after a while but the parameter estimates have been damaged and do not recover. The moral of this example is that noise or loads $d(t)$ affect *both* the output $y(t)$ and the estimates $\hat{G}(t)$, so we must ensure:

1. The control law is *insensitive* or *robust* with respect to incorrectly estimated parameters, and:
2. the estimator is insensitive with respect to noise.

This second point can be seen more clearly with the following analysis. The model error $\tilde{G}$ can be written as:

$$\begin{aligned}
\tilde{G}(t) &= G - \hat{G}(t) = \tilde{G}(t-1) - \lambda\frac{\epsilon(t)}{u(t-1)} \\
&= \tilde{G}(t-1) - \lambda\frac{\tilde{G}(t-1)u(t-1) + d(t)}{u(t-1)} \\
&= (1-\lambda)\tilde{G}(t-1) - \lambda\frac{d(t)}{u(t-1)}.
\end{aligned}$$

We obviously want $\tilde{G}(t)$ to be as small as possible. If there is no noise ($d = 0$) then putting $\lambda = 1$ is correct, but if there is noise it is rejected completely only if λ is 0. Hence in the noisy case we must choose λ to be a number less than 1, thereby *trading speed of adaptation against sensitivity to noise*. In a 'full' RPE algorithm such as recursive-least-squares with a 'variable forgetting-factor' (Fortescue *et al*, 1981) this trade-off is done automatically, giving fast convergence initially yet less sensitivity in the steady-state, though even there they may be problems associated with certain patterns of noise. Note that a large value of $u(t-1)$ – i.e. a *good signal/noise ratio* – also improves matters. Hence realistic estimators must take account of current S/N ratios and possibly a test-signal might have to be injected in the worst case if adaptation is required at all times.

1.4 Plant models for general self-tuners

A model is intended to crystalize the plant behaviour in terms of a finite set of parameters so that good predictions can be made and the effect of current and future controls can be optimized. Hence the proper choice of model structure is vitally important in the implementation of a practical self-tuning algorithm; inappropriate assumptions can lead to poor or even unstable performance. Here we assume that the underlying process is the continuous-time single input/single output system:

$$y(t) = \frac{B(s)}{A(s)}u(t - T_d) + d(t),$$

in which $A(s)$ and $B(s)$ are polynomials in the differential operator s. There is a dead-time T_d before the plant input $u(t)$ affects the output $y(t)$, and $d(t)$ is a general disturbance term. Although $d(t)$ is often modelled in terms of a stationary process with rational spectral density, most practical plants are characterized by disturbances including:

1. a (possibly *slowly varying*) constant reflecting the fact that A and B are local linearizations of the process in which the incremental gain differs from the static gain;
2. significant *quantization* levels produced by the output transducer;
3. *load-disturbances* better modelled by random steps at random times;
4. *unmodelled dynamics* as there are high-frequency modes (higher-order dynamics) in the plant not captured by the lower-order model. Rohrs *et al* (1982), show how this can effect badly-designed adaptive controllers.

As self-tuners are implemented digitally, the process model as seen by the controller is the Z-transform of the plant after the usual ZOH has been inserted in the forward path and a sampler placed on the output. Considering q^{-1} to be the backward-shift operator ($qx(t) \equiv x(t+1)$), the general noise-free input-output relationship can therefore be written (with a slight abuse of notation) where t is now a sample index:

$$y(t) = \frac{B^*(q^{-1})}{A^*(q^{-1})} u(t-k).$$

Here the dead-time of k samples is $\mathrm{INT}(T_d/h)+1$. If k is unknown one procedure is to assume some minimum value (such as 1) and to have an 'extended B polynomial' for which the leading $k-1$ terms should be zero, giving (dropping the *) the Deterministic AutoRegressive and Moving Average or DARMA form:

$$A(q^{-1})y(t) = B(q^{-1})u(t-1),$$

which, if na and nb are the degrees of the respective polynomials, gives the difference equation:

$$y(t) + a_1 y(t-1) + \cdots + a_{na} y(t-na) = b_0 u(t-1) + \cdots + b_{nb} u(t-nb-1).$$

For a correctly chosen structure na should equal the degree of $A(s)$ and nb should be $n+k$ (-1 if there is no fractional delay). In practice, however, na could be set equal to only the number of *significant* poles, the effect of the others being presumed to be swamped by the low-pass property of the Nyquist filter in the ADC subsystem. Choosing a rather larger value for nb not only allows for variations in dead-time, but the extra terms give additional flexibility in fitting the plant data. Some algorithms even go as far as assuming an all-zero model given by the weighting-sequence $H(q^{-1}) = A^{-1}(q^{-1})B(q^{-1})$. In principle the weighting-sequence (which cannot be applied with unstable plant) involves an infinite numbers of parameters, but on the other hand simple *convolution* is used to develop the output:

$$y(t) = \sum_{j=1}^{\infty} h_j u(t-j).$$

In practice the plant is assumed to 'settle' with $h_j = 0$ after a finite number N of samples, giving an FIR model. For some plant (e.g. with lightly-damped modes) N can be large (some 50 or so) and the problem of how to choose or to analyse the effect of truncation is difficult. However a weighting-sequence (or the related step-response) is easy to find experimentally, such as by injecting a unit-pulse (or step) or by using a PRBS test-signal and performing cross-correlation. Some claim that $H(q^{-1})$ is a 'non-parametric' model: this is clearly false as it in fact requires many more parameters than the DARMA form, particularly if there is a spread of time-constants in the plant which have to be captured.

Some early self-tuning designs proved to be sensitive to the existence of zeros of $B(q^{-1})$ outside the stability region (so-called *nonminimum-phase* zeros) – see Clarke (1984b). These happen to be surprisingly pervasive, even for normal continuous-time plant, because of the sampling process:

- a finite NMP zero of $B(s)$ maps into a finite NMP zero of $B^*(q^{-1})$;
- if $G(s)$ has j more poles than zeros, then $G^*(q^{-1})$ becomes, as $h \to 0$, proportional to that obtained if $G(s)$ were simply j integrators. In particular for $j \geq 2$ at least one zero of the DARMA model is then on or outside the stability region;
- as the fractional delay $\delta \to h$ at least one zero of G^* will become NMP: a well-known example is the simple plant $\exp(-sT_d)/s$ whose discrete-time model has a zero outside the unit disc if $h/2 < T_d < h$. Note that T_d here includes *computational delay.*

There are three approaches to this problem: sampling at a slow enough rate to avoid NMP zeros (though at what cost in control performance?); using continuous-time self-tuners (but this does not avoid NMP zeros due to dead-time); or providing an algorithm which is insensitive to the zero positions. The latter is the preferred choice.

A further possibility is to employ a *state-space* model so that the extensive machinery for optimal LQ control (Kucera, 1979; Clarke *et al*,1985) can be used. Chapter 11 describes an application of state-space self-tuning, whilst Chapter 12 looks at autopilots from an adaptive LQG point of view, based on the polynomial developments of Chapter 3. A minimal representation of the DARMA plant with $n = \max(na, nb)$ is (in observable canonical form):

$$
\begin{aligned}
\mathbf{x}(t+1) &= \mathbf{A}\mathbf{x}(t) + \mathbf{b}u(t) \\
y(t) &= \mathbf{c}'\mathbf{x}(t), \quad \text{where:} \\
\mathbf{A} &= \begin{bmatrix} -a_1 & 1 & 0 & 0\ldots0 \\ -a_2 & 0 & 1 & 0\ldots0 \\ \vdots & & & \\ . & & & 1 \\ -a_n & 0 & 0 & 0\ldots0 \end{bmatrix} \\
\mathbf{b} &= [b_0, b_1, \ldots, b_n]' \\
\text{and} \quad \mathbf{c} &= [1, 0, 0, \ldots, 0]'.
\end{aligned}
$$

One use of this model is in providing extra insight such as stabilizing results in predictive control; as it stands the state $\mathbf{x}(t)$ is not directly measurable and must be reconstructed by an observer. One interesting *nonminimal* state representation defines:

$$\mathbf{x}_1(t) \equiv [y(t), y(t-1), \ldots, y(t-na); u(t-1), u(t-2), \ldots, u(t-nb)]'.$$

The advantage of this formulation is that the state is now completely known at time t without use of an observer and that state-feedback of the form $u(t) = -\mathbf{k}'\mathbf{x}(t)$ can be interpreted immediately in the form of a difference-equation.

1.4.1 Incorporation of disturbances

For process control in particular, regulation against disturbances is a principal requirement so that it is surprising that several methods include neither deterministic nor stochastic disturbances in their models. Other approaches consider disturbances of the moving-average form $C(q^{-1})\xi(t)$ where:

$$C(q^{-1}) = 1 + c_1 q^{-1} + \cdots + c_{nc} q^{-nc}$$

and $\xi(t)$ is an uncorrelated $(0, \sigma^2)$ random sequence, leading to the Controlled AutoRegressive and Moving Average or CARMA model:

$$A(q^{-1})y(t) = B(q^{-1})u(t-1) + C(q^{-1})\xi(t). \tag{1.4}$$

This, however, is insufficient to characterize offsets for which a zero control signal is typically accompanied by a nonzero output mean. A popular modification is to model this class of 'disturbance' by an unknown constant d, giving:

$$A(q^{-1})y(t) = B(q^{-1})u(t-1) + C(q^{-1})\xi(t) + d.$$

It is possible then in self-tuning control to *estimate* d along with the other unknown parameters. However in practice (Latawiec and Chyra, 1983) it has been found better to assume step-like or Brownian motion disturbances using a CARIMA (*Integrated*) representation (Tuffs and Clarke, 1985):

$$A(q^{-1})y(t) = B(q^{-1})u(t-1) + \frac{C(q^{-1})}{\Delta}\xi(t). \tag{1.5}$$

This structure leads to inherent integral control action in a natural way as a consequence of the internal model principle – a prerequisite for realistic algorithms.

1.4.2 Predictive models

Many self-tuners are based on predictive control laws which can be considered as generalizations of the classical Smith-predictor (1959) method. This is not just so that they can give tighter control of dead-time plant than PID regulators, but additional features such as *feedforward* and *preprogrammed set-points* are easily included, and strong stabilization theorems based on LQ ideas (Kwakernaak and Sivan, 1972) can be derived. Early designs used *k-step-ahead* prediction, arguing that $y(t+k)$ is the first output influenced by the current control choice $u(t)$. More recently however *long-range predictive control*, in which a whole set of future outputs are predicted based on assumptions concerning current and future controls, has been found to be very robust and effective (Peterka, 1984; de Keyser and van Cauwenberghe, 1985; Clarke *et al*, 1987).

As an example of deriving a predictive model, consider eqn.4 for which it is known that the first $k-1$ coefficients of B are zero so that the control term becomes $Bu(t-k)$, and the ***Diophantine equation***:

$$C(q^{-1}) = E(q^{-1})A(q^{-1}) + q^{-k}F(q^{-1}), \tag{1.6}$$

from which the polynomials $E(q^{-1})$ and $F(q^{-1})$ can be deduced based on the known $A(q^{-1})$, $C(q^{-1})$ and k. Multiplying eqn.4 by $q^kE(q^{-1})$ gives:

$$\begin{aligned} EAy(t+k) &= EBu(t) + EC\xi(t+k). \\ \text{So by eqn.6:}\quad Cy(t+k) - Fy(t) &= Gu(t) + EC\xi(t+k), \\ \text{where:}\quad G(q^{-1}) &= E(q^{-1})B(q^{-1}). \\ \text{This gives:}\quad y(t+k) &= \frac{F(q^{-1})y(t) + G(q^{-1})u(t)}{C(q^{-1})} + E(q^{-1})\xi(t+k). \end{aligned}$$

The optimal predictor of $y(t+k)$, given data up to time t, is then:

$$C(q^{-1})\hat{y}(t+k|t) = F(q^{-1})y(t) + G(q^{-1})u(t), \tag{1.7}$$

$$\text{with error}\quad \tilde{y}(t+k|t) = E(q^{-1})\xi(t+k). \tag{1.8}$$

It is seen that the prediction depends on current and past I/O data and that, as E is a polynomial of degree $k-1$, the prediction error is orthogonal to $\hat{y}(t+k|t)$ for it contains only future error terms $\xi(t+i)$. Note from eqn.7 that this is a *positional* predictor, in which full value data u and y are used to predict the full value $y(t+k)$. It is found that this form is sensitive to offset errors and so *incremental* predictors, in which *changes* in the output are forecast, are to be preferred.

To derive a long-range predictor consider eqn.5 with $C = 1$ and the Diophantine equation:

$$1 = E_j(q^{-1})A\Delta + q^{-j}F_j^1(q^{-1}),$$

and multiply eqn.5 by $q^jE_j\Delta$ to obtain:

$$E_jA\Delta y(t+j) = E_jB\Delta u(t+j-1) + E_j\xi(t+j).$$

Replacing $E_jA\Delta$ from the Diophantine equation gives:

$$y(t+j) = G_j\Delta u(t+j-1) + F_j^1 y(t) + E_j\xi(t+j).$$

But (by substituting $q = 1$) we note that $F_j^1(1) = 1$ so that we can write:

$$F_j^1(q^{-1}) = 1 + F_j(q^{-1})\Delta \quad \text{and hence:}$$

$$y(t+j) = G_j\Delta u(t+j-1) + y(t) + F_j\Delta y(t) + E_j\xi(t+j). \tag{1.9}$$

This is in the desired incremental form, for note that if u and y are constant then $\Delta y(t)$ and $\Delta u(t)$ are zero and $y(t+j)$ is (on average) equal to $y(t)$.

In eqn.9 it is seen that the control term includes both current and future elements as well as previous known values. The equation can then be split into those signals available at time t and those which are either unpredictable – $\xi(t+j)$ – or yet to be determined: $u(t+i); i \geq 0$. Hence we can write:

$$\hat{y}(t+j|t) = \sum_{i=0}^{j} g_{ji}\Delta u(t+j-i) + p_j,$$

where p_j is the signal whose components are known at time t:

$$p_j = \sum_{i=j+1}^{ng} g_{ji}\Delta u(t+j-i) + y(t) + \sum_{i=0}^{nf} f_{ji}\Delta y(t-i).$$

There are many ways of simplifying the calculations: useful recursive relations between E_j, F_j and E_{j+1}, F_{j+1} can be deduced; the first j terms of G_j are just the ordinates of the plant's step response; and p_j is computed for all j simply by iterating the plant model assuming that $\xi(t+i) = 0$ and that future controls equal the previous control $u(t-1)$: i.e. zero control increments.

By combining all the predictions $\{\hat{y}(t+j); j = 1 \ldots N_2\}$ into a vector $\hat{\mathbf{y}}$ we arrive at the following *key equation*:

$$\hat{\mathbf{y}} = \mathbf{G}\tilde{\mathbf{u}} + \mathbf{p}, \tag{1.10}$$

$$\text{where: } \begin{aligned} \hat{\mathbf{y}} &= [\hat{y}(t+1|t), \hat{y}(t+2|t), \ldots, \hat{y}(t+N_2|t]' \\ \tilde{\mathbf{u}} &= [\Delta u(t), \Delta u(t+1), \ldots, \Delta u(t+N_2-1)]' \\ \mathbf{p} &= [p_1, p_2, \ldots, p_{N_2}]'. \end{aligned}$$

The matrix $\mathbf{G}$, of dimension $N_2 \times N_2$, is highly significant in determining the properties of LRPC algorithms. It is of banded lower-triangular structure:

$$\mathbf{G} = \begin{bmatrix} g_1 & 0 & \ldots & 0 \\ g_2 & g_1 & \ldots & 0 \\ \vdots & . & \ldots & . \\ g_N & g_{N-1} & \ldots & g_1 \end{bmatrix}. \tag{1.11}$$

Note in particular if $k > 1$ the first row(s) of $\mathbf{G}$ consist entirely of zeros so that the matrix is not invertible: this is relevent when formulating self-tuning algorithms for plant with unknown dead-time.

1.5 Predictive control laws

Consider the k-step-ahead predictor of eqn.7 in which all that there is to determine is the current control $u(t)$. The objective of regulating y around the value 0 with *minimum variance* is achieved by the choice of feedback which sets $\hat{y}(t+k|t)$ to 0:

$$u(t) = -F(q^{-1})/G(q^{-1}),$$

for then $y(t+k)$ will consist of only the remnant $\tilde{y}(t+k|t)$.

A self-tuned version of the above defines a regression model of the same form as eqn.7 but including only data known at time t:

$$y(t) = F(q^{-1})y(t-k) + G(q^{-1})u(t-k) + \epsilon(t),$$

whose unknown parameters are organised into the vector:

$$\theta \equiv [f_0, f_1, \ldots; g_0, g_1, \ldots]'$$

and with corresponding input/output data:

$$\mathbf{x}(t) \equiv [y(t-k), y(t-k-1), \ldots; u(t-k), u(t-k-1), \ldots]'.$$

Updating of estimates $\hat{\theta}$ can be achieved by, say, a recursive-least-squares or other RPE algorithm as described in Chapter 2. It can be shown that, under certain conditions, a certainty-equivalent control law:

$$\hat{F}(q^{-1})y(t) + \hat{G}(q^{-1})u(t) = 0,$$

which accepts the latest parameter estimates, leads asymptotically to the desired MV law: it possesses the self-tuning property.

Astrom and Wittenmark's (1973) implicit *self-tuning regulator* follows the above ideas (see also Peterka, 1970) and was a major stimulus to the area as it rapidly lead to applications (Astrom *et al*, 1977; Astrom, 1983). However the minimum-variance objective is not often useful as it leads to a *cancellation* (Isermann, 1981) scheme (the plant zeros become poles of the controller and hence lead to instability for NMP B polynomials) and there is no trade-off between control and output variances. These objections were overcome by the *self-tuning controller* of Clarke and Gawthrop (1975, 1979) which introduced the *Generalized Minimum Variance* or GMV cost-function:

$$J_{GMV} \equiv E\{[y(t+k) - w(t)]^2 + \lambda u^2(t)|t\},$$

in which the control-weight λ is used both to moderate control effort and to allow for stabilizing certain NMP plant. Design polynomials can be added to the GMV scheme which Gawthrop (1977) interpreted as providing Smith-predictive and (de-tunable) model-reference 'knobs'. The general development is described by Clarke (1984a) and was implemented in a portable microcontroller 'SESAME' and applied to a variety of plants (Clarke and Gawthrop, 1981). ASEA (1983) incorporate a GMV self-tuner in a commercial multi-loop controller, and Fjeld and Wilhelm (1981) describe how a conventional DDC system with additional software for self-tuning can enhance the control of paper-making plant.

The major objection to MV and GMV laws is that they depend on (fairly) good knowledge of the dead-time k. Whilst reaction-curve tests and on-line 'jacketting' of estimators can provide reasonably good values it is better to derive algorithms which are less sensitive to this prior knowledge. One such is *Generalized Predictive Control* by Clarke *et al* (1987) which defined a cost-function:

$$J_{GPC}(N_1, N_2, NU, \lambda) \equiv \sum_{j=N_1}^{N_2} [\hat{y}(t+k|t) - w(t+k)]^2 + \lambda \sum_{j=1}^{NU} \Delta u^2(t+j-1),$$

where:

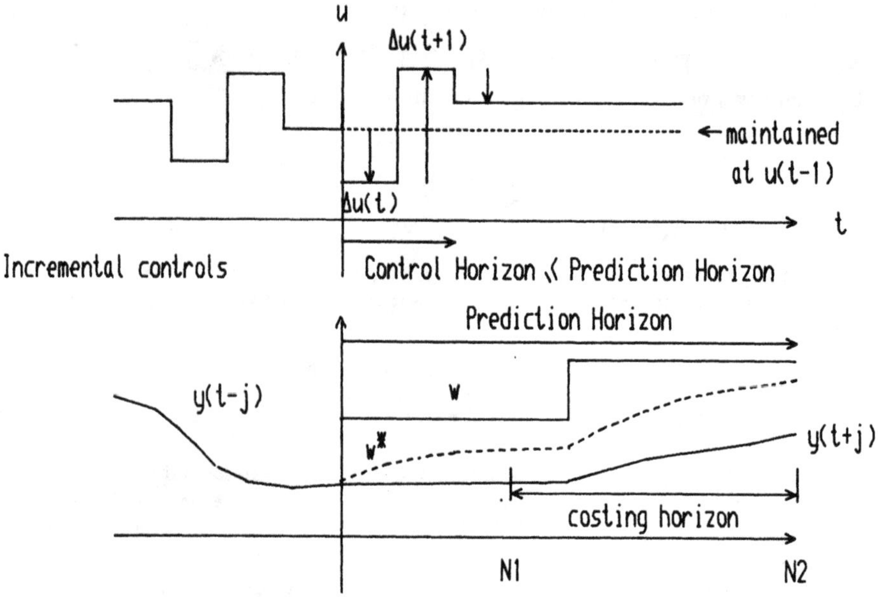

Figure 1.7: The general LRPC set-up

- N_1 is the *minimum costing horizon*,
- N_2 is the *maximum costing horizon*, and
- NU is the *control horizon*.

subject to the important constraint that *control increments* $\Delta u(t+j)$ *for* $j \geq NU$ *are zero*. It is found that particular choices of the various horizons lead to well-understood and viable control laws, and in particular $N_1 > 1$ and $NU < N_2$ are especially important for achieving useful control behaviour.

GPC, as other explicit LRPC approaches, uses a *receding-horizon* strategy, as shown in Fig.7, where for each sample-instant t:

1. RPE is used to provide a suitable plant model such as eqn.5.
2. Given $y(t)$ and previous values of $[y; u; \Delta u]$ the predictions of the freely responding plant are computed and compared with the vector of *future* set-points $w(t+j)$ – which may be known *a priori* or assumed equal to $w(t)$.
3. For the given $[N_1, N_2, NU, \lambda]$ the optimal future control vector $\tilde{\mathbf{u}}$ is computed by minimizing J_{GPC}.
4. The first element $u(t) = u(t-1) + \tilde{\mathbf{u}}_1$ is asserted and sequences are shifted ready for the next sample interval.

With the assumptions about the GPC horizons, eqn.10 becomes:

$$\hat{\mathbf{y}} = \mathbf{G}\tilde{\mathbf{u}} + \mathbf{p}, \text{ where } \tilde{\mathbf{u}} = [\Delta u(t), \ldots, \Delta u(t+NU-1)]'$$

$$\text{and: } \mathbf{G} = \begin{bmatrix} g_{N_1} & g_{N_1-1} & \cdots & 0 & 0 \\ g_{N_1+1} & g_{N_1} & g_{N_1-1} & \cdots & 0 \\ \vdots & & & & \\ g_{N_2-1} & g_{N_2-2} & \cdots & \cdots & g_{N_2-NU} \\ g_{N_2} & g_{N_2-1} & g_{N_2-2} & \cdots & g_{N_2-NU+1} \end{bmatrix}$$

. Here $\mathbf{G}$ is a $(N_2 - N_1 + 1) \times NU$ matrix with zero entries G_{ij} for $j - i > N_1$. With this the GPC cost-function can be written:

$$J_{GPC} = (\mathbf{w} - \mathbf{G}\tilde{\mathbf{u}} - \mathbf{p})'(\mathbf{w} - \mathbf{G}\tilde{\mathbf{u}} - \mathbf{p}) + \lambda\tilde{\mathbf{u}}'\tilde{\mathbf{u}}$$

and the solution giving the minimum cost is then:

$$\tilde{\mathbf{u}}_{opt} = (\mathbf{G}'\mathbf{G} + \lambda\mathbf{I})^{-1}\mathbf{G}'(\mathbf{w} - \mathbf{p}),$$

where $\mathbf{G}'\mathbf{G}$ is a $NU \times NU$ matrix. In particular if $NU = 1$ (as in the EPSAC algorithm of de Keyser and van Cauwenberghe, 1985) this reduces to a scalar which is always invertible provided that at least one g_i is nonzero. In general the inversion (i.e. for the case where λ is 0) is possible when rank$(\mathbf{G}'\mathbf{G}) = NU$; this is clearly easier to achieve when $NU \ll N_2$ and moreover the computational burden is then much reduced (see Chapter 4).

As GPC is based on optimization (see Lawson and Hansen, 1974) it is possible to add *constraints* to the control signals. Future reference signals, as in robotics or batch applications, are catered for automatically. Fig.8 shows the overall structure of the algorithm as a feedback loop.

The behaviour of GPC depend on the values chosen for the three horizons $[N_1, N_2, NU]$ and the control weighting λ. This gives the approach considerable versatility but might arouse concern about the plethora of choices involved. Space precludes a full discussion here, but Mohtadi and Clarke (1986), Clarke and Mohtadi (1987) give guidelines based on theorems showing that standard, well-understood control laws can be deduced as special cases of GPC. For example even a pole-placement law (Wellstead *et al*, 1979), which is popular in applications, can be derived using the settings $[n, 2n-1, n, 0]$. In general there are the following guidelines:

- $N_1 = nb$: the first costing-horizon at least exceeds the maximum expected plant dead-time and avoids NMP zeros.
- N_2 large: at least $\geq 2nb - 1$ but up to the rise-time of the plant may be better.
- $NU \leq na$: NU determines the degrees of freedom in future control increments. For process plant $NU = 1$ is generally satisfactory whilst 'difficult' plant require $NU \approx$ number of unstable/ underdamped poles.
- $\lambda = 0$: the easiest choice, though sometimes $\lambda = \epsilon$ where ϵ is small helps numerical robustness. If desired λ can be considered to be a fine 'tuning-knob', whereas NU is a coarse knob for which a drop in its value can induce a marked reduction in control variance.

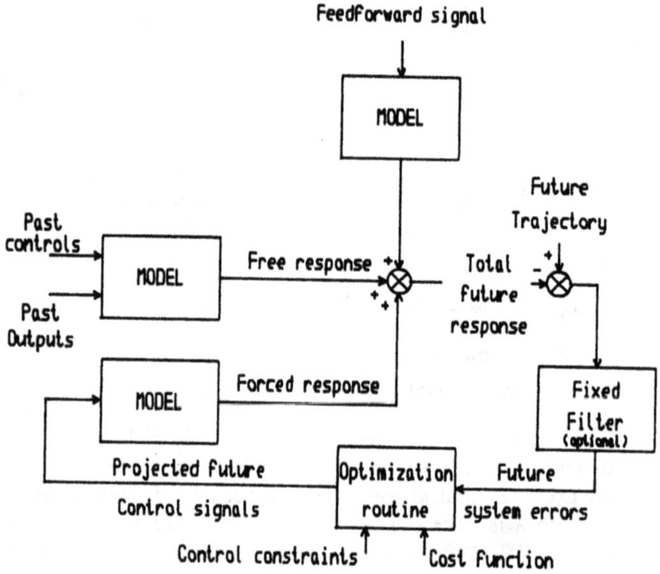

Figure 1.8: Optimization in GPC

Multivariable self-tuning applications of GMV and GPC are described in Chapter 13. Chapters 9 and 10 discuss other LRPC algorithms and show their potential. Unbehauen (1980) and Seborg *et al* (1983) give further prectical self-tuning examples.

1.6 Conclusions

Self-tuning has matured over the last few years from a theoretical subject to a serious industrial tool. Practical experience is growing rapidly and it appears that the methods are becoming a standard part of a control engineer's armoury. Microprocessors have played a crucial role in achieving acceptance of self-tuning for they allow the computationally demanding algorithms to be implemented cheaply. In some cases, however, cost or bandwidth requirements dictate simplified methods (see Chapter 5) which still provide self-tuning performance.

Self-tuners do not remove the need for an engineer's skill but allow him to think more positively about the real control needs and constraints of his plant rather than simply hoping that manually-tuned and fixed PID laws will solve all his problems. At present customers are demanding that PID manufacturers add tuning features to their product range, though there are signs that the more advanced ideas discussed in this book are becoming accepted as case-studies and industrial trials demonstrate the applicability of the algorithms.

The real need now is to add *engineering features* to the methods. Self-tuning relies on the ability to estimate an effective process model from input/output data. This task is made harder by actuator and sensor nonlinearities, by unmodelled dynamics, and by certain types of disturbance. The aim is to provide RPE methods which are reliable in these circumstances: often 'jacketting' software must be added to detect periods of poor data. The 'user-interface' should be designed with care: how much choice is available to the engineer? can default values be sufficient for most plant? is the algorithm understandable enough so that setting-up is easy? should a standard initial test sequence such as 'bumping' the plant be used? should adaptation be automatic or on demand? how is a self-tuner integrated into a network of microcontrollers? etc. These questions have been answered for PID algorithms in a commonly-accepted way. One nice feature of model-based methods is that the operator's display can include *future* predictions as well as historical data. Moreover rapid changes in the current model compared with the nominal may be used to provide a certain level of fault-detection which enhances the traditional limits and rate-limits. Finally, the software needs to be engineered (Chapter 8): reliable and reusable code which clearly expresses the appropriate algorithms and data structures, and which can be easily embedded into a novel self-tuning product.

Bibliography

ASEA (1983). Adaptive control - a NEW age in process control. *ASEA Innovation.*

Astrom,K.J.(1983). Theory and applications of adaptive control - a survey. *Automatica*, **19**, 471-486.

Astrom,K.J. and T.Hagglund (1984). Automatic tuning of simple regulators with specifications on phase and amplitude margins. *Automatica*, **20**, 645-651.

Astrom,K.J. and B.Wittenmark (1973). On self-tuning regulators. *Automatica*, **9**, 185-199.

Astrom,K.J., U.Borisson, L.Ljung and B.Wittenmark(1977). Theory and applications of self-tuning regulators. *Automatica*, **13**, 457-476.

Bristol,E.H. (1977). Pattern recognition: an alternative to parameter identification in adaptive control. *Automatica*, **13**, 197-202.

Clarke,D.W. (1981). Introduction to self-tuning controllers. in *Self-tuning and Adaptive Control* (ed. Harris and Billings) Peter Perigrinus.

Clarke,D.W. (1984a). Self-tuning control design and implementation. in *Real-time Computer Control*, (ed. Bennett and Linkens), Peter Perigrinus.

Clarke,D.W. (1984b). Self-tuning control of nonminimum-phase systems. *Automatica*, **20**, 501-517.

Clarke,D.W. (1985). PID algorithms and their computer implementation. *Trans.Inst.M.C.*, **6**, 305-316.

Clarke,D.W. and P.J.Gawthrop (1975). Self-tuning controller. *Proc.IEE*, **122**, 929-934.

Clarke,D.W. and P.J.Gawthrop (1979). Self-tuning control. *Proc.IEE*, **126**, 633-640.

Clarke,D.W. and P.J.Gawthrop (1981). Implementation and application of microprocessor-based self-tuners. *Automatica*, **17**, 233-244.

Clarke,D.W., P.P.Kanjilal and C.Mohtadi (1985). A generalized LQG approach to self-tuning control (2 Parts). *Int.J.Control*, **41**, 1509-1544.

Clarke,D.W. and C.Mohtadi (1987). Properties of generalized predictive control. *IFAC World Congress*, Munich.

Clarke,D.W., C.Mohtadi and P.S.Tuffs (1987). Generalized predictive control (2 Parts). *Automatica*, **23**, 137-160.

Fjeld,M. and R.G.Wilhelm (1981). Self-tuning regulators - the software way. *Control Engineering*, October, 99-102.

Fortescue,T.R., L.S.Kershenbaum and B.E.Ydstie (1981). Implementation of self-tuning regulators with variable forgetting factors. *Automatica*, **17**, 831-835.
Francis,B.A. (1987). *A Course in H^∞ Control Theory*. Springer.
Gawthrop,P.J. (1977). Some interpretations of the self-tuning controller. *Proc.IEE*, **124**, 889-894.
Gawthrop,P.J. (1980). Hybrid self-tuning control. *Proc.IEE*, **127**, Pt.D, 229-236.
Gawthrop,P.J. (1982). A continuous-time approach to discrete-time self-tuning control. *Optimal Control App. and Methods*, **3**, 399-414.
Goodwin,G.C. and K.S.Sin (1984). *Adaptive Filtering, Prediction and Control.* Prentice-Hall.
Harris,C.J. and S.A.Billings (eds) (1981). *Self-tuning and Adaptive Control: Theory and Applications.* Peter Perigrinus.
Horowitz,I.M. (1963). *Synthesis of Feedback Systems.* Academic Press.
Isermann,R. (1981). *Digital control systems.* Springer-Verlag.
Kalman,R.E. (1958). Design of a self-optimising control system. *Trans.ASME*, **80**, 468-478.
de Keyser,R.M.C. and A.R.van Cauwenberghe (1985). Extended prediction self-adaptive control. *IFAC Symp.Ident.Syst.Param.Est.*, York.
Kraus,T.W. and T.J.Myron (1984). Self-tuning PID controller uses pattern recognition approach. *Control Engineering*, June, 106-111.
Kucera,V.(1979). *Discrete Linear Control.* Wiley.
Kwakernaak,H. and R.Sivan (1972). *Linear Optimal Control Systems.* Wiley.
Latawiec,K. and M.Chyra (1983). On low frequency and long-run effects in self-tuning control. *Automatica*, **19**, 419-424.
Lawson,C.L. and R.J.Hanson (1974). *Solving Least-squares Problems.* Prentice-Hall.
Mishkin,E. and L.Braun (1961). *Adaptive Control Systems.* McGraw-Hill.
Mohtadi,C. and D.W.Clarke (1986). Generalized predictive control, LQ, or pole-placement: A unified approach. *CDC Conf.*, Athens.
Narendra,K.S. and R.V.Monopoli (eds) (1980). *Applications of Adaptive Control.* Academic Press.
Peterka,V. (1970). Adaptive digital regulation of noisy systems. *IFAC Symp.Ident.Proc.Param.Est.*, Prague.
Peterka,V. (1984). Predictor-based self-tuning control. *Automatica*, **20**, 39-50.
Rohrs,C.E., L.Valavani, M.Athans and G.Stein (1982). Robustness of adaptive control algorithms in the presence of unmodelled dynamics. *21st IEEE Conf.Dec.Cont.*.
Seborg,D.E., S.L.Shah and T.F.Edgar (1983). Adaptive control strategies for process control: a survey. *AIChE Diamond Jubilee Mtg.*, Washington,DC.
Smith,O.J.M. (1959). A controller to overcome dead-time. *ISA.Journal*, **6**, 28-33.
Tuffs,P.S. and D.W.Clarke (1985). Self-tuning control of offset: a unified approach. *Proc.IEE*, **132**, Pt.D, 100-110.
Unbehauen,H. (ed.) (1980). *Methods and Applications in Adaptive Control.* Springer-Verlag.
Wellstead,P.E., J.M.Edmunds, D.Prager and P.Zanker (1979). Self-tuning pole/zero assignment regulators. *Int.J.Control*, **30**, 1-26.
Westcott,J.H. (ed.) (1962). *An Exposition of Adaptive Control.* Pergamon Press.
Ziegler,J.G. and N.B.Nichols (1942). Optimum settings for automatic controllers. *Trans.ASME*, **64**, 759-768.

Chapter 2

RLS based estimation schemes for self-tuning control

S. L. Shah and W. R. Cluett

2.1 INTRODUCTION

Broadly speaking, an adaptive or a self-tuning controller is a combination of an on-line, recursive estimation law with a suitable control strategy. The performance and convergence properties play an important role in the overall stability and performance of an adaptive controller. Recursive least squares (RLS) is probably the most popular parametric identification method in adaptive control.

RLS is one member of a family of prediction error identification methods that are based on the minimization of prediction error functions. The basic RLS method is the result of minimization of a quadratic cost function of the prediction error, e(t), of the following form:

$$J_N(\theta) = \frac{1}{N}\sum_{t=1}^{N} \alpha_t e(t)^2 = \frac{1}{N}\sum_{t=1}^{N} \alpha_t \left(y(t) - \theta^T \cdot \phi(t)\right)^2 \qquad (1)$$

Here we pose a problem where a parameter vector, θ, is to be estimated from measurements of y(t), u(t); t=1,2,...N; α_t is a sequence of positive numbers (usually chosen as 1) that allow us the flexibility of giving different weights to different observations.

The origin of the batch least squares algorithm, according to Ljung and Soderstrom (23), can be traced back to Gauss (7), whereas the recursive version appears to have been derived by Plackett (18) in 1950. Since then numerous papers have appeared on different aspects of the least squares method. At the present time, the RLS algorithm in its various modified forms is the most widely used and studied estimation algorithm in adaptive control. This paper provides a brief introduction to the basic RLS algorithm and its convergence properties, and then surveys the various modifications of this algorithm.

2.1.1 Preliminaries

Suppose the system to be identified is described by the following linear discrete model

$$y(t) = \theta_0^T \cdot \phi(t) \qquad (2)$$

where

$$\theta^T = [a_1, a_2, \ldots, b_0, b_1, \ldots]; \quad \phi(t) = [y(t-1), y(t-2), \ldots, u(t-1),\ u(t-2), \ldots\ ]$$

This equation represents the usual dynamic system:

$$A(q^{-1})y(t) = B(q^{-1})\ u(t-1) \tag{3}$$

Recursive estimation methods are concerned with computation of an estimate $\hat{\theta}(t)$ of θ based on present and past input and output observations.

In the following sections where we discuss the performance of the RLS algorithm and its many variants, some performance analysis will also be supplemented by simulation results on the following example.

$$y(t) = a\ y(t-1) + b\ u(t-1) + e(t)$$

where true value of $a = 0.9\ \forall\ t$

$b = 0.1\ \forall\ t \quad \{\ 1,\ 499\ \}$

$= 0.5\ \forall\ t \quad \{500,\ 600\ \}$

The process is excited by a zero-mean PRBS sequence of magnitude $\pm$ 1.0 and period 19. The corresponding output is shown in figure 1. Notice that for $t \geq 500$, the process gain is increased by a factor of 5 by changing the value of b; and the signal to noise ratio also increases by the same order of magnitude. $\{\ e(t)\ \}$ is a zero-mean noise sequence of variance =0.02. The input and output sequences so generated for 600 sample intervals are used to illustrate various features of the RLS based algorithms for a 2 parameter estimation problem. In each illustration, the following initial conditons are used:

$$\hat{\theta}(0) = 0 \quad \text{and } P(0) = I \text{ or } \begin{bmatrix} 1 & 0 \\ 0 & 1 \end{bmatrix}$$

2.2 THE ORDINARY RECURSIVE LEAST SQUARES ALGORITHM

The ordinary RLS algorithm in its simplest form can be obtained by minimizing the quadratic cost function of the prediction error shown in eqn. (1). The resulting estimation law in recursive form, with α_t=1, is

$$P(t)^{-1} = P(t-1)^{-1} + \phi(t)\ \phi(t)^T \tag{4}$$

$$\hat{\theta}(t) = \hat{\theta}(t-1) + P(t)\ \phi(t)\ [y(t) - \hat{\theta}(t-1)^T\ \phi(t)] \tag{5}$$

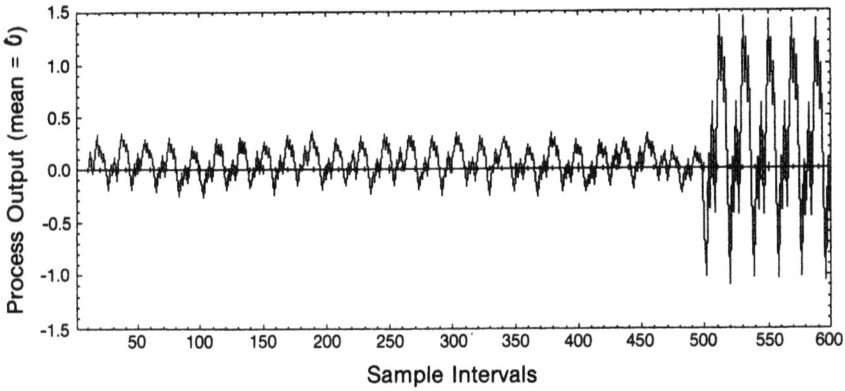

Figure 1: Process Output Under PRBS Input Excitation

This algorithm is computationally costly since it requires the inversion of the matrix P(•) at each sample interval. With the use of the matrix inversion lemma the algorithm can be re-expressed in a form so that the inversion of P(•) is now replaced by inversion of a scalar.

Parameter vector update law:

$$\hat{\theta}(t) = \hat{\theta}(t-1) + K(t)\ [y(t) - \hat{\theta}^T(t-1)\ \phi(t)] \qquad (6)$$

Gain (vector) update:

$$K(t) = P(t-1)\ \phi(t)/[\lambda + \phi(t)^T\ P(t-1)\ \phi(t)] \qquad (7)$$

Covariance matrix update

$$P(t) = \left[P(t-1) - \frac{P(t-1)\phi(t)\phi(t)^T P(t-1)}{\lambda+\phi(t)^T P(t-1)\phi(t)}\right]\frac{1}{\lambda} \qquad (8)$$

(Note $\lambda = 1$ for ordinary RLS)

This algorithm with λ=1 has a number of interesting convergence properties (Goodwin and Sin, pg. 60(9)). The key ones of interest to us here are:

i) $$\frac{\|\tilde{\theta}(t)\|^2}{\|\tilde{\theta}(0)\|^2} \le K_0 = \text{condition no. of } P(0)^{-1} = \frac{\lambda_{max}\ P(0)^{-1}}{\lambda_{min}\ P(0)^{-1}}$$

ii) $$\frac{\|\tilde{\theta}(t)\|^2}{\|\tilde{\theta}(t-1)\|^2} \le K_{t-1} = \frac{\lambda_{max} P(t-1)^{-1}}{\lambda_{min} P(t-1)^{-1}}\ ;\ \tilde{\theta}(t) = \hat{\theta}(t) - \theta_0$$

Note that $\hat{\theta}(t)$ does not necessarily coverge to θ_o. The first, property implies that $\hat{\theta}(t)$ is <u>never</u> further from θ_o than κ_o times $\|\tilde{\theta}(0)\|^2$. (Note that $\kappa \geq 1$). In contrast, the projection algorithm (which is equivalent to the RLS algorithm with a constant P(.)) guarantees that:

$$\|\tilde{\theta}(t)\|^2 \leq \|\tilde{\theta}(t-1)\|^2 \ \forall\ t$$

However, the RLS algorithm initially has much faster convergence than the projection algorithm. The basic difficulty with the ordinary RLS is that the covariance matrix gradually decays to a small value and therefore the algorithm does not retain its alertness or adaptivity. This is easily seen from the correction term on the right hand side of eqn. (8). This term is always positive or zero. If there is sufficient new information it is positive and consequently P(t) becomes smaller, and if there is no new information $P(t-1)\phi(t) \simeq 0$. This property is clearly illustrated in figures 2 and 3. The lack of adaptivity or alertness is evident as the process cannot track time-varying parameters (cf. figure 2) This figure shows plots in the time domain of the estimated parameter divided by the true value of the parameter. An alert estimator would maintain this ratio of $\hat{\theta}_i/\theta_{io}$ close to or at unity at all times. Notice the rapid decay of the trace of P as shown in figure 3. The trace of P is a measure of the "magnitude" of P and hence of the estimator gain vector, K(t-1).

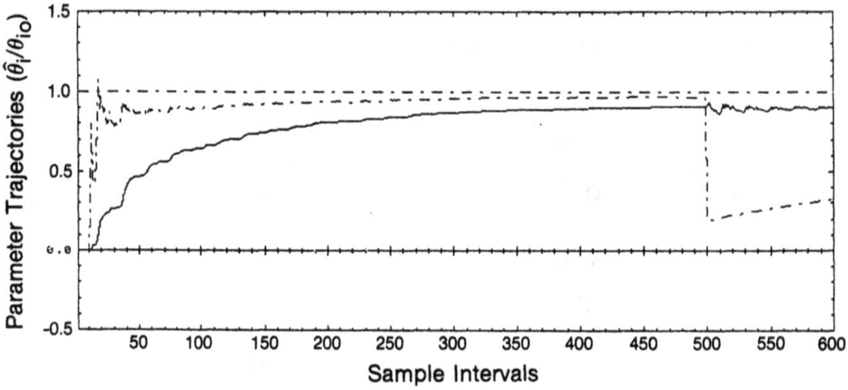

Figure 2: Parameter Trajectories for the Ordinary RLS Algorithm

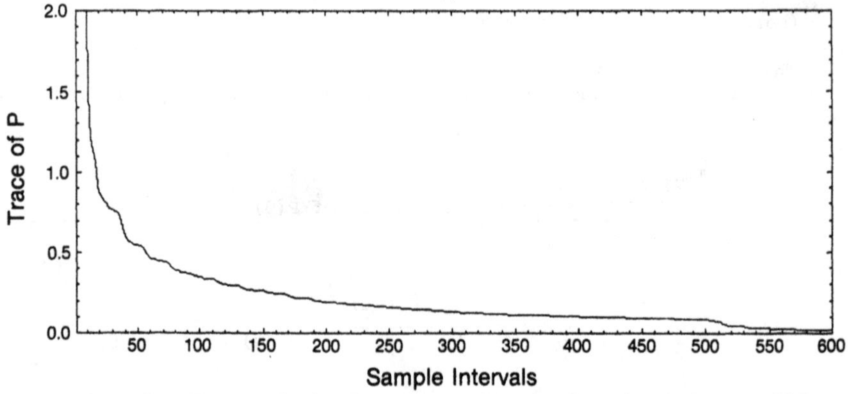

Figure 3: Trace of the Covariance Matrix for the Ordinary RLS Algorithm

It is important to emphasize that the covariance matrix update (eqn. 4 or 8) occurs in the direction and with the magnitude that minimizes the sum of squares of the prediction errors. In doing so, the covariance matrix automatically takes into account the new information content in the regressor vector and also accommodates the units of the output, y, and the input, u. To perturb the update equations in an ad-hoc manner would be to ruin this property. Yet this is routinely done in many applications without any regard for the geometry and direction of the updates. An interesting insight into the geometry of least squares with respect to eigenvectors and eigenvalues of the $P(\cdot)^{-1}$ matrix has been obtained by Rogers (20) by examining the ellipsoids described by the Lyapunov function $V(t) = \tilde{\theta}(t)^T P(t-1)\tilde{\theta}(t)$, where $\tilde{\theta}$ (t) denotes the parameter error vector: $\tilde{\theta}(\cdot) = \hat{\theta}(\cdot) - \theta_0$.

2.3 RLS WITH EXPONENTIAL DATA WEIGHTING

One method of ensuring that the algorithm retains its alertness is to introduce a 'forgetting' or discounting factor, λ ($\lambda < 1$ for exponential data weighting; $\lambda = 1$ for ordinary RLS). Effectively this means minimization of the following function where the most recent data has highest weighting.

$$J = \sum_{t=0}^{N} \lambda^{N-t} \left[y(t) - \phi(t)^T \hat{\theta}(t-1)\right]^2$$

The algorithm remains essentially unchanged with respect to the ordinary RLS except for the forgetting term $\lambda < 1$. The choice of λ also involves compromises: too fast discounting (small λ) will make the estimates uncertain (P and hence K become large) and too slow discounting will make it difficult to track fast parameter variations. The parameter trajectory to illustrate good tracking with exponentially weighted observations, $\lambda = 0.9$, is shown in figure 4. The corresponding trajectory for trace P is shown in figure 5. Notice that relatively large trace P values leads to higher gains which in turn lead to large parameter estimate variances. Under rich excitation (in the interval $t \geq 500$), the right-hand side of eqn. 8 inside the square brackets decays at a faster rate than it can be inflated by the multiplier $1/\lambda$. Also this forgetting works well only if the process has excitation. Otherwise exponential forgetting leads to covariance windup or what is commonly referred to as 'blow-up'. This can be easily seen as follows: If $P(t-1)\phi(t) = 0$ then $P(t) = P(t-1)/\lambda$. Thus $P(\cdot)$ will grow exponentially and any small change in the future $\phi(\cdot)$ will lead to large parameter changes and corresponding oscillations in the control and even possible instability.
The latter phenomena is illustrated in figures 6 and 7. In the absence of any excitation in the interval 200 T_s to 400 T_s, the covariance matrix blows up or trace $P(\cdot)$ increases rapidly (cf. fig.6). The parameters remain bounded, however their variances are larger in the same interval (cf. figure 7).

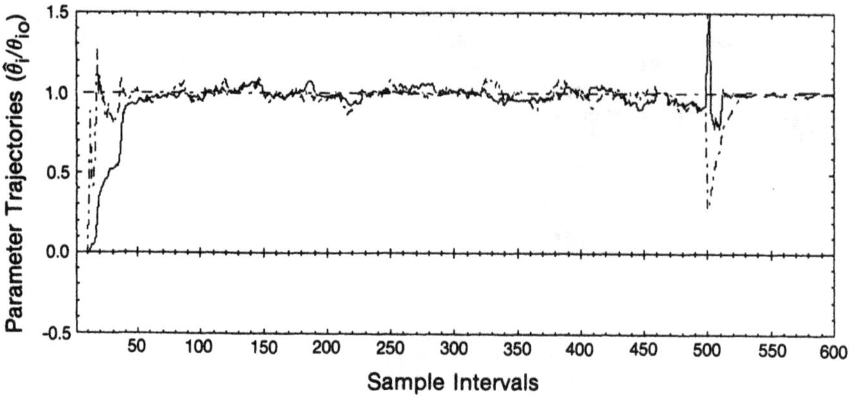

Figure 4: Parameter Trajectories for an Exponentially Weighted RLS Algorithm ($\lambda = 0.90 \; \forall \; t$)

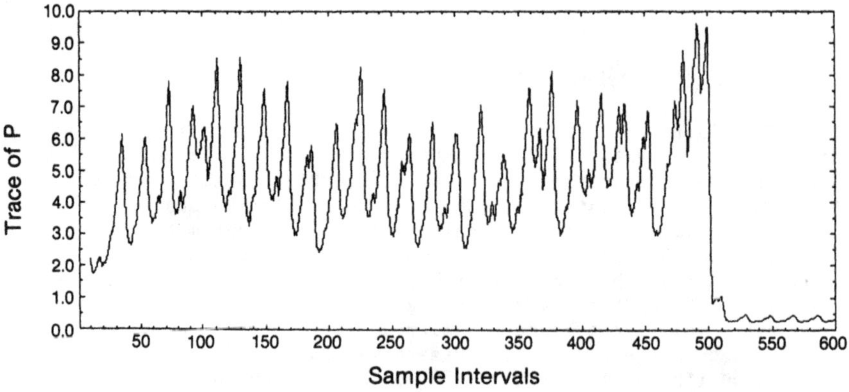

Figure 5: Trace of the Covariance Matrix for the Exponentially Weighted RLS Algorithm

2.4 RLS WITH VARIABLE FORGETTING

This idea due to Fortescue et al.(6) relies on a time varying $\lambda(t)$ which is automatically set so that $\lambda(t) \rightarrow 1$ when the prediction error is small and $\lambda(t)$ is set to a small value ($\lambda < 1$) if the prediction error is large. However, an additional mechanism to ensure that $P(\cdot)$ remains bounded has to be imposed or else even with a variable forgetting factor the covariance matrix can grow exponentially (Cordero and Mayne, (3)). For the variable forgetting factor algorithm eqns. (7) and (8) are replaced by the following steps:

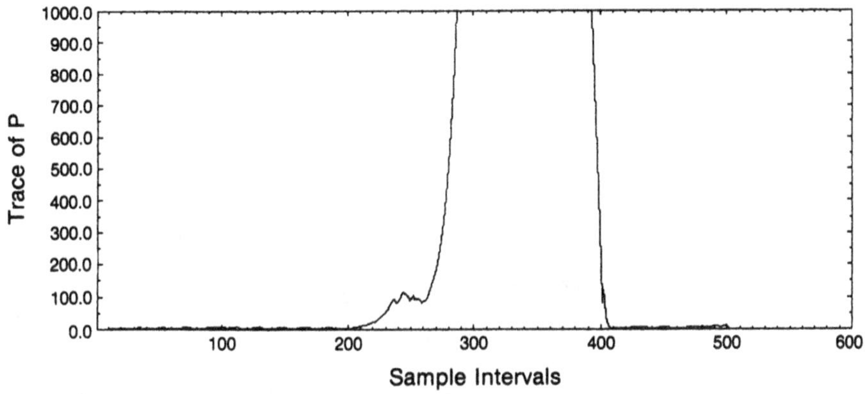

Figure 6: Trace of the Covariance Matrix for the Exponentially Weighted RLS Algorithm in the Absence of Process Excitation in the Interval $200T_s$ to $400T_s$ (cf.Figure 5)

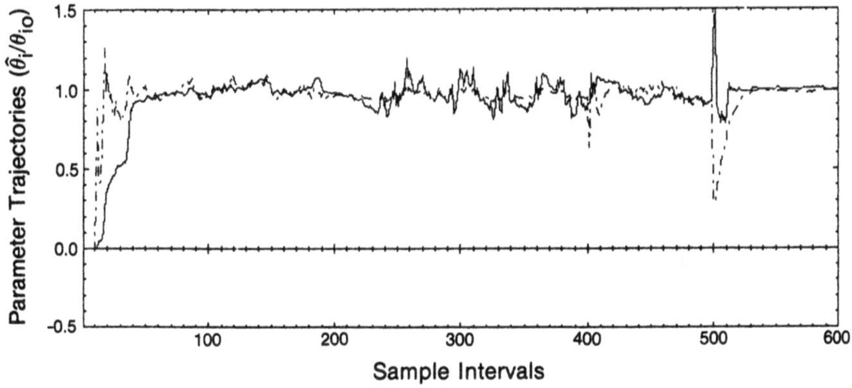

Figure 7: Parameter Trajectories for an Exponentially Weighted RLS Algorithm ($\lambda = 0.90\ \forall\ t$) in the Absence of Process Excitation in the interval $200T_s$ to $400T_s$ (cf.Figure 4)

$$\lambda(t) = 1 - \frac{[y(t) - \phi(t)^T\hat{\theta}(t-1)]^2}{[1+\phi(t)^T P(t-1)\phi(t)]\sigma} \qquad (9)$$

$$K(t) = P(t-1)\phi(t) / \left[1+\phi(t)^T P(t-1)\phi(t)\right] \tag{10}$$

$$W(t) = P(t-1) - K(t)\phi(t)^T P(t-1) \tag{11}$$

To ensure an upperbound on P(t-1), we update it as follows:

$$P(t) = W(t)/\lambda(t) \text{ if trace } W(t)/\lambda(t) \leq C$$

$$\text{else } P(t) = W(t) \quad (\text{i.e. } \lambda(t) = 1) \tag{12}$$

(σ/σ_w is selected to be a large number $\simeq 1000$, σ_w is the variance of a zero mean noise sequence in the process described by eqn (2). Note that $\sigma/\sigma_w = 1000$ means that at steady state, $\lambda(t) = 0.999$). This scheme has already been tried successfully on a chip refiner and a paper machine box (Ydstie et al. (22)). Figures 8 and 9 show the excellent performance of this variable forgetting factor scheme. The variation of λ with respect to t is shown in figure 9. In practical applications one would set a lower bound on $\lambda(t)$ to be something other than zero, e.g. 0.9.

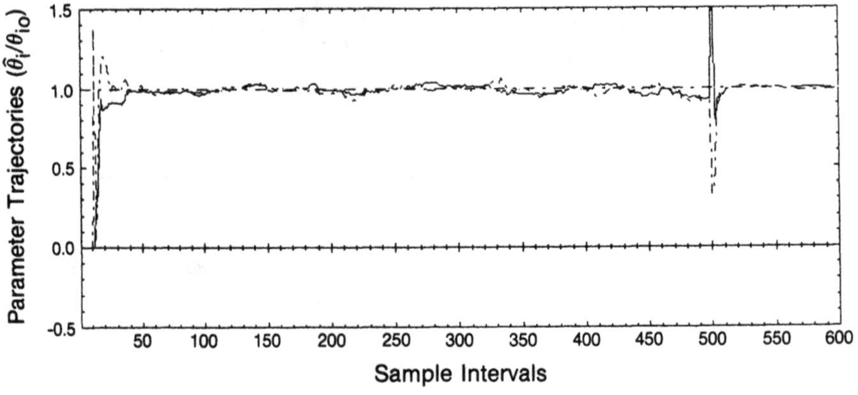

Figure 8: Parameter Trajectories for an RLS Algorithm with a Varia ble Forgetting Factor

2.5 RLS WITH COVARIANCE RESETTING

An alternative mechanism to stop P(•) from decaying to zero is to update it by a non-zero value Q (e.g. $Q = K_o I$) at various times. For this algorithm eqn. (8) would be replaced by:

$$\bar{P}(t) = P(t) - \frac{P(t-1)\phi(t)\phi(t)^T P(t-1)}{1+\phi(t)^T P(t-1)\phi(t)} \tag{13}$$

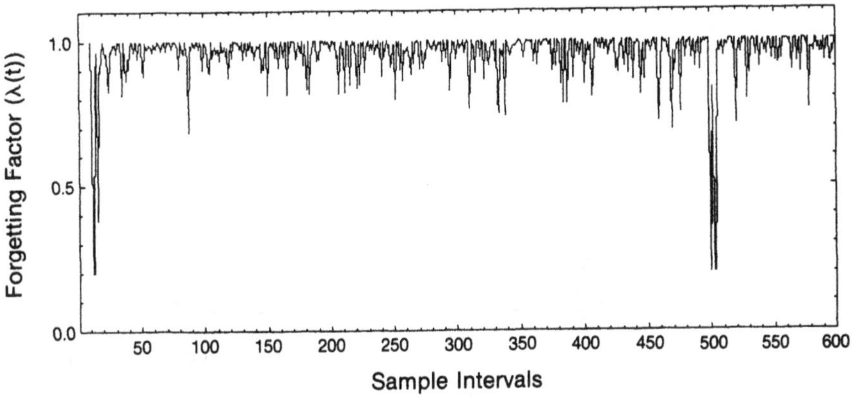

Figure 9: Variation of the Forgetting Factor ($\lambda(t)$) for the Variable Forgetting Factor RLS Algorithm

$$P(t) = \bar{P}(t) + Q \qquad (14)$$

(Q is added at various times to keep the algorithm alert, e.g. if trace $\bar{P}(t) < K_{min}$ or otherwise when the prediction error is large). Note that P(t) is not reset if

$$\text{trace } \bar{P}(t) > k_{max} \quad \text{in which case } P(t) = \bar{P}(t) \qquad (15)$$

A convergence analysis of this algorithm for use in adaptive control is contained in Goodwin et al. (8).

2.6 RLS WITH MATRIX REGULARIZATION

This mechanism for updating the covariance matrix was first proposed by Praly (19) and modified by Ortega et al. (17) for use in their work on robust adaptive control. The basic idea is a combination of a covariance resetting feature and a guarantee of lower and upperbounds on the covariance matrix. In its unnormalized form this algorithm replaces eqn. (8) by:

$$P(t) = (1 - \frac{\lambda_0}{\lambda_1}) \left[P(t-1) - \frac{P(t-1)\phi(t)\phi(t)^T P(t-1)}{1+\phi^T(t)P(t-1)\phi(t)} \right] + \lambda_0 I \qquad (16)$$

where $\lambda_0 < \lambda_1$, and both are strictly positive scalars. It can be shown that $\lambda_{max}\, P(t) \leq \lambda_1$, and $\lambda_{min}\, P(t) \geq \lambda_0$. The algorithm is especially designed to have an upperbound on the maximum eigenvalue for robustness. In its normalized form (by including a normalized regressor in place of $\phi(t)$) the algorithm can guarantee stability and robustness in the presence of unmodelled dynamics. Notice that if λ_0 is chosen equal to λ_1 then the algorithm reduces to the constant gain or a projection type, i.e. $P(\cdot) = \lambda_0 I$, and if $\lambda_0 \approx 0$ then this eqn. (16) becomes similar to the standard covariance update for the ordinary RLS (cf. eqn. 8).

2.7 RLS WITH CONSTANT TRACE

This interesting variant of the basic RLS has been suggested by Irving (12) and more recently by Goodwin et al. (10) in a different form. The form in (10) requires that the covariance update (eqn. (8)) be followed by the additional steps:

Let C_0, C_1 denote two positive constants, $C_1 > C_0$ Initialize $P(\cdot) = (C_1/m)I$ with m as the number of parameters to be estimated; also let $\tau =$ trace $P(\cdot)$. Then the final covariance matrix, $P'(t)$, to be used in eqns. (6) and (7) is given by:

If $\tau > C_0$ update $P(t-1)$ as follows:

$$P'(t-1) = P(t-1) + \frac{(C_1-\tau)}{m} I \tag{17}$$

If $\tau \leq C_0$ then update $P'(t-1)$ as:

$$P'(t-1) = \left[\frac{C_0}{\tau}\right] P(t) + \left[\frac{C_1-C_0}{m}\right] I \tag{18}$$

The algorithm ensures a constant trace of value C_1 and therefore naturally an upperbound on $\lambda_{max} P'(\cdot)$, as well as the following lower bound on $\lambda_{min} P'(\cdot)$:

$$(\lambda_{min} P'(\cdot) \geq (C_1-C_0)/m; \ \lambda_{max} P'(\cdot) \leq C_1$$

2.8 RLS WITH DEAD-ZONE AND A NORMALIZED REGRESSOR

Robustness of adaptive controllers in the presence of unmodelled dynamics has been a subject of much current research. A necessary prerequisite in the theoretical treatment of this problem is the normalization factor. The role of normalization in theory is to ensure that the unmodelled terms contributing to the output can be considered bounded. Normalization as first proposed by Egardt (5) is a key requirement in the treatment of robust adaptive stability in the presence of model-plant mismatch. Cluett et al. (1) have defined the following normalization factor:

$$x(t) = \phi(t)/n(t); \ y^n(t) = y(t)/n(t)$$

where $n(t) = \max\left(\max_{1 \le i \le m} |\Phi_i(t)|, C\right), C>0$

Note that dimension of Φ is greater than the dimension of the regressor vector of the real process; and $\phi(\cdot)$ denotes a regressor vector for a reduced order model. Various other forms of normalization factors have also been proposed. The justification in proposing a particular form of normalization is to facilitate stability and convergence analysis of the algorithm. (See Cluett et al. (2) for a detailed discussion on the choice of normalization factors).

The following algorithm guarantees global stability and convergence of an adaptive system in the presence of unmodelled dynamics.

$$\hat{\theta}_r(t) = \hat{\theta}_r(t-1)+\psi(t)^2K(t)[\, y^n(t) - \hat{\theta}_r(t-1)^T \; x(t)] \tag{19}$$

$$K(t) = P(t-1)x(t)/(1+\psi(t)^2x(t)^TP(t-1)x(t)) \tag{20}$$

$$P(t) = P(t-1) - \frac{\psi(t)^2P(t-1)x(t)x(t)^TP(t-1)}{1+\psi(t)^2x(t)^TP(t-1)x(t)} \tag{21}$$

The subscript 'r' denotes parameter estimates for a reduced order model. The scalar term $\psi(t)^2$ is selected to turn adaptation on or off depending on the size of the 'dead-band' around the normalized prediction error, i.e. roughly speaking:

$$\psi(t)^2 = 1 \quad \text{if } |y^n(t) - \hat{\theta}_r(t-1)^Tx(t)| > \Delta'_b \qquad \text{(adaptation ON)}$$

$$= 0 \quad \text{otherwise} \qquad \text{(adaptation OFF)}$$

where Δ'_b is a function of the class of unmodelled dynamics being considered. This algorithm may also be modified to include a variable forgetting factor and a dead zone to retain alertness. This algorithm in its unnormalized form, may be used if only bounded disturbances and noise are present.

2.9 RLS WITH CONSTANT TRACE AND SCALING

In a recent paper Sripada and Fisher (21) have proposed the following four modifications to the basic least squares algorithm.

i) Normalization
ii) Scaling
iii) Constant trace through a variable forgetting factor, and
iv) An "information content" based criterion for turning adaptation on or off.

The importance of normalization and a variable forgetting factor has already been discussed earlier.

The modification with respect to scaling is concerned with improving the numerical properties of the algorithm but has no effect on the convergence properties of the algorithm. Specifically, a diagonal scaling matrix, S, is calculated according to a procedure due to Noble (16) such that the condition number of the matrix (SQ) is minimized. Note that $P = QQ^T$, i.e. the square root of P is scaled.

Property iii) concerns updating of the covariance matrix. The forgetting factor $\lambda(t)$ is selected so that the trace of the covariance matrix is constant. The following choice of $\lambda(t)$ ensures that

$0 < \lambda(t) \leq 1$ and that trace $P_s(\cdot)$ is constant.

$$\lambda(t) = 1 - \left[r(t) - \left\{ r(t)^2 - \frac{4\| P_s(t-1)\phi_{ns}(t) \|^2}{\mathrm{tr}\, P_s(t-1)} \right\}^{\frac{1}{2}} \right] / 2 \qquad (22)$$

where $r(t) = 1 + \phi_{ns}(t)\, P_s(t-1)\, \phi_{ns}(t)$; $P_s(\cdot)$ corresponds to the scaled covariance matrix; and $\phi_{ns}(\cdot)$ corresponds to the normalized and scaled regressor. The constant trace of $P_s(\cdot)$, as chosen by the user, effectively ensures an upper bound on the maximum eigenvalue. Notice also that unlike other modifications to the covariance update eqn.8, this modification determines the extent of discounting of old information in the current update of $P_s(t)$. In contrast to this, the covariance resetting modification adds constant diagonal elements to the P(t) update causing changes in their relative magnitudes and thus affects the direction and geometry of the overall parameter update.

The fourth and last property checks for information content in the new regressor vector and if through this measure the input $u(\cdot)$ and therefore $\phi(\cdot)$ is found to be sufficiently rich the adaptation is turned ON. Sripada and Fisher have shown that an approximate but good measure of the persistence of excitation of ϕ is the condition number of $P_s(\cdot)$, an upperbound for which is easily computed by computing $\| P_s(\cdot) \|_\infty \cdot \| P_s(\cdot) \|_\infty^{-1}$. For other computation details of the algorithm, the reader is referred to (21).

Figures 10 and 11 illustrate the excellent performance of this algorithm. The trace of the covariance matrix was maintained constant at 2 for all t, through the variable forgetting factor $\lambda(t)$ (cf. eqn. (22) and figure 11)).

2.10 RLS WITH DATA-DEPENDENT UPDATING AND FORGETTING FACTOR

In a very recent paper DasGupta and Huang (4) have proposed that for a realistic estimation problem one should minimize J_1 subject to the restriction that $(y-\hat{y})^2 \leq \nu^2$, where ν^2 is an upperbound on external disturbance sequence $\{\xi^2\}$ that enters the process, i.e.

$$y(t) = \theta^T \cdot \phi(t) + \xi(t); \quad \xi(t)^2 \leq \nu^2 \ \forall \ t \qquad (23)$$

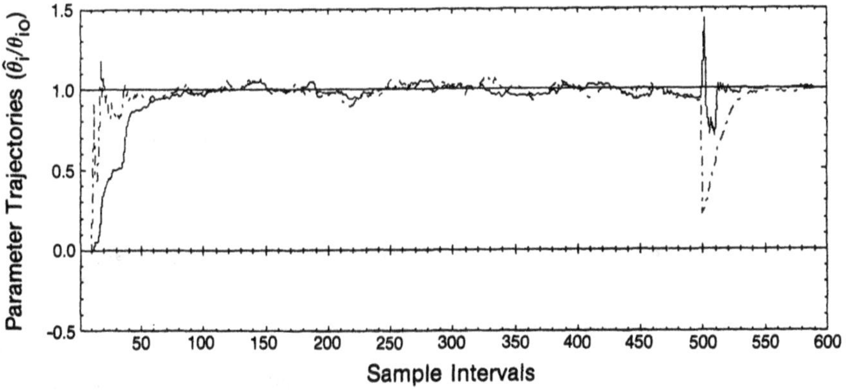

Figure 10: Parameter Trajectories for a Constant Trace RLS Algorithm Achieved through a variable forgetting factor.

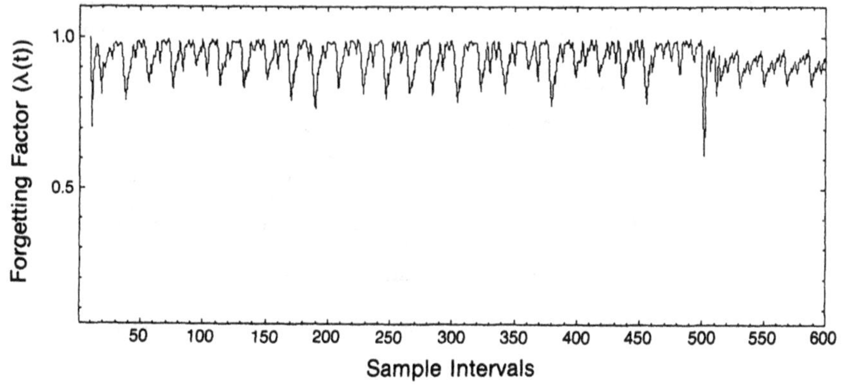

Figure 11: Variation of the Forgetting Factor for the Constant Trace RLS Algorithm

Because of the particular type of minimization function, a set of $\{\hat{\theta}(t)\}$ is obtained at each step rather than a unique $\hat{\theta}(\cdot)$. The set of $\{\hat{\theta}(t)\}$ obtained corresponds to an ellipsoid and the new estimate $\hat{\theta}(t)$ obtained is the centre of this ellipsoid. The key equations of their algorithm are:

$$P(t)^{-1} = (1-\lambda(t))P(t-1)^{-1} + \lambda(t)\phi(t)\cdot\phi(t)^T \tag{24}$$

$$\hat{\theta}(t) = \hat{\theta}(t-1) + \lambda(t)P(t)\phi(t)[\,y(t) - \hat{\theta}(t-1)^T\phi(t)\,] \tag{25}$$

In the above equation $\lambda(t)$ is determined on the basis of the information content in the data or the regressor. Notice also the similarity between $(1-\lambda(t))$ which is here derived as an information-dependent forgetting factor and the conventional forgetting factor defined earlier. The term, $\lambda(t)$, plays a dual role here. It affects the forgetting factor and it also acts as a gain update. Another algorithm based on a similar type of objective function has also been derived independently by Lozano-leal and Ortega (14).

The most important contribution of these two papers is that unlike other algorithms where dead-zones are introduced to facilitate convergence analysis in the presence of external disturbances, the dead-zone or the parameter stopping criterion arises naturally out of the form of the minimization function of the prediction errors.

2.11 RLS WITH LEAKAGE

The basic RLS algorithm in the presence of disturbances leads to parameter drift or bias. In fact a reoccurring disturbance may eventually lead to unbounded $\hat{\theta}(\cdot)$. To prevent this and to ensure bounded estimation parameters, Ioannou and Kokotovic (11) have introduced the idea of "σ-modification". This modification as explained below is also known by the term "leakage" in the signal processing literature. The parameter update equation with leakage is:

$$\begin{aligned} \hat{\theta}(t) &= (1-\nu)\hat{\theta}(t-1) + P(t)\phi(t)[\,y(t) - \hat{\theta}(t-1)^T\phi(t)\,] \\ 1-\nu &= \sigma;\ 0 \leq \nu < 1 \end{aligned} \tag{26}$$

The term $-\nu\,\hat{\theta}(t-1)$, corresponds to the amount of update that is "leaked". This term now ensures a stable parameter trajectory even in the presence of persistently occurring disturbances. One drawback of this approach is that the modified parameter update cannot guarantee convergence of estimated parameters to their "true" values even in the absence of disturbances. An additional modification due to M'Saad (15) is to make ν a function of time and to ensure parameter convergence to a user specified set of parameters, θ^*, based on a priori information. Alternately θ^* can be the last good set of estimated parameters which gave satisfactory control.

$$\hat{\theta}(t) = (1-\nu)\hat{\theta}(t-1)+P(t)\phi(t)[\ y(t)\ -\hat{\theta}(t-1)^T\ \phi(t)\ ]\ +\nu\ \theta^* \qquad (27)$$

where ν = 0 in the absence of disturbance or parameter drift

$0 < \nu < 1$ otherwise

2.12 RLS WITH PARAMETER WEIGHTING

In another recent study, Lambert (13) has introduced a novel and practical modification to classical RLS with exponential forgetting. His basic idea is to exercise control over the amount of parameter variation in one sample interval: $\|\hat{\theta}(t)-\hat{\theta}(t-1)\|$ or the variation of $\|\hat{\theta}(t)-\theta^*\|$ where θ^* is a user-specified parameter set which gives satisfactory control. Lambert achieves this through minimization of the following objective functions:

$$J_1 = \sum_{k=0}^{N} \lambda^{N-t}\ [y(t)\ -\ \theta^T\cdot\phi(t)]^2\ +\ \lambda_c(t)\,\|\ \hat{\theta}(t)\ -\ \theta^*\|^2 \qquad (28)$$

or

$$J_2 = \sum_{k=0}^{N} \lambda^{N-t}\ [y(t)\ -\ \theta^T\cdot\phi(t)]^2\ +\ \lambda_d(t)\,\|\ \hat{\theta}(t)\ -\ \hat{\theta}(t-1)\|^2 \qquad (29)$$

J_1 takes deviation of $\hat{\theta}(t)$ from the nominal value θ^* into account and effectively bounds it in a region the size of which depends on the weighting term $\lambda_c(t)$. J_2 on the other hand allows one to trade-off RLS with the projection algorithm and is certainly useful in a noisy or stochastic process. The corresponding parameter and covariance update equations are:

For J_1:

$$\hat{\theta}(t) = \hat{\theta}(t-1)+P(t)\phi(t)\varepsilon(t)\ +\ P(t)[\lambda_c(t)-\lambda\lambda_c(t-1)][\theta^* - \hat{\theta}\ (t-1)]$$

$$(\varepsilon(t)\ =\ y(t)\ -\ \hat{y}(t)) \qquad (30)$$

$$P(t)^{-1} = \lambda P(t-1)^{-1}\ +\phi(t)\phi(t)^T\ +[\lambda_c(t)\ -\lambda\lambda_c(t-1)]I \qquad (31)$$

For J_2:

$$\hat{\theta}(t)\ =\ \hat{\theta}(t-1)+P(t)\phi(t)\varepsilon(t)\ +\ P(t)\lambda\lambda_d(t)[\ \hat{\theta}(t-1)\ -\ \hat{\theta}(t-2)] \qquad (32)$$

$$P(t)^{-1} = \lambda P(t-1)^{-1}\ +\phi(t)\phi(t)^T\ +\ [\lambda_d(t)\ -\lambda\lambda_d(t-1)]\ I \qquad (33)$$

2.13 OTHER AD-HOC MODIFICATIONS

In addition to the above mentioned RLS estimation schemes there are many ad-hoc modifications that are in use in simulation and practical work in adaptive control. One example of such a modification is the following: a common concern in estimation is a method for bounding parameter estimates or to confine estimates into a small region given the necessary apriori information. One such scheme due to Egardt (5) is to project parameters into a sphere of centre θ_c and radius R.

$$\bar{\theta}(t) = \theta_c + (\hat{\theta}(t) - \theta_c) \max \left[1, \frac{R}{\| \hat{\theta}(t) - \theta_c \|} \right] \tag{34}$$

($\bar{\theta}(t)$ is the parameter update that would be used in subsequent control and estimation error calculations).

This mechanism ensures that

$$\bar{\theta}(t) = \hat{\theta}(t) \quad \text{whenever} \quad \| \hat{\theta}(t) - \theta_c \| \leq R$$

$$\text{and} \quad \bar{\theta}(t) = \theta_c + \frac{[\hat{\theta}(t) - \theta_c]}{\| \hat{\theta}(t) - \theta_c \|} R \quad \text{otherwise} \tag{35}$$

i.e. the parameters are projected onto the surface of the sphere if $\| \hat{\theta}(t) - \theta_c \| > R$. This is often a useful remedy in situations where a plant is subjected to disturbances that would cause large estimation jumps. The use of this 'projection' mechanism requires apriori knowledge of θ_c and R.

2.14 CONCLUSIONS

Many variations of the RLS algorithm exist in the literature and many more ad-hoc modifications are in use in simulation and practical work in adaptive control. This paper has surveyed some of the more useful variants of the RLS algorithm that retain the alertness or adaptivity of the algorithm in the presence of time-varying parameters, and improve the robustness of the algorithm in the presence of bounded disturbances and unmodelled dynamics.

However, this survey does not suggest the idea that there exists a single general purpose RLS algorithm. The choice of a particular set of modifications is usually problem specific.

REFERENCES

(1) Cluett, W.R., J.M. Martin-Sanchez, S.L. Shah, and D.G. Fisher, (1987). Stable discrete-time adaptive control in the presence of unmodelled dynamics. IEEE Trans of Auto Control. (In Press).

(2) Cluett, W.R., S.L. Shah and D.G. Fisher (1987). Robustness analysis of discrete-time adaptive control systems using input-output stability theory: a tutorial. Accepted for publication in Proc. of IEE, part D.

(3) Cordero, A.O. and D.Q. Mayne (1981). Deterministic convergence of a self-tuning regulator with a variable forgetting factor. Proc. of IEE, 128, Part D, 19-23.

(4) Dasgupta, S., and Y.F. Huang (1987). Asymptotically convergent modified recursive least-squares with data-dependent updating and forgetting factor for systems with bounded noise. IEEE Trans. on Information Theory. 33, 383-392.

(5) Egardt, B. (1979). Stability of Adaptive Controllers (Lecture notes in Control and Information Sciences, No. 20). Springer-Verlag, Berlin.

(6) Fortescue, T.R., L.S. Kershenbaum and B.E. Ydstie (1981). Implementation of self-tuning regulators with variable forgetting factors. Automatica, 17, 831-835.

(7) Gauss, K.F. (1809). Theoria motus corporum coelestium. English translation: Theory of the motion of the heavenly bodies. Dover, New York (1963).

(8) Goodwin, G.C., H. Elliott and E.K. Teoh (1983). Deterministic convergence of a self-tuning regulator with covariance resetting. Proc. of IEE, 130, Part D, 6-8.

(9) Goodwin, G.C. and K.S. Sin (1984). Adaptive filtering prediction and control. Prentice-Hall, New Jersey.

(10) Goodwin, G.C., D.J. Hill and M. Palaniswami (1985). Towards an adaptive robust controller. Proc. of IFAC Conf. on Identification and system parameter estimation, York, U.K. 997-1002.

(11) Ioannou, P.A. and P.V. Kokotovic (1983). Adaptive systems with reduced models (Lecture notes in control and information sciences, No. 47). Springer-Verlag, Berlin.

(12) Irving, E. (1979) Improving power network stability and unit stress with adaptive generator control. Automatica, 15, 31-46.

(13) Lambert, E. (1987). Process control applications of long-range prediction. D. Phil thesis submitted to Dept. of Engineering Science, Oxford University.

(14) Lozano-leal, R. and R. Ortega (1987) Reformulation of the parameter identification problem for systems with bounded disturbances. Automatica, 23, 247-251.

(15) M'Saad M., M. Duque and I.D. Landau (1986). Practical implications of recent results in robustness of adaptive control schemes. 1986 CDC Proc., Athens, Greece, 477-481 and private communication (1987).

(16) Noble, B. (1969). Applied Linear algebra. Englewood Cliffs, Prentice Hall.

(17) Ortega, R., L. Praly and I.D. Landau (1985). Robustness of discrete-time adaptive controllers, IEEE Trans. of Auto. Control, 30, 1179-1187.

(18) Plackett, R.L. (1950). Some theorems in Least Squares. ometrika, Vol. 37, p. 149.

(19) Praly, L. (1983) Robustness of Model reference adaptive control. Proc. of 3rd Yale Workshop on Adaptive Control, New Haven, CT. June 15-17.

(20) Rogers, M. (1987). Process model representation and parameter identification for adaptive control. M.Sc. thesis (in preparation), Dept. of Chemical Engineering, U. of Alberta.

(21) Sripada, R. and D.G. Fisher (1987). Improved Least Squares Identification. In Press - Int. J. of Control.

(22) Ydstie, B.E., L.S. Kershenbaum and R.W.H. Sargent (1985). Theory and application of an extended horizon self-tuning controller. AIChE J., 31, 1771-1780.

(23) Ljung, L. and T. Soderstrom (1983). Theory and practice of recursive identification. M.I.T. press, Cambridge, MA.

Chapter 3

LQG based self-tuning controllers

K. J. Hunt and M. J. Grimble

3.1 INTRODUCTION

There are several approaches to the design of control systems using stochastic optimal control theory ('LQG' control). The method followed in this chapter is the polynomial equation approach developed by Kučera (1979) in which the design procedure reduces to the solution of polynomial equations. These equations can be solved using fast and efficient numerical algorithms.

The polynomial approach to LQG control offers a flexible design method which can be readily used as the basis of a self-tuning control algorithm. A comprehensive account of the LQG self-tuning control technique is given in this chapter. The new algorithm presented contains some refinements and important extensions of the earlier work by Grimble (1984) and Hunt et al (1986). In particular, the optimal tracking problem in the presence of a measurable disturbance has recently been solved by Grimble (1986) and Šebek et al (1987). The solution naturally involves the use of a feedforward compensator and is described here in the self-tuning control framework for the first time.

A further feature of the design method presented is the use of frequency-dependent weighting elements in the cost function. The dynamic weights allow the frequency-response of the closed-loop system to be shaped in a straightforward manner.

The optimal controller consists of three parts (feedback, reference and feedforward) which process the system output, reference and measurable disturbance signals separately. Three possible design strategies are proposed:

(i) The complete general solution of the optimal control problem which involves three couples of polynomial equations, one couple being associated with each part of the controller.

(ii) Each couple of polynomial equations can be reduced to a single 'implied' equation. Under certain stated conditions the three implied polynomial equations can be solved to obtain the unique optimal controller. Solution of the implied equations is computationally simpler than solution of the original couples.

(iii) In the optimal control design the feedback part of the controller is independent of the reference and feedforward parts. A third strategy is proposed in which the optimal feedback controller is calculated in the normal way and then the reference and feedforward parts are calculated non-optimally using steady-state considerations. The method is useful in situations where the available computation time is short, as the reference and feedforward polynomial equations no longer need to be solved.

The robustness properties of the LQG self-tuner are discussed by summarising the important features of the control design and various techniques which have been used to achieve robust parameter estimation. The main results of a recent convergence analysis are also presented.

The chapter concludes with a discussion of practical issues relating to control law implementation, cost-function weight selection and computational issues.

A complete treatment of the theory and application of the proposed LQG self-tuner can be found in Hunt (1987) from which this chapter is drawn. This reference contains an investigation of the application of LQG self-tuning control in power systems. An extensive simulation study and comparison with other methods is given in Jones (1987).

Notation

All systems considered are assumed to be linear, time-invariant and discrete-time. The systems are described by means of real polynomials in the delay operator d.

For simplicity the arguments of polynomials in d are often ommitted such that X(d) is denoted by X. The conjugate of a polynomial X(d) is written as $X^*(d) \triangleq X(d^{-1})$, or simply X^*. (Editor's note $d \equiv q^{-1}$)

The absolute coefficient of X is denoted by <X>. Stable (strictly Hurwitz) polynomials are those with all their zeros strictly outside the unit circle of the d-plane. The power spectrum of a signal x(t) is denoted by ϕ_x.

3.2 MODEL STRUCTURE

The open-loop model for the single-input single-output plant under consideration is shown in Figure 3.1 The plant is governed by the equation:

$$y(t) = p(t) + x(t) + d(t) \tag{3.1}$$

$$= W_p u(t) + W_x \ell(t) + W_d \phi_d(t) \tag{3.2}$$

where y(t) is the output to be controlled, u(t) is the plant control input, $\phi_d(t)$ is an unmeasurable disturbance, and $\ell(t)$ is a disturbance which is available for measurement. Denoting the least-common-denominator of W_p, W_x and W_d as A, these sub-systems may be expressed as:

$$W_p = A^{-1}B \tag{3.3}$$

$$W_d = A^{-1}C \tag{3.4}$$

$$W_x = A^{-1}D \tag{3.5}$$

where A,B,C and D are polynomials in the delay operator d.

Reference generator

The system output y(t) is required to follow as closely as possible a reference signal r(t). The signal r(t) is represented as the output of a generating sub-system W_r which is driven by an external stochastic signal $\phi_r(t)$:

$$r(t) = W_r\phi_r(t) \tag{3.6}$$

The sub-system W_r is represented in polynomial form as:

$$W_r = A_e^{-1}E_r \tag{3.7}$$

where A_e and E_r are polynomials in d.

The tracking error e(t) is defined as:

$$e(t) \triangleq r(t) - y(t) \tag{3.8}$$

Any common factors of A_e and A are denoted by D_e such that:

$$A_e = D_eA_{ec}' \text{ , } A = D_eA' \tag{3.9}$$

Measurable disturbance generator

The measurable disturbance signal $\ell(t)$ may be represented as the output of a generating sub-system W_ℓ driven by an external stochastic signal $\phi_\ell(t)$;

$$\ell(t) = W_\ell\phi_\ell(t) \tag{3.10}$$

The sub-system W_ℓ is represented in polynomial form as:

$$W_\ell = A_\ell^{-1}E_\ell \tag{3.11}$$

where A_ℓ and E_ℓ are polynomials in d.

Assumptions

1. Each sub-system is free of unstable hidden modes.

2. The plant input-output transfer-function W_p is assumed strictly causal i.e. $\langle B\rangle = 0$.

3. The polynomials A and B must have no unstable common factors.

4. Any unstable factors of A_e must also be factors of A.

5. Any unstable factors of A_ℓ must also be factors of both A and D.

6. The polynomials C, E_r and E_ℓ may, without loss of generality, be assumed stable.

These assumptions, together with the assumptions on the cost-function weighting elements given in the following section, amount to the solvability conditions for the optimal control problem. By making these assumptions, therefore, we ensure that a solution to any given problem exists. The question of artificially ensuring problem solvability is discussed in Section 3.7.3.

Different types of reference and disturbance signals (r(t), $\ell(t)$ and d(t) in Figure 3.1) may be admitted by appropriate definition of the external stochastic signals, $\psi_r(t)$, $\psi_\ell(t)$ and $\psi_d(t)$, namely:

(i) Coloured zero-mean signals are generated when the driving source (ψ_r, ψ_ℓ or ψ_d) is a zero-mean white noise sequence and the filter (W_r, W_ℓ or W_d) is asymptotically stable.

(ii) Random walk sequences are generated when the driving source is a zero-mean white noise sequence and the filter has a denominator 1-d.

(iii) Step-like sequences consisting of random steps at random times are generated when the driving source is a Poisson process and the filter has a denominator 1-d.

(iv) Deterministic sequences (such as steps, ramps or sinusoids) are generated when the driving source is a unit pulse sequence and the filter has poles on the unit circle of the d-plane.

3.3 CONTROLLER DESIGN

In the closed-loop system a two-degrees-of-freedom (2DF) control structure is used. In addition, a feedforward compensator is employed to counter the effect of the measurable disturbance $\ell(t)$.

Controller structure

The closed-loop system is shown in Figure 3.2. The control law is given by:

$$u(t) = -C_{fb}y(t) + C_r r(t) - C_{ff}\ell(t) \tag{3.12}$$

where the feedback controller C_{fb}, the reference controller C_r, and the feedforward controller C_{ff} may be expressed a ratios of polynomials in the delay operator d as:

$$C_{fb} = C_{fbd}^{-1}C_{fbn} \tag{3.13}$$

$$C_r = C_{rd}^{-1}C_{rn} \tag{3.14}$$

$$C_{ff} = C_{ffd}^{-1}C_{ffn} \tag{3.15}$$

Cost function

The desired optimal controller evolves from minimisation of the cost-function:

$$J = E\{(H_q e)^2(t) + (H_r u)^2(t)\} \quad (3.16)$$

where H_q and H_r are dynamic (i.e. frequency dependent) weighting elements which may be realised by rational transfer-functions.

Using Parseval's Theorem the cost-function may be transformed to the frequency domain and expressed as:

$$J = \frac{1}{2\pi j} \oint_{|z|=1} \{Q_c \phi_e + R_c \phi_u\} \frac{dz}{z} \quad (3.17)$$

where ϕ_e and ϕ_u are the tracking error and control input spectral densities, respectively, and:

$$Q_c = H_q H_q^* \ , \ R_c = H_r H_r^* \quad (3.18)$$

The weighting elements Q_c and R_c may be expressed as ratios of polynomials in the delay operator d as:

$$Q_c \triangleq \frac{B_q^* B_q}{A_q^* A_q} \ , \ R_c \triangleq \frac{B_r^* B_r}{A_r^* A_r} \quad (3.19)$$

Assumptions

1. The weighting elements Q_c and R_c are strictly positive on $|d|=1$.

2. A_q, B_q, B_r and A_r are stable polynomials.

3.3.1 Optimal Control Law

The stable spectral factor D_c is defined by:

$$D_c^* D_c = B^* A_r^* B_q^* B_q A_r B + A^* A_q^* B_r^* B_r A_q A \quad (3.20)$$

The feedback, reference and feedforward parts of the control law (3.12) which minimises the cost-function (3.17) are as follows:

(i) Optimal feedback controller

$$C_{fb} = \frac{GA_r}{HA_q} \quad (3.21)$$

where G, H (along with F) is the solution having the property:

$(D_c^* z^{-g})^{-1}F$ strictly proper

of the polynomial equations:

$$D_c^* z^{-g} G + FAA_q = B^* A_r^* B_q^* B_q C z^{-g} \tag{3.22}$$

$$D_c^* z^{-g} H - FBA_r = A^* A_q^* B_r^* B_r C z^{-g} \tag{3.23}$$

where $g > 0$ is the smallest integer which makes the equations (3.22)-(3.23) polynomial in d.

(ii) Optimal reference controller

$$C_r = \frac{MA_r C}{E_r HA_q} \tag{3.24}$$

where M (along with N and Q) is the solution having the property:

$(D_c^* z^{-g})^{-1}N$ strictly proper

of the polynomial equations:

$$D_c^* z^{-g} M + NA_q A_e = B^* A_r^* B_q^* B_q E_r z^{-g} \tag{3.25}$$

$$D_c^* z^{-g} Q - NBA_r A_{ec}' = A^* A_q^* B_r^* B_r A' E_r z^{-g} \tag{3.26}$$

(iii) Optimal feedforward controller

$$C_{ff} = \frac{A_r(XC - GE_\ell D)}{HA_q E_\ell A} \tag{3.27}$$

where X (along with Z and Y) is the solution having the property:

$(D_c^* z^{-g})^{-1} Z$ strictly proper

of the polynomial equations:

$$D_c^* z^{-g} X + ZAA_q A_\ell = B^* A_r^* B_q^* B_q DE_\ell z^{-g} \tag{3.28}$$

$$D_c^* z^{-g} Y - ZBA_r A_\ell = A^* A_q^* B_r^* B_r DE_\ell z^{-g} \tag{3.29}$$

3.3.2 Implied Diophantine Equations

The general solution of the optimal control problem involves three couples of polynomial equations, one couple for each part of the control law. However, under certain conditions each couple can be replaced by a single, related, equation. Optimality of the implied diophantine equations requires the following additional assumptions:

Assumptions

1. The disturbance sub-systems $A^{-1}C$, $A^{-1}D$, $A_e^{-1}E_r$ and $A_\ell^{-1}E_\ell$ are assumed to be proper rational transfer-functions.

2. The cost-function terms $A_q^{-1}B_q$ and $A_r^{-1}B_r$ are assumed to be proper rational transfer-functions.

3. The polynomial pairs A_q, A_r, A_q, B and A_r, A each assumed to be coprime.

We can easily ensure that Assumptions 2 and 3 are satisfied by appropriate choice of the cost-function weights.

(i) Implied feedback equation

The polynomials G and H in equations (3.22)-(3.23) also satisfy the polynomial equation:

$$AA_qH + BA_rG = D_cC \tag{3.30}$$

The optimal feedback controller polynomials G and H are determined uniquely by the solution of equation (3.30) having the property:

$(AA_q)^{-1}G$ strictly proper

iff the polynomials A and B are coprime.

(ii) Implied reference equation

The polynomials M and Q in equations (3.25)-(3.26) also satisfy the polynomial equation:

$$D_eA_qQ + BA_rM = D_cE_r \tag{3.31}$$

The optimal reference controller polynomial M is determined uniquely by the solution of equation (3.31) having the property:

$(A_eA_q)^{-1}M$ strictly proper

iff A_e is a divisor of the denominator of the plant input-output transfer-function, W_p. When this condition holds then, from equation (3.9), $D_e = A_e$ and the implied equation (3.31) becomes:

$$A_eA_qQ + BA_rM = D_cE_r \tag{3.32}$$

(iii) Implied feedforward equation

The polynomials X and Y in equations (3.28)-(3.29) also satisfy the polynomial equation:

$$AA_qY + BA_rX = D_cDE_\ell \tag{3.33}$$

Assume now that A_ℓ is a divisor of both A and D. From equations (3.28)-(3.29) A_ℓ must then also divide both X and Y. The implied feedforward diophantine equation (3.33) becomes:

$$AA_qY' + BA_rX' = D_cD'E_\ell \tag{3.34}$$

where:

$$Y \underline{\underline{\Delta}} A_\ell Y', \; X \underline{\underline{\Delta}} A_\ell X', \; D \underline{\underline{\Delta}} A_\ell D' \tag{3.35}$$

The conditions for optimality of the implied feedforward diophantine equation may now be stated as follows; the optimal feedforward controller polynomial X is given by
$X = A_\ell X'$ where X' is determined uniquely by the solution of equation (3.34) having the property:

$$(AA_q)^{-1}X' \quad \text{strictly proper}$$

iff the polynomials A and B are coprime and the polynomial A_ℓ is a divisor of both A and D.

Discussion

The condition that the A and B polynomials must be coprime is required in order that the implied feedback and feedforward diophantine equations uniquely determine the optimal feedback and feedforward parts of the controller. This condition means that all the poles of the disturbance sub-systems W_d and W_x must also be poles of the plant input-output transfer-function W_p.

The condition required for optimality of the implied reference equation is that A_e, the reference generator denominator, must divide the plant input-output transfer-function W_p. In the case of the unstable reference generators of greatest practical interest (such as steps, ramps etc), this condition corresponds to one of the optimal control problem solvability conditions (see Assumption 4 in Section 3.2). If, therefore, the reference generator is unstable and the optimal control problem is solvable (by satisfying assumptions 1-6 in Section 3.2 and Assumptions 1-2 in Section 3.3 we can ensure that the problem is solvable) then the implied reference diophantine equation will uniquely determine the optimal reference controller.

The conditions required for optimality of the feedforward diophantine equation are that A and B must be coprime and that A_ℓ, the measurable disturbance generator denominator, must divide both A and D. For the unstable disturbance generators of practical importance the condition that A_ℓ must divide A and D again corresponds to one of the optimal control problem solvability conditions (see Assumption 5 in Section 3.2). If, therefore, the measurable disturbance generator is unstable and the optimal control problem is solvable then the condition for optimality of the implied feedforward diophantine equation reduces to the condition that A and B must be coprime (this is the same condition required for optimality of the implied feedback diophantine equation).

The conditions relating to optimality of the implied feedback and reference equations in the scalar case have been known for some time (Šebek and Kučera 1982, Kučera 1984). The corresponding result for the feedback equation in the multivariable case has recently been derived by Hunt et al (1987). A proof of all three results (the feedforward result is new) can be found in Hunt (1987).

The question of ensuring problem solvability is discussed in Section 3.7.3.

3.3.3 Simplifed Design

The control structure used allows flexibility in the design of the reference and feedforward parts of the controller. Having designed the optimal feedback controller the reference and feedforward parts can then, if desired, be designed independently of the feedback properties of the system. This option may be important in cases where the full LQG design cannot be implemented due to computational constraints. In such cases the reference and feedforward parts of the controller can be designed to ensure proper steady-state performance as follows:

(i) Reference controller

The reference controller may be defined as:

$$C_r = \frac{\gamma A_r C}{H A_q} \tag{3.36}$$

where the scalar γ replaces the polynomial M in equation (3.24). From the closed-loop system structure shown in Figure 3.2 the transfer-function between the reference signal r(t) and the controlled output y(t) may be calculated as:

$$T_{y/r} = \frac{\gamma B}{D_c} \tag{3.37}$$

The scalar γ is chosen so as to ensure unity steady-state gain between r(t) and y(t) as:

$$\gamma = \frac{D_c(1)}{B(1)} \tag{3.38}$$

Shaping of the command response may also be achieved by cascading a unity-gain shaping filter with the reference controller.

(ii) Feedforward controller

From the closed-loop system structure shown in Figure 3.2 the transfer-function between the measurable disturbance $\ell(t)$ and the controlled output y(t) may be calculated as:

$$T_{y/\ell} = \frac{C_{fbd}(DC_{ffd} - BC_{ffn})}{C_{ffd}(AC_{fbd} + BC_{fbn})} \tag{3.39}$$

The effect of $\ell(t)$ on the output is eliminated when:

$$DC_{ffd} - BC_{ffn} = 0 \tag{3.40}$$

or, when the feedforward controller is defined as:

$$C_{ff} = \frac{D}{B} \tag{3.41}$$

There are, however, two major problems with this design:

(i) Whenever the delay associated with the measurable disturbance filter D/A is less than the delay in the plant input-output transfer-function B/A a non-causal controller results.

(ii) The feedforward controller is unstable whenever the plant input-output transfer-function B/A is inverse unstable i.e. when B(d) has zeros inside the unit circle of the d-plane.

A solution to these problems is to sacrifice the transient properties of the feedforward controller in favour of a static feedforward design which is calculated to ensure the elimination of the measurable disturbance in steady-state:

$$C_{ff} = \frac{D(1)}{B(1)} \tag{3.42}$$

The two problems of the non-optimal design outlined above demonstrate a clear advantage of the optimal feedforward design since a causal and stable feedforward controller will always result in the optimal design regardless of the relative delays in the B and D polynomials, and regardless of the positions of the zeros of B.

3.4 LQG SELF-TUNING CONTROL ALGORITHM

An explicit LQG self-tuning controller may be constructed using the certainty equivalence argument, where the A,B,C and D polynomials in the LQG design of Section 3.3 are replaced by their estimated values $\hat{A},\hat{B},\hat{C}$ and $\hat{D}$.

The explicit LQG self-tuning control algorithm may be summarised as follows:

Step 1 : Choose cost-function weights

Step 2 : Update estimates of A,B,C and D polynomials.

Step 3 : Solve spectral factorisation (3.20) for D_c.

Step 4 : Solve equations (3.22)-(3.23) for G and H, and form feedback controller according to equation (3.21)

Step 5 : Solve equations (3.25)-(3.26) for M, and form reference controller according to equation (3.24).

Step 6 : Solve equations (3.28)-(3.29) for X, and form feedforward controller according to equation (3.27)

Step 7 : Calculate and implement new control signal.

Step 8 : Goto Step 1 at next sample instant.

In steps 4-6 it may be possible to solve the implied diophantine equations to obtain the optimal controller polynomials, as discussed in Section 3.3.2. Similarly, in steps 5-6 it may be desirable to use the steady-state designs outlined in Section 3.3.3

3.5 ROBUSTNESS OF THE LQG SELF-TUNER

The subject of robustness of self-tuning controllers is one which has generated a great deal of discussion and controversy in recent years. This discussion has largely been stimulated by the paper of Rohrs et al (1982) which analysed the robustness properties of the self-tuning regulator (STR) and of model reference adaptive controllers (MRAC) in the presence of unmodelled dynamics and disturbances. These authors concluded that adaptive controllers are inherently non-robust and this stimulated a very active debate leading to some useful insights into the robustness question.

Åström (1983b) directly challenged the allegations of Rohrs, while Goodwin et al (1985) pointed out that the approach of analysing existing high performance adaptive controllers would almost certainly reveal poor robustness properties.

Åström (1983a) and Goodwin et al (1985) take a more pragmatic approach to the robustness issue by attempting to obtain a robust adaptive controller by combining a robust control law with a robust identification algorithm.

In the following discussion the robustness properties of the LQG design are examined and methods of achieving robust parameter estimation are discussed.

3.5.1 Robustness of the LQG Design

The performance and properties of feedback control systems have long been understood by control engineers in terms of the frequency responses of the various system transfer-functions (Truxal 1955, Horowitz, 1963). The main ideas of conventional frequency-response design methods have recently been supported by theoretical analysis (Doyle and Stein, 1981).

These ideas can be summarised with reference to a typical Bode plot of the loop gain as shown in Figure 3.3. It is well understood that there are three important frequency regions:

(i) The low-frequency region where the loop gain should be high to achieve good command response, disturbance rejection and robust performance properties.

(ii) The crossover region where the stability margins should be adequate.

(iii) The high-frequency region where the loop gain should fall off rapidly to achieve robust stability (i.e. insensitivity to unmodelled high-frequency dynamics) and insensitivity to measurement noise.

Any competent design of a digital control system should include anti-aliasing filters to eliminate signal transmission above the Nyquist frequency (Aström and Wittenmark, 1984). The high-frequency properties of the controller will therefore depend critically upon the sampling period.

The relevant features of the LQG controller may be investigated by summarising the properties of the design presented in Section 3.3:

(i) The feedback, reference and feedforward parts of the controller each have poles due to the A_q weighting term, and zeros due to the A_r weighting term (see equations (3.21), (3.24) and (3.27). Thus, loop-shaping may easily be achieved by manipulation of the cost function weights. In particular, the desired high gain at low-frequency can be achieved by introducing integral action when the A_q term is chosen as $A_q = 1-\nabla d$, $\nabla \to 1$.

A particularly simple formulation of the weighting terms Q_c and R_c which requires the selection of only two parameters is presented in Section 3.7.2

(ii) The stability margins for the closed-loop system can be examined using the implied feedback diophantine equation (3.30):

$$AA_qH + BA_rG = D_cC \triangleq T \tag{3.43}$$

From the feedback controller equation (3.21) and the closed-loop system model in Figure 3.2 it may easily be verified that equation (3.43) is the characteristic equation of the closed-loop system, where T is defined as the characteristic polynomial. This shows that the nominal closed-loop system is guaranteed to be stable, since the polynomials D_c and C are by definition stable.

This result should be contrasted with the stability properties of the Self-Tuning Regulator (Åström and Wittenmark, 1973 and the Self-Tuning Controller (Clarke and Gawthrop, 1975, 1979). It is possible that the closed-loop system for these control laws will be nominally unstable, particularly when the controlled process has zeros inside the unit circle in the d-plane.

Equation (3.43) also shows that standard pole-assignment may be obtained as a by-product of the LQG algorithm by solving the equation:

$$AA_qH + BA_rG = T_cC \tag{3.44}$$

where T_c is chosen as the desired closed-loop pole polynomial. Notice that in this formulation of the pole-assignment algorithm the loop-shaping properties of the LQG design are retained since the A_q and A_r polynomials remain in equation (3.44). Use of the pole-assignment algorithm introduces computational savings since the spectral factorisation (3.20) is no longer required, and in subsequent calculations the polynomial D_c is replaced by T_c.

3.5.2 Robust Parameter Estimation

In this section the recursive Extended-Least-Squares (ELS) estimation algorithm is described and the methods which can be used to achieve robust parameter tracking are briefly reviewed. The plant model (3.2) is re-written in the approximate form:

$$y(t) = \underline{\phi}^T(t)\underline{\theta}(t) + \psi_d(t) \tag{3.45}$$

where the parameter vector $\underline{\theta}(t)$ and regression vector $\underline{\phi}(t)$ are defined by:

$$\underline{\theta}^T(t) = [a_1 \ldots a_{na};\ b_o \ldots b_{nb};\ d_o \ldots d_{nd}; c_1 \ldots c_{nc}] \tag{3.46}$$

$$\underline{\phi}^T(t) = [-y(t-1) \ldots -y(t-na); u(t-k_1) \ldots u(t-k_1-nb);$$
$$\ell(t-k_2) \ldots \ell(t-k_2-nd); v(t-1) \ldots v(t-nc)] \tag{3.47}$$

The parameters, $a_\cdot, b_\cdot, c_\cdot$ and $d_\cdot$ are the coefficients of the polynomials A, B, C and D. k_1 and k_2 represent the time-delays in the sub-systems W_p and W_x as integer multiples of the sample period. The signal v(t) is a proxy to the unmeasurable signal $\psi_d(t)$, defined by:

$$v(t) = y(t) - \underline{\phi}^T(t)\underline{\hat{\theta}}(t) \tag{3.48}$$

where $\underline{\hat{\theta}}(t)$ is the vector of estimated parameters.

The recursive ELS algorithm may be summarized as follows:

$$P(t) = \frac{1}{\lambda(t)}\left(P(t-1) - \frac{P(t-1)\underline{\phi}(t)\underline{\phi}^T(t)P(t-1)}{\lambda(t) + \underline{\phi}^T(t)P(t-1)\underline{\phi}(t)}\right) \tag{3.49}$$

$$\underline{k}(t) = P(t)\underline{\phi}(t) = \frac{P(t-1)\underline{\phi}(t)}{\lambda(t) + \underline{\phi}^T(t)P(t-1)\underline{\phi}(t)} \tag{3.50}$$

$$v(t) = y(t) - \underline{\phi}^T(t)\underline{\hat{\theta}}(t) \tag{3.51}$$

$$\underline{\hat{\theta}}(t) = \underline{\hat{\theta}}(t-1) + \underline{k}(t)v(t) \tag{3.52}$$

In the above algorithm the 'forgetting factor' $\lambda(t)$ (where $0 < \lambda(t) \leq 1$) weights the measurements, such that a measurement received n samples ago will have a weighting proportional to λ^n (assuming a constant forgetting factor $\lambda(t) = \lambda$).

The constant forgetting factor technique for parameter tracking has frequently been used in self-tuning control algorithms. The method, however, has some potential implementation difficulties:

(i) If the algorithm is to remain capable of tracking sudden parameter changes the updating gain $\underline{k}$ must be prevented from becoming too small as the parameter estimates improve. Moreover, if good data is arriving and $\underline{k}$ becomes small, equation (3.49) implies that P is near singular and roundoff error over many updating steps may cause the computed P to become indefinite and the algorithm to break down (the UD factorisation technique (Bierman, 1977) is normally used to alleviate this problem). Thus, λ must not stay too close to unity.

(ii) On the other hand, when λ is less than 1 and little new information on $\underline{\theta}$ is being brought in by the observations, equation (3.49) shows that P may increase as λ^{-1} (the well known 'burst phenomenon', sometimes also known as 'estimator wind-up'). If P becomes large in this way then observation noise, or a sudden increase in information, may induce large spurious variations in $\hat{\underline{\theta}}$.

When using the constant forgetting factor method, therefore, choice of λ is a difficult and often unsatisfactory compromise. Many methods of adjusting $\lambda(t)$ automatically in the recursion have been devised (Åström 1980, Fortescue et al 1981, Wellstead and Sanoff, 1981). Alternatively, P may be adjusted directly. For instance, a constant matrix, which can be interpreted as the covariance of random increments in the parameters, may be added at each step then some upper bound applied to P, or the new P may be formed as a weighted sum of the old P and the identity matrix I, the weights being chosen to keep trace P constant (Goodwin et al, 1985). Other methods are described by Egardt (1979), Hagglund (1983), Kulhavy and Karny (1984) and Andersson (1985).

Chen and Norton (1987) have recently described a new parameter tracking method which enables the recursive ELS algorithm to track both abrupt and smooth parameter changes. The method differs from methods based on a scalar forgetting factor by its use of vectors to detect parameter variation, which then results in the relevant element in the updating gain vector being incremented. It also incorporates a test to determine when parameter updating should be suspended so as to avoid divergence when little new information about the parameters is arriving.

The algorithm does not, therefore, have the drawbacks associated with the constant forgetting factor method. The algorithm also embodies one of the key ideas behind robust estimation, namely that estimation should only be performed when 'good' data is arriving.

The method has been described in the context of LQG self-tuning control by Hunt et al (1986).

3.6 CONVERGENCE PROPERTIES

One of the key theoretical problems in self-tuning control which has received growing attention in recent years is convergence analysis. Chen and Caines (1984) and Moore (1984) have tackled the problem for state-equation based LQG algorithms.

A global convergence result for explicit polynomial based LQG self-tuning control algorithms of the type under discussion in this chapter has recently been derived and will be summarised in the following. The result relates to the regulator case (i.e. $r(t) = \ell(t) = 0$) and to a stochastic approximation type identification algorithm. The result is due to Grimble (1986).

To guarantee the convergence properties of the algorithm the following assumptions must be made:

Assumptions

1. Upper bounds $n_a = n$, $n_b = m$ and $n_c = q$ on the polynomials A,B and C are known.

2. The polynomial $C - k'/2$ is input strictly passive (strictly positive real) for some real positive constant k'.

3. There exists a finite T_2 such that $\hat{C}(t;d)$ remains stable for all $t < T_2$.

4. Any common roots of $\hat{A}(t;d)$ and $\hat{B}(t;d)$ are strictly and uniformly outside the unit circle of the d-plane as $t \to \infty$, with probability one.

Remarks

(a) Assumption 1 is generally valid and has the useful property that the transport delay need not be known exactly.

(b) A positive real condition such as Assumption 2 often arises as a sufficient condition for the convergence of recursive parameter estimation schemes.

The global convergence theorem for the explict LQG self-tuner may now be stated:

Theorem

Subject to Assumptions 1-4 the explicit LQG self-tuning regulator, using a stochastic approximation identification algorithm, is globally convergent in the following sense:

$$\text{(i)} \quad \lim_{T\to\infty} \sup \frac{1}{T} \sum_{t=1}^{T} y^2(t) < \infty \quad \text{a.s.} \tag{3.53}$$

$$\text{(ii)} \quad \lim_{T\to\infty} \sup \frac{1}{T} \sum_{t=1}^{T} u^2(t) < \infty \quad \text{a.s.} \tag{3.54}$$

(iii) $$\lim_{T\to\infty} \frac{1}{T} \sum_{t=1}^{T} E[(y(t)-\hat{y}(t/t-1))^2/F_{t-1}] = 1 \quad a.s. \tag{3.55}$$

(iv) The closed-loop characteristic polynomial T(t;d) converges in the sense:

$$\lim_{T\to\infty} \frac{1}{T} \sum_{t=1}^{T} E[T(t;d)y(t)-H(t;d)\hat{C}(t-1;d)\varepsilon(t)]^2 = 0 \quad a.s. \tag{3.56}$$

Proof: Grimble (1986)

Remarks

(a) Parts (i) and (ii) of the theorem ensure that the output and control signals are bounded.

(b) If the system parameters and past values of $\varepsilon(t)$ are known the prediction error is obtained as 1. Part (iii) of the theorem shows that this is obtained asympotically.

(c) From the system model shown in Figure 3.2, and from equations (3.21) and (3.43), the transfer-function between the disturbance $\psi_d(t)$ and the output y(t) may be written:

$$y(t) = \frac{CH}{T} \psi_d(t) \tag{3.57}$$

which may be re-written in the form:

$$Ty(t) - CH\psi_d(t) = 0 \tag{3.58}$$

This equation allows the convergence result in part (iv) of the proof (equation (3.56)) to be more easily interpreted.

(d) The theorem remains substantially unchanged if Assumption 4 is replaced by a weaker assumption and the identification algorithm is replaced by extended least-squares (see Grimble, 1986).

3.7 PRACTICAL ISSUES

3.7.1 Control Law Implementation

When implementing the control law (3.12) it is necessary to include in the forward path the term A_r/HA_q since this term, which is common to each part of the controller (see equations (3.21), (3.24) and (3.27)), is not necessarily stable. Thus, the control law should be implemented using the equivalent structure shown in Figure 3.4 where:

$$C_{fb} \triangleq C'_{fb} \frac{A_r}{HA_q} \tag{3.59}$$

$$C_r \triangleq C'_r \frac{A_r}{HA_q} \tag{3.60}$$

$$C_{ff} \triangleq C'_{ff} \frac{A_r}{HA_q} \tag{3.61}$$

From equations (3.21), (3.24) and (3.27)

$$C'_{fb} = G \tag{3.62}$$

$$C'_r = \frac{MC}{E_r} \tag{3.63}$$

$$C'_{ff} = \frac{XC - GE_\ell D}{E_\ell A} \tag{3.64}$$

It is also necessary to show that C'_{ff} in equation (3.64) is stable. To this end, multiply equations (3.22)-(3.23) by DE_ℓ and equations (3.28)-(3.29) by C. When the resulting equations are compared the following equations, after some algebraic manipulation, are obtained;

$$\frac{CX - GE_\ell D}{A} = \frac{A_q(FDE_\ell - ZA_\ell C)}{D_c^* z^{-g}} = \frac{A_q(HDE_\ell - YC)}{BA_r} \tag{3.65}$$

By assumption the pairs A,B and A,A_r can have no unstable common factors. Since D_c^* is unstable (due to equation (3.20)) all the fractions in (3.65) are, in fact, polynomials. Thus C'_{ff} in equation (3.64) is stable (E_ℓ is stable by definition).

3.7.2 Cost-function Weight Selection

In the LQG optimal controller design the cost-function weighting elements, Q_c and R_c, are the major design parameters which must be selected by the system user. Perhaps one of the key practical objectives in any self-tuning control algorithm is to reduce the number of design parameters (the 'on-line tuning knobs') to a minimum, and to give these parameters a clear physical interpretation.

In the cost-function of equation (3.17) there are many ways to choose the weighting elements, allowing various performance characteristics and loop-shaping properties to be achieved. However, a straightforward technique appropriate for self-tuning systems which involves only two design parameters, each with a simple interpretation, is described in the following.

As previously mentioned, the controller will have integral action when the error weighting denominator has the form $A_q = 1 - d$. Since in the majority of practical problems it is desirable to include integral action in the controller the following definition for the cost-function weights is appropriate:

$$Q_c = \frac{B_q^* B_q}{A_q^* A_q} = \frac{(1-\beta d)^*(1-\beta d)}{(1-\beta)^2(1-d)^*(1-d)} \tag{3.66}$$

$$R_c = \frac{B_r^* B_r}{A_r^* A_r} = \frac{1}{\rho} \tag{3.67}$$

These definitions correspond to the following:

$$B_q = (1-\beta d)/(1-\beta) \tag{3.68}$$

$$A_q = 1 - d \tag{3.69}$$

$$B_r = 1 \tag{3.70}$$

$$A_r = \rho^{1/2} \tag{3.71}$$

In this formulation the scalars ρ and β effectively become the on-line tuning parameters. The interpretation of their effect is straightforward : the control weighting $1/\rho$ varies the relative magnitude of error and control penalty while β determines the amount of integral action. As ρ is increased (i.e. the control weight is decreased) the error signal will be decreased at the expense of increased control effort, an effect analogous to increasing the proportional gain of a PI controller, since the term $A_r = \rho^{1/2}$ appears in the controller numerator (see equations (3.21), (3.24) and (3.27)). As $\beta \to 1$ the integral action is removed since the term (1-d) becomes cancelled in (3.66) (in practice β is never allowed to come too close to 1). Conversely, decreasing the value of β increases the effect of the integral action, which is analogous to decreasing the integral time constant in a PI controller.

Athough a strict application of the theory requires that A_q is stable the use of unstable A_q, such as $A_q = 1 - d$, can be justified using the argument in the following section.

3.7.3 Solvability Conditions and Unstable Weighting Terms

Solvability of the optimal control problem is dependent upon the conditions (see Section 3.2):

1. The polynomials A and B must have no unstable common factors. This condition is equivalent to the requirement that any unstable terms in A_d and A_x (where A_d and A_x are the denominators of W_d

and W_x, respectively) must also be factors of the denominator of the plant input-output transfer-function W_p.

2. Any unstable factors of A_e must also be factors of A.

3. Any unstable factors of A_ℓ must also be factors of A and D. This condition is equivalent to the requirement that any unstable terms in $A_x A_\ell$ must also be factors of the denominator of the plant input-output transfer-function W_p.

To summarise, any unstable terms in A_e, A_d or $A_x A_\ell$ must also be factors of the denominator of the plant input-output transfer-function W_p.

In any practical situation where an unstable term in A_e, A_d or $A_x A_\ell$ does not appear in the denominator of W_p then this term must be artificially introduced into the forward path using A_q. Similarly, in other situations it may be desirable to use an unstable A_r weighting term. In some situations, therefore, the plant conditions may dictate that the use of unstable weighting terms is desirable.

A strict application of the theory, however, requires that A_q and A_r should be stable. The use of unstable A_q and A_r in practice can be justified using the following argument; let us formally define the plant as that part of the system which is known a priori in advance of controller design. Assume now that the given data is such that we know unstable weighting terms are desirable. Denote A_q and A_r as follows:

$$A_q = A_q^+ A_q^- \tag{3.72}$$

$$A_r = A_r^+ A_r^- \tag{3.73}$$

where + indicates a stable polynomial and - indicates an unstable polynomial. The forward path is then as shown in Figure 3.5. Since the given problem data tells us a priori that the terms A_q^- and A_r^- are necessary let us consider the transfer-function A_r^-/A_q^- as part of the plant, as shown in Figure 3.6. The controller is then designed for the new plant in Figure 3.6 where the solvability conditions have now been satisfied by appropriate choice of A_q^-. This approach is equivalent to minimising the cost-function:

$$J = \frac{1}{2\pi j} \oint_{|z|=1} \left\{ \frac{B_q^* B_q}{A_q^{+*} A_q^+} \phi_e + \frac{B_r^* B_r}{A_r^{+*} A_r^+} \phi_{u'} \right\} \frac{dz}{z} \tag{3.74}$$

Finally, the control signal u is implemented as follows:

$$u = \frac{A_r^-}{A_q^-} u' \tag{3.75}$$

This argument allows us to ensure that the problem solvability conditions are always satisfied without violating the theoretical conditions on the cost-function weights. The original approach of directly using unstable weights will, nevertheless, result in precisely the same closed-loop system. To illustrate the point, consider a system which has a drifting disturbance due to a factor $1 - d$ in the denominator of the disturbance transfer-function W_d. Assume that this term is <u>not</u> present in the denominator of the plant. It is immediately apparent that integral action is needed to counter the effect of the drifting disturbance. A_q is then defined as $A_q = 1 - d$ and the controller is calculated. However, that this problem violates the theoretical solvability conditions is apparent for three reasons:

(i) The polynomials A and B have an unstable common factor $1-d$.

(ii) A_q is unstable.

(iii) A drifting control signal u will result in order to counteract the drifting disturbance. The cost-function will, therefore, be infinite.

The above argument can, however, be used to justify the design; let us assume that the term $1/A_q^-$ is included in the plant and then calculate a controller based on this newly defined plant to minimise the cost-function (3.74). This design will result in the same closed-loop system as the original design. The new design will, however, have the following properties.

(i) The newly defined A and B polynomials will not have any unstable common factors.

(ii) The effective weighting terms A_q^+ and A_r^+ are stable.

(iii) The pseudo control signal u' will be stable.

From equation (3.75) $u' = A_q^- u$. In this example $A_q^- = 1 - d$ which means that <u>changes</u> in the control signal are, effectively, being weighted.

3.7.4 <u>Computational Algorithms</u>

Efficient computational routines for diophantine equation solution have been derived by Kučera (1979) and Ježek (1982). The spectral factorisation can be performed using the iterative algorithms proposed by Kučera (1979) or Ježek and Kučera (1985) . Iteration of the spectral factorisation routine is terminated either when a specified tolerance is reached or when a specified maximum number of iterations have been performed. These algorithms have the necessary property that the solution obtained after each iteration is guaranteed to be stable.

3.7.5 <u>Common Factors in A and B</u>

The implied feedback diophantine equation (3.30) uniquely determines the optimal feedback controller when the plant A and B polynomials are coprime. If A and B have a stable common factor then the couple of equations (3.22)-(3.23) must be solved to obtain the optimal feedback controller.

If, however, it is deemed necessary in a particular application to always solve the implied equation (since this is computationally simpler than solving the original couple) regardless of any possible stable common factors in A and B then this equation will still be solvable since any stable common factors of A and B will also divide D_c (see equation (3.20)). Such a solution will lead to a closed-loop system with optimal pole positions but sub-optimal zero positions.

This property of the optimal design should be contrasted with standard pole-assignment algorithms (Wellstead et al, 1979) where the diophantine equation (3.44) must be solved. Since the arbitrary polynomial T_c appears on the right-hand side of this equation (in place of D_c) any common factors in A and B will render this equation unsolvable (unless, coincidentally, this factor also divides T_cC).

3.8 REFERENCES

Andersson, P. (1985). Adaptive forgetting in recursive identification through multiple models. International Journal of Control, 42, 1175.

Åström, K.J. (1980). Self-tuning regulator - design principles and applications. In K.S. Narendra and R.V. Monopoli (Eds), Applications of Adaptive Control, Academic Press, New York.

Åström, K.J., (1983a). LQG self-tuners. Proceedings IFAC workshop on adaptive control, San Francisco.

Åström, K.J., (1983b). Analysis of Rohrs counterexamples to adaptive control. Proceedings IEEE Conference on decision and control, San Antonio.

Åström, K.J., and B. Wittenmark (1973). On self-tuning regulators. Automatica, 9, 185-199.

Åström, K.J., and B. Wittenmark (1984). Computer controlled systems. Prentice Hall, Englewood Cliffs.

Biermann, G.J., (1977). Factorisation methods for discrete sequential estimation. Academic Press, New York.

Chen, H.F., and P.E. Caines (1984). Adaptive linear quadratic control for discrete time systems. Proceedings IFAC world congress, Budapest.

Chen, M.J., and J.P. Norton (1987). Estimation technique for tracking rapid parameter changes. International Journal of Control, 45, 1387-1398.

Clarke, D.W. and P.J. Gawthrop (1975). Self-tuning controller. Proc. IEE, 122, 929-934.

Clarke, D.W. and P.J. Gawthrop (1979). Self-tuning control. Proc. IEE, 16, 633-640.

Doyle, J.C. and G. Stein (1981). Multivariable feedback design : concepts for a classical/modern synthesis. Trans IEEE on automatic control, AC-26, 4-16.

Egardt, B. (1979). Stability of adaptive controllers. Springer-Verlag, Berlin.

Fortescue, T.R., L.S. Kershenbaum and B.E. Ydstie (1981). Implementation of self-tuning regulators with variable forgetting factors. Automatica, 17, 831.

Goodwin, G.C., D.J. Hill and M. Palaniswami (1985). Towards an adaptive robust controller, IFAC Symposium and Identification and System Parameter Estimation, York.

Grimble, M.J. (1984). Implicit and explicit LQG self-tuning controllers. Automatica, 20, 661-669.

Grimble, M.J. (1986). Convergence of explicit LQG self-tuning controllers. To appear, Proc. IEE.

Hagglund, T. (1983). New estimation techniques for adaptive control. PhD. thesis, Lund.

Horowitz, I.M. (1963). Synthesis of feedback systems. Academic Press, New York.

Hunt, K.J. (1987). Stochastic optimal control theory with application in self-tuning control. PhD thesis, University of Strathclyde, Glasgow.

Hunt, K.J., M.J. Grimble, M.J. Chen and R.W. Jones (1986). Industrial LQG self-tuning controller design. Proceedings IEEE conference on decision and control, Athens.

Hunt, K.J., M. Sebek and M.J. Grimble (1987). Optimal multivariable LQG control using a single diophantine equation. To appear, International Journal of Control.

Ježek, J., (1982). New algorithm for minimal solution of linear polynomial equations. Kybernetika, 18, 505.

Ježek, J., and V. Kučera (1985). Efficient algorithm for matrix spectral factorisation. Automatica, 21, 663.

Jones, R.W. (1987). PhD Thesis, University of Strathclyde, Glasgow.

Kučera, V. (1979). Discrete Linear Control. John Wiley, Chichester.

Kučera, V. (1984). The LQG control problem : A study of common factors. Problems of control and information theory, 13, 239-251.

Kulhavy, R. and M. Karny (1984). Tracking of slowly varying parameters by directional forgetting. Proceedings IFAC World Congress, Budapest.

Moore, J.B. (1984). A globally convergent recursive adaptive LQG regulator. Proceedings IFAC world congress, Budapest.

Rohrs, C.E., L. Valavani, M. Athans and G. Stein (1982). Robustness of adaptive control algorithms in the presence of unmodelled dynamics. Proceedings IEEE Conference on decision and control, Orlando.

Šebek, M., K.J. Hunt and M.J. Grimble (1987). LQG regulation with disturbance measurement feedforward. To appear, International Journal of Control.

Šebek, M., and V. Kucera (1982). Polynomial approach to quadratic tracking in discrete linear systems. Trans. IEEE on automatic control, AC-27, 1248-1250.

Truxal, J.G., (1955). Control system synthesis. McGraw Hill, New York.

Wellstead, P.E., D. Prager and P. Zanker (1979). Pole assignment self-tuning regulator. Proceedings IEE, 126, 781-787.

Wellstead, P.E. and S.P. Sanoff (1981). Extended self-tuning algorithm. International Journal of Control, 34, 433-455.

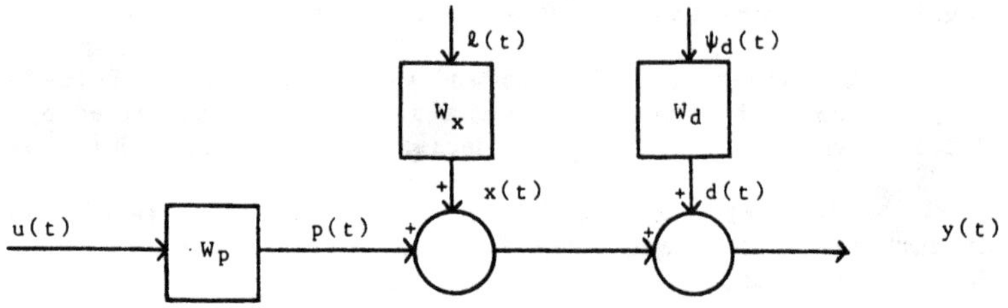

Figure 3.1 Plant model

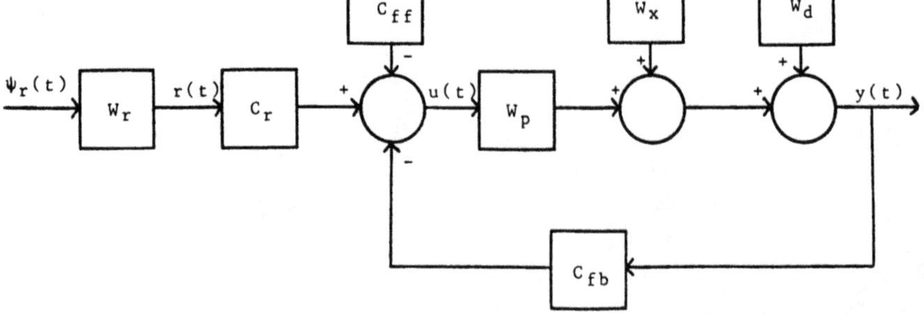

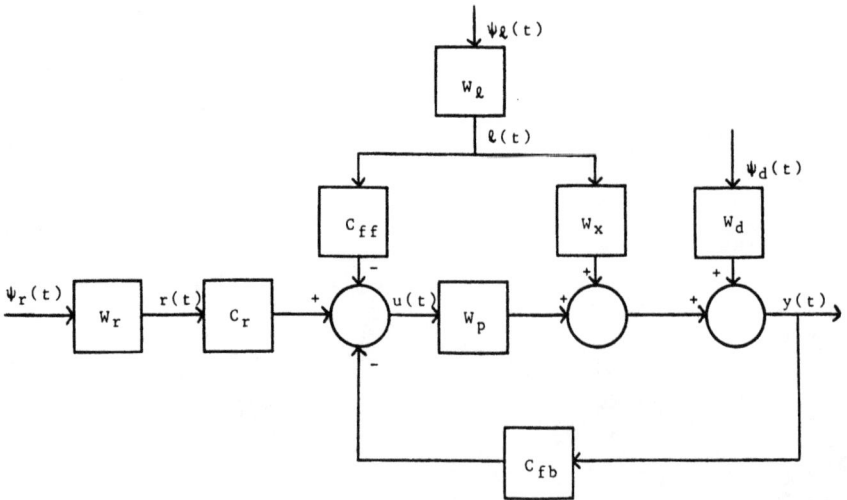

Figure 3.2 Closed-loop system

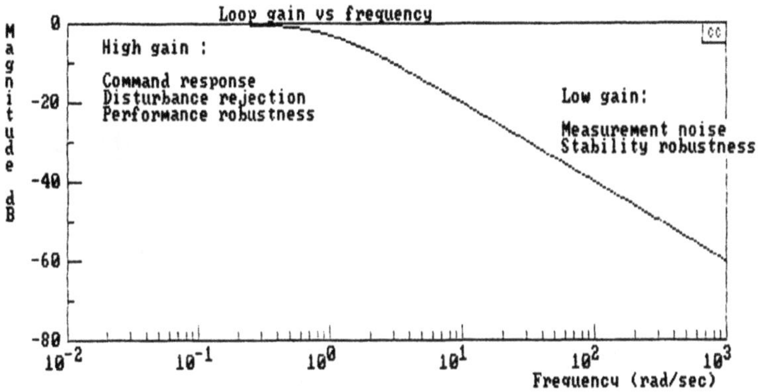

Figure 3.3 Loop gain

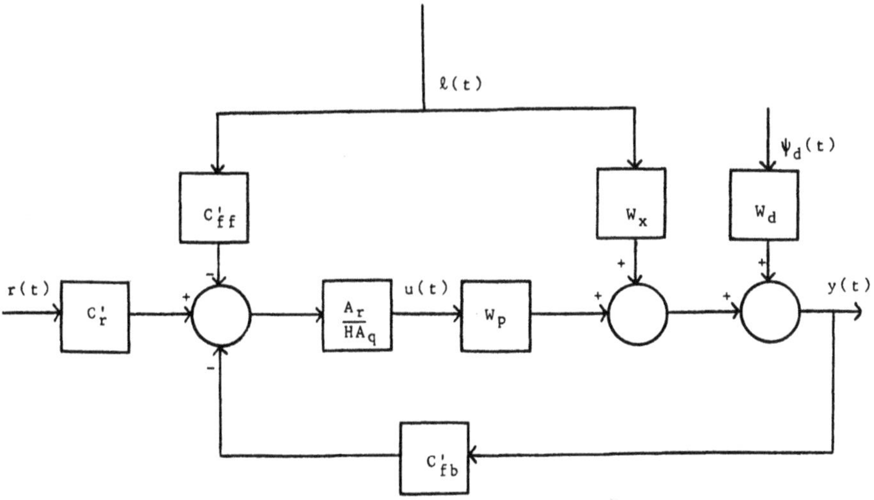

Figure 3.4 Equivalent control structure

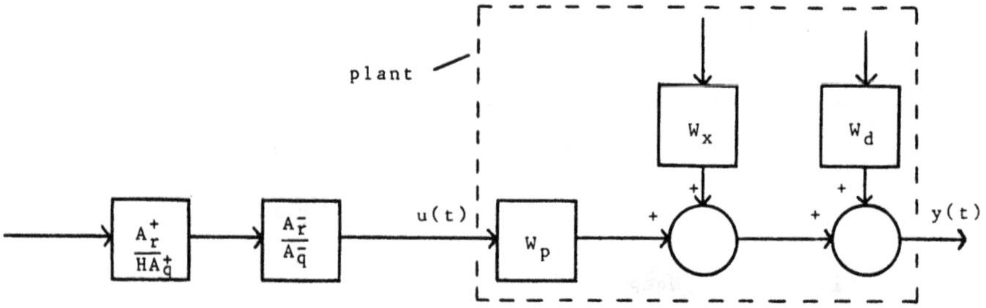

Figure 3.5 Forward path

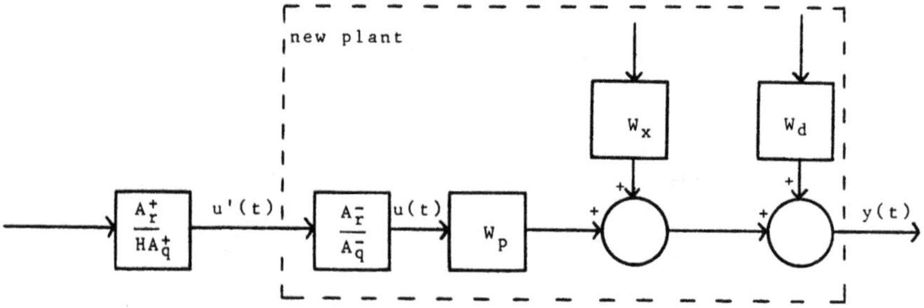

Figure 3.6 Artificial plant

Chapter 4

Numerical algorithms in self-tuning control

C. Mohtadi

4.1 Introduction

Encouraged by current advances in microprocessor technology, researchers in control system design and self-tuning have developed many different and useful estimation and control algorithms. The field of self-tuning has come a long way since the inception of the first Minimum-Variance self-tuning regulator of Peterka (1970) and Åström and Wittenmark (1973) and the first implementation of the so-called Generalized Minimum-Variance approach of Clarke and Gawthrop (1975) on the Intel 8080 microprocessor (Clarke, Cope and Gawthrop 1975) written entirely in Assembler using 3-byte floating point software. Later implementations in intermediate-level languages required substantial modifications to the interpreters and compilers to include real-time features such as interrupts and watch-dog-timers (e.g. Control Basic of Clarke and Frost (1979) as an extension of Li-Chen Wang's 'Palo Alto Tiny Basic' interpreter). Modern hardware and software have opened new avenues for application of self-tuners (e.g. Lambert, M. (1987) has applied a multi-tasking/multi-processor Motorola 68020/68010 system for full self-tuning control estimating 4th order models of a flexible single-link robot arm at sample rates in excess of 100Hz; Jota (1987) used Distributed Pascal Plus (DPP) for implementing multi-loop and multi-variable self-tuning and supervisory control of a HVAC system, on a bank of distributed microprocessors). New high level languages such as Ada[1], Modula-2 and occam[2] (Jones 1987) provide a novel and exciting view of the software engineering and design aspects of multi-tasking multi-processor real-time digital control. More recently the current advances in VLSI chip technology have made it possible to design specific fine-grain parallel architecture (e.g. systolic type arrays, see Kung, Whitehouse and Kailath 1985) to improve computational efficiency of suitably structured algorithms.

Despite the recent growth of activity in modern algorithm design many issues are still fully embedded in their traditional roots. Although consideration of memory space, computation time or round-off errors with fixed-point arithmetic are *not*

[1] Ada is a registered trademark of the U.S. Government, Ada Joint Program Office
[2] occam is the trademark of the INMOS Group of Companies

as crucial as they used to be – except in implementations using present generation signal processing chips (e.g. TMS32010) – other concepts such as numerical stability and long term integrity remain as important as ever.

This chapter attempts to give a brief review of some of the commonly used schemes studied in the literature. The chapter is organised as follows:

- Section 4.2.1 deals with the parameter estimation and the so-called "covariance update" in recursive least squares type algorithms and reiterates some of the recent results in the field.
- Section 4.2.2 outlines the two basic fast algorithms for recursive least squares: the fast RLS (transversal filter estimation) and the fast lattice filter estimation.
- Section 4.3.1 concentrates on the solution of the well known Diophantine equation where various approaches are compared and examined.
- Section 4.3.2 deals with the solution of multivariable Diophantine identities.
- Section 4.3.3 examines the algorithms relating to state estimation and state feedback type controllers, mainly using the observable canonical form.
- Section 4.3.4 relates the earlier designs to the input/output formulation using finite stage minimisation techniques, namely the Peterka's predictive controller (1984) and Generalized Predictive Control approach (Clarke *et al* 1987).
- Section 4.3.5 examines the infinite stage LQG solutions using the so-called polynomial spectral factorization techniques.

First we deal with some preliminaries.

4.1.1 Models and operators

The algorithms described in the forthcoming sections exclusively deal with linear plant models of the type:

$$A(q^{-1})y(t) = B(q^{-1})u(t-1) + \mathcal{X}(t)$$

where $A(q^{-1})$ and $B(q^{-1})$ are polynomials in the backward shift operator q^{-1}. $y(t)$, $u(t)$ and $\mathcal{X}(t)$ are the output, control signal and the disturbance respectively. Different assumptions about $\mathcal{X}$ yields several classes of models: CARMA (Controlled Autoregressive and Moving Average $\mathcal{X}(t) = C(q^{-1})\xi(t)$) and CARIMA (Controlled Autoregressive and Integrated Moving Average $\mathcal{X}(t) = \frac{C(q^{-1})\xi(t)}{\Delta}$) models being the most extensively used (see Tuffs and Clarke 1985 for a general overview). The algorithms in this chapter are explained for the simplest case but are usually easily extendable to the more complicated cases.

Some of the methods discussed are not exclusive to the backward shift operator representation and can be used with other operators such as δ where $\delta = (q-1)/h$ and h is the sample-time (Agarwal and Burrus 1975, Goodwin *et al* 1986) and s the Laplace operator. In other cases, as for example with some of the spectral factorization algorithms the mapping, $z = (1+s)/(1-s)$ or $z = h\delta + 1$ may be used.

4.1.2 Gentleman's Algorithm for dyadic reduction

Self-tuning controllers are currently realized through numerical methods on digital computers. Frequently, the algorithms are coded up in a form fairly close to the original mathematical developments. They are subsequently tested using a variety of simulated experiments of approximately 1000 samples or so. It is then conjectured that "if the method operates for 1000 samples it will work for *ever*!"

The real-life situation is somewhat different. On real industrial plant the minimum expected continuous duration of operation for a prospective self-tuner is perhaps a few hours on batch processes, and maybe upto 5 years say on a continuous process. At a typical sample interval of 10 seconds this means that the self-tuner is expected to operate reliably and without *failure* for anything between 5000 samples to 15,000,000 samples: four orders of magnitude longer than the simulated runs!

With a standard Recursive Least Squares[3] estimator, implemented exactly as explained in most text books (Ljung and Soderstrom 1983) the typical length of experiment to the onset of instability of the algorithm is about 10,000 samples – just long enough to work correctly in the batch processing example, say, but sufficiently long to give the commissioning engineer of the continuous process the impression of working albeit with disasterous consequences! An important lesson in the development of algorithms for reliable operation is the "folk theorem"[4] of Laub (1985):

Theorem 4.1 *If an algorithm is amenable to "easy" hand calculation, it's probably a poor method if implemented in finite floating point arithmetic of a digital computer.*

Fortunately, in case of most self-tuning algorithms it is usually possible to find a substitute method which is at least one to two orders of magnitude better than the simple-minded algorithms. A majority of designs are based on linear algebra which has been extensively studied by numerical analysts. Foremost among these reliable algorithms are the so-called orthogonal transformations used in solution of quadratic minimizations.

The method described here is due to Gentleman (1973) and is a basic primtive for orthogonal transformations (see Lawson and Hanson 1974). It is the square root free version of Givens' rotation. This algorithm is extensively used by Peterka (1986a,b) where a Pascal coding is also given. Consider a matrix **M** (capital bold lettering refers to matrices and lower-case bold lettering to vectors) of rank at most 2:

$$\begin{aligned} \mathbf{M} &= \mathbf{f} d_f \mathbf{f}^T + \mathbf{r} d_r \mathbf{r}^T \\ \mathbf{f} &= [1, f_1, f_2, \cdots, f_n]^T \\ \mathbf{r} &= [r_0, r_1, r_2, \cdots, r_n]^T \end{aligned}$$

Consider a rotation such that:

$$\begin{aligned} \mathbf{M} &= \tilde{\mathbf{f}} \tilde{d}_f \tilde{\mathbf{f}}^T + \tilde{\mathbf{r}} \tilde{d}_r \tilde{\mathbf{r}}^T \\ \tilde{f}_0 &= 1 \\ \tilde{r}_0 &= 0 \end{aligned}$$

[3]RLS is chosen as an example because it constitutes one of the central components of most self-tuners

[4]as a corollary of Murphy's law

The algorithm is given in Fig.4.1.

$\tilde{d_f} = d_f + d_r r_0^2$
IF $\tilde{d_f} = 0$ **skip** the update
$\tilde{d_r} = (d_f/\tilde{d_f})d_r$
$k_r = (d_r/\tilde{d_f})r_0$
For $j = 1 \rightarrow n$;
$\quad \tilde{r_j} = r_j - r_0 f_j$
$\quad \tilde{f_j} = f_j + k_r \tilde{r_j}$

Figure 4.1: Gentleman's dyadic reduction

Repeated application of the dyadic reduction to a matrix gives the final orthogonal transformation required. The following numerical example from Jover and Kailath (1986) explains the use of Gentleman's rotation to obtain a lower triangular matrix. This is extensively used in parameter estimation and Kalman filtering. Consider the set of dyads below (dyadic matrix of $[\mathbf{f}_1, \mathbf{r}_2, \mathbf{r}_3, \mathbf{r}_4]$, $[\text{ diag } d_i]$)

$$\begin{bmatrix} 1 & 65 & 103 & 107 \\ 0 & 1 & 0 & 0 \\ 0 & 3 & 1 & 0 \\ 0 & -2 & 3 & 1 \end{bmatrix}, \begin{bmatrix} 571 & & & \\ & 1 & & \\ & & 3 & \\ & & & 2 \end{bmatrix} \xrightarrow{\Theta_D(1,4)}$$

$$\begin{bmatrix} 1 & 65 & 103 & 0 \\ 0 & 1 & 0 & 0 \\ 0 & 3 & 1 & 0 \\ 0.9118E-1 & -2 & 3 & 1 \end{bmatrix}, \begin{bmatrix} 23469 & & & \\ & 1 & & \\ & & 3 & \\ & & & 0.00487 \end{bmatrix} \xrightarrow{\Theta_D(1,3)}$$

$$\begin{bmatrix} 1 & 65 & 0 & 0 \\ 0 & 1 & 0 & 0 \\ 0.5588E-2 & 3 & 1 & 0 \\ 0.2063E-1 & -2 & 2.0608 & 1 \end{bmatrix}, \begin{bmatrix} 55296 & & & \\ & 1 & & \\ & & 1.2733 & \\ & & & 0.0487 \end{bmatrix} \xrightarrow{\Theta_D(1,2)}$$

$$\begin{bmatrix} 1 & 0 & 0 & 0 \\ 0.1092E-2 & 1 & 0 & 0 \\ 0.8468E-2 & 2.6368 & 1 & 0 \\ 0.1699E-1 & -3.3412 & 2.0608 & 1 \end{bmatrix}, \begin{bmatrix} 59521 & & & \\ & 0.9290 & & \\ & & 1.2733 & \\ & & & 0.0487 \end{bmatrix}$$

4.1.3 Polynomial Multiplication and Division

Polynomial multiplication and division are the elementary operations required for any polynomial based controller design. The polynomial division algorithm described here is also known as the Diophantine recursion algorithm (Mohtadi 1986) and Markov recursion algorithm (Gawthrop 1987).

Given
$A(s) = a_0 s^n + a_1 s^{n-1} + \cdots + a_n$
$B(s) = b_0 s^m + b_1 s^{m-1} + \cdots + b_m$
$R(s) = 0$
For $i = 0 \to n$
  For $j = 0 \to m$
    $r_{i+j} = r_{i+j} + a_i b_j$

Figure 4.2: Polynomial multiplication

Given
$A(s) = s^n[a_0 + a_1 s^{-1} + \cdots + a_n s^{-n}] = s^n A'(s)$
$B(s) = s^m[b_0 + b_1 s^{-1} + \cdots + b_m s^{-m}] = s^m B'(s)$
Check A' to ensure that a_0 is *nonzero*
$Q_1 = 0; \quad R_0 = B'$
For $k = 1 \to m - n - 1$
  $h_{k+1} = r_{k0}/a_0$
  $Q_{k+1} = sQ_k + h_{k+1}$
  $R_{k+1} = sQ_{k+1} - h_{k+1}A'$
Q_{m-n} is the quotient and R_{m-n} the remainder.

Figure 4.3: Polynomial division

4.2 Parameter Estimation

4.2.1 Covariance Update

Shah (Chapter 2 of this book) discusses some of the theoretical as well as practical aspects of identification and recursive parameter estimation in adaptive control. Here we concentrate on the algorithmic implementation of the variants of the Recursive Least Squares (RLS) algorithm, in particular the so-called covariance update:

$$\mathbf{P}(t) = \frac{1}{\beta(t)}\left\{\mathbf{P}(t-1) - \frac{\mathbf{P}(t-1)\mathbf{x}(t)\mathbf{x}^T(t)\mathbf{P}(t-1)}{\alpha(t) + \mathbf{x}^T(t)\mathbf{P}(t-1)\mathbf{x}(t)}\right\} + \mathbf{Q}(t) \tag{4.1}$$

where $\mathbf{P}(t)$ is the unscaled covariance matrix, $\mathbf{x}(t)$ is the data vector, $\alpha(t)$, $\beta(t)$ and $\mathbf{Q}(t)$ are given weights associated with the appropriate forgetting mechanism.

Although a direct implementation of eqn. 4.1 using standard matrix/vector multiplication and addition is a relatively simple task it is *not* recommended as it is not possible to guarantee positive definiteness of $\mathbf{P}(t)$: $\mathbf{P}(t)$ *must* be positive definite to ensure stability of the parameter adaptation algorithm. In order to remedy this shortcoming it is possible to factorize $\mathbf{P}(t)$ as follows:

$$\mathbf{P}(t) = \mathbf{L}(t)\mathbf{D}(t)\mathbf{L}^T(t) \tag{4.2}$$

where $\mathbf{L}(t)$ is a lower triangular matrix with ones along the diagonal. If instead of $\mathbf{P}(t)$ the pair $\{\mathbf{L}(t), \mathbf{D}(t)\}$ are updated then the covariance is guaranteed *positive definite*. For simplicity consider the case where $\mathbf{Q}(t) = 0$ and $\beta(t) = 1$. In order to derive one such algorithm consider the Schur's complement:

$$\begin{bmatrix} \alpha(t) + \mathbf{x}^T(t)\mathbf{P}(t-1)\mathbf{x}(t) & \mathbf{x}^T(t)\mathbf{P}(t-1) \\ \mathbf{P}(t-1)\mathbf{x}(t) & \mathbf{P}(t-1) \end{bmatrix}$$

which can be factorized to either of the forms below:

$$\begin{bmatrix} 1 & \mathbf{x}^T(t)\mathbf{L}(t-1) \\ 0 & \mathbf{L}(t-1) \end{bmatrix} \begin{bmatrix} \alpha(t) & 0 \\ 0 & \mathbf{D}(t-1) \end{bmatrix} \begin{bmatrix} 1 & 0 \\ \mathbf{L}^T(t-1)\mathbf{x}(t) & \mathbf{L}^T(t-1) \end{bmatrix} \tag{4.3}$$

or:

$$\begin{bmatrix} 1 & 0 \\ \mathbf{k}(t) & \tilde{\mathbf{L}}(t) \end{bmatrix} \begin{bmatrix} \tilde{d}(t) & 0 \\ 0 & \tilde{\mathbf{D}}(t) \end{bmatrix} \begin{bmatrix} 1 & \mathbf{k}^T(t) \\ 0 & \tilde{\mathbf{L}}^T(t) \end{bmatrix} \tag{4.4}$$

where

$$\frac{\mathbf{k}(t)}{\tilde{d}(t)} = \frac{\mathbf{P}(t-1)\mathbf{x}(t)}{\alpha(t) + \mathbf{x}^T\mathbf{P}(t-1)\mathbf{x}(t)}.$$

The algorithm will therefore consist of forming the prearray of 4.3 and using any orthogonal transformation (e.g. the dyadic reduction Fig.4.1) to obtain the upper-triangular post-array 4.4. For non-unity $\beta(t)$ simply the division $\tilde{\mathbf{D}}/\beta(t)$ is performed after the transformation. For cases where $\mathbf{Q}(t) = \mathbf{L}_Q\mathbf{D}_Q\mathbf{L}_Q^T$ is non-zero then an additional upgrade has to be performed whereby the prearray

$$\begin{bmatrix} \tilde{\mathbf{L}}(t) & \mathbf{L}_Q \end{bmatrix} \begin{bmatrix} \tilde{\mathbf{D}}(t) & 0 \\ 0 & \mathbf{D}_Q \end{bmatrix} \begin{bmatrix} \tilde{\mathbf{L}}^T(t) \\ \mathbf{L}_Q^T \end{bmatrix} \tag{4.5}$$

is transformed to the post array

$$\begin{bmatrix} \mathbf{L}(t) & 0 \end{bmatrix} \begin{bmatrix} \mathbf{D}(t) & 0 \\ 0 & 0 \end{bmatrix} \begin{bmatrix} \mathbf{L}^T(t) \\ 0 \end{bmatrix} \tag{4.6}$$

by repeatedly applying Gentleman's dyadic reduction Fig.4.1.

The method outlined here has excellent numerical properties (Verhaegen and Van Dooren 1986) and more importantly the same algorithm is used for all updates unlike the approach of (Bierman 1977) where Agee-Turner $\mathbf{UDU}^T$ factorization is used for the measurement update (i.e. the case where $\mathbf{Q}(t) = 0$) and Modified Weighted Gram Schmidt is used for addition of $\mathbf{Q}(t)$. Jover and Kailath (1986) have described a systolic-type array architecture for implementation of the above algorithm.

Another important aspect of the **LD** factorization not commonly considered in the literature is in the ability to perform on-line model order selection. Consider the case where:

$$\mathbf{x}(t) = [y(t-n), u(t-n), y(t-n+1), u(t-n+1), \cdots, y(t-1), u(t-1)]^T$$

and the parameter vector of the nth order model $\hat{\theta}_n(t)$ is available. The orthogonal basis parameters are given by:

$$\hat{\tilde{\theta}}_n(t) = \mathbf{L}^{-1}(t)\hat{\theta}_n(t)$$

In order to obtain the mth order least sqaures estimate say $(m < n)$, simply set the first $2(n-m)$ elements in $\hat{\tilde{\theta}}_n$ equal to zero and map the parameters back using the inverse tramsformation.

$$\begin{bmatrix} 0 \\ \vdots \\ 0 \\ \hat{\theta}_m(t) \end{bmatrix} = \mathbf{L}(t)diag\{\overbrace{0,\cdots,0}^{2(n-m)}1,\cdots,1\}\hat{\tilde{\theta}}_n(t) \tag{4.7}$$

Thus it is quite straightforward to obtain an efficient algorithm for computing the mth order least squares estimate using $2l(l+2m+1)$ floating point operations where $l = n - m$.

In many applications of least squares estimators or some of its variants (e.g. "constrained" least sqaures and "damped" least squares see Lambert, E. 1987) some form of covariance regularization may be necessary especially in applications to non-linear systems (Bard 1974). The regularization takes the form of adding a matrix $\mathbf{\Lambda}$ (usually diagonal or $\lambda\mathbf{I}$) to the inverse of the so-called covariance matrix $\mathbf{P}(t)$. λ is commonly known as the Levenberg-Marquardt parameter (Levenberg 1944, Marquardt 1963).

$$[\mathbf{P}^*(t)]^{-1} = [\mathbf{P}(t)]^{-1} + \mathbf{\Lambda}$$

The update can be either performed by iterative calls to the basic **LD** filter above setting $\mathbf{x} = \sqrt{\lambda_i}\mathbf{e}_i$ where $\mathbf{e}_i$ is the ith basis vector or through using the two factorizations of the Schur's complement:

$$\begin{bmatrix} \mathbf{\Lambda}+\mathbf{P}(t) & \mathbf{P}(t) \\ \mathbf{P}(t) & \mathbf{P}(t) \end{bmatrix} = \begin{bmatrix} \mathbf{I} & \mathbf{L}(t) \\ 0 & \mathbf{L}(t) \end{bmatrix} \begin{bmatrix} \mathbf{\Lambda} & 0 \\ 0 & \mathbf{D}(t) \end{bmatrix} \begin{bmatrix} \mathbf{I} & 0 \\ \mathbf{L}^T(t) & \mathbf{L}^T(t) \end{bmatrix} \tag{4.8}$$
$$= \begin{bmatrix} \mathbf{L}_1(t) & 0 \\ \mathbf{K}(t) & \mathbf{L}^*(t) \end{bmatrix} \begin{bmatrix} \mathbf{\Delta}^*(t) & 0 \\ 0 & \mathbf{D}^*(t) \end{bmatrix} \begin{bmatrix} \mathbf{L}_1^T(t) & \mathbf{K}^T(t) \\ 0 & [\mathbf{L}^*(t)]^T \end{bmatrix}$$

As before the algorithm simply consists of forming the prearray and computing the postarray via *any* orthogonal transformation (e.g. Gentleman's dyadic reduction).

Another form of regularization quite popular and sometimes necessary in applications of the **LD** filter is to put hard constraints on the **D** parameters when they are computed. This ensures both positive definiteness and boundedness of the matrix elements (Ljung and Soderstrom 1983, Tuffs 1984).

4.2.2 Fast Least Squares and Ladder Algorithms

Development of fast and efficient algorithms for least squares prediction has been an area of active reasearch during the past decade. Much of the work has been conducted for applications in adaptive speech processing, design of equalizers and noise cancellation devices. A brief examination of the properties of such algorithms, their relevance to self-tuning controllers and their possible applications in the field is of some importance.

The generic recursive least squares estimator of the previous section requires $O(n^2)$ floating point operations for the covariance update and parameter estimation where n is the number of estimated parameters. In some applications such as adaptive speech processing or self-tuning control using impulse-response models (Bruijn *et al* 1980, Clarke and Zhang (1984)) n can be as large as large as 15 to 50 (in adaptive equalization n might be upto 4000!) in order to obtain reasonable models. Clearly, employing the standard algorithm can become computationally prohibitive even with fast processors. Recall that the derivation of the basic algorithm above does *not* employ the sequential nature of the assimilated data. Fast algorithms (so-called because they use $O(n)$ floating point operations to compute the estimated parameters) use this sequential characteristics of the data or more precisely the shift-invariance property of the appropriate correlation matrix. There are two basic fast algorithms: fixed order methods whereby the Kalman gains are computed using a fast algorithm, and ladder or lattice algorithms where the reflection coefficients of an orthogonalized linear system are estimated and where unlike the other methods recursions in the order of the model are very simple as higher order components do not affect the lower coefficients. In general the usual trade-offs apply and these algorithms have worse numerical properties with respect to round-off and error propagation than the equivalent full **LD** factorization techniques described above.

4.2.2.1 Fixed order methods

The fast RLS or fast Kalman algorithms were first derived by (Ljung, Morf and Falconer 1978). The basic concept of the scheme revolves round the two prediction and smoothing equations:

$$\begin{aligned}\hat{y}(t|t-1) &= -\mathcal{A}^T(t-1)\psi(t)\\ \hat{y}(t-n-1|t-1) &= -\mathcal{B}^T(t-1)\psi(t)\end{aligned}$$

where $\psi(t)$ is the vector of measurements (i.e. $[x(t-1), x(t-2), \cdots, x(t-n)]^T$), and $\mathcal{A}$ and $\mathcal{B}$ are the associated vector of parameters. The orthogonality of the predicted and smoothed outputs to the appropriate regressors subspace lead to the required algorithms (see Ljung and Soderstrom 1983, Goodwin and Sin 1984, Alexander 1986 and Marple 1987 for some excellent tutorial discussions). Some variants of the method (e.g. normalized form), their geometrical derivation and interpretations as well as the associated numerical properties (e.g. operation count and numerical stability to round-off errors) are discussed in some detail in (Cioffi and Kailath 1984). The bare bones of the algorithm are given in Fig.4.4. $x(t)$ is set equal to $y(t)$ for Auto-Regressive parameter estimation and to $\{y(t), u(t-1), \epsilon(t-1)\}$ for CARMA parameter estimation where $\epsilon(t)$ is the estimation error representing a proxy for the noise term.

The total number of floating point operations varies between $7n \rightarrow 11n$ depending on the specific implementation and as such the method is only useful if n the number of estimated parameters exceeds 10. Unfortunately the numerical properties of the method are not as favourable as they may seem initially. A thorough study of the error propagation properties of the fast RLS algorithm is performed by Ljung and Ljung (1986). They conclude that as far as error propagation is concerned the algorithm is *unstable* for $\beta < 1$. Normalization and rescue devices may be

1. $\mathcal{A}(0) = 0; \mathcal{B}(0) = 0; R^e(0) = \delta\mathbf{I}; k(1) = 0$
2. Given $\mathcal{A}(t-1); \mathcal{B}(t-1); R^e(t-1)$ and $k(t)$
$e(t) = x(t) + \mathcal{A}^T(t-1)\psi(t)$
$\mathcal{A}(t) = \mathcal{A}(t-1) - k(t)e^T(t)$
$\lambda(t) = k^T(t)\psi(t)$
$\bar{e}(t) = [1 - \lambda(t)]e(t)$
$R^e(t) = \beta R^e(t-1) + \bar{e}(t)e^T(t)$
$k^*(t) = [R^e(t)^{-1}\bar{e}(t), k(t) + \mathcal{A}(t)R^e(t)^{-1}\bar{e}(t)]^T$
$k^*(t) = [M(t), \cdots, \mu(t)]^T$
$r(t) = x(t-n) + \mathcal{B}^T(t-1)\psi(t+1)$
$\mathcal{B}(t) = [\mathcal{B}(t-1) - M(t)r^T(t)][\mathbf{I} - \mu(t)r^T(t)]^{-1}$
$k(t+1) = M(t) - \mathcal{B}(t)\mu(t)$

Figure 4.4: The fast RLS algorithm

used to alleviate this problem (Cioffi and Kailath 1984) at the expense of increased numerical complexity (i.e. $2n$ extra flops, square-root extraction, etc).

4.2.2.2 Ladder or Lattice filters

Ladder or lattice filters of Itakura and Saito (1971) are somewhat more complicated than the fast RLS method pointed out above but have superior numerical properties. The algorithm is very closely related to AR and ARMA estimators but instead of estimating the coefficients of the ARMA filter a set of reflection coefficients are estimated. For control algorithms where the coefficients of the CARMA model are required a subsequent mapping is required. The main importance of the algorithm is in the fact that models with variable orders can be easily estimated without any requirements for extra computations (unlike the **LD** factorization method). Moreover, as pointed out earlier the resulting model parameters (i.e. reflection coefficients) are more robust towards numerical round-off and finite word-length representation than the equivalent ARMA filter. An excellent derivation of the algorithm is given in Ljung and Soderstrom (1983). See also Lee *et al* (1982) for the geometric interpretations. The algorithm is depicted in Fig.4.5.

4.3 Control Algorithms

This section examines the implementation of various algorithms used in control synthesis of self-tuning strategies. These in short include solution of Diophantine equations, solution of Riccati equations, state-estimation, spectral factorization and solutions to finite-stage receding horizon optimizations. A majority of the control schemes proposed in the literature use one or more of the algorithms discussed here.

1. Initialize at time $t = 0$;
$R_n^e(0) = \delta \mathbf{I}, R_n^r(-1) = \delta \mathbf{I}$
$F_n(0) = 0, r_n(0) = 0, \quad \text{for } n = 0 \rightarrow M - 1$
2. At time $t - 1$ store;
$R_n^e(t-1), F_n(t-1), R_n^r(t-1), r_n(t-1) for n = 0 \rightarrow M - 1$
3. At time $t - 1$ compute;
$K_n(t-1) = F_n^T(t-1)[R_n^r(t-2)]^{-1}$
$K_n^*(t-1) = F_n(t-1)[R_n^e(t-2)]^{-1}$
$K_n^y(t-1) = [F_n^y(t-1)]^T[R_n^r(t-2)]^{-1}$
4. Compute $\hat{x}_n(t)$ for $n = 1 \rightarrow M$
$\hat{x}_n(t) = \hat{x}_{n-1}(t) + K_{n-1}r_{n-1}(t-1)$
$\hat{x}_0(t) = 0$
$\hat{y}_n(t) = \hat{y}_{n-1}(t) + K_{n-1}^y r_{n-1}(t-1)$
5. Compute for $n = 0 \rightarrow M - 1$;
$R_n^r(t-1) = \beta R_n^r(t-2) + [1 - \lambda_n(t)]r_n(t-1)r_n^T(t-1)$
$\lambda_{n+1}(t) = \lambda_n(t) + [1 - \lambda_n(t)]^2 r_n^T(t-1)[R_n^r(t-1)]^{-1} r_n(t-1)$
$\lambda_0(t) = 0$
6. $x(t)$ is recieved and compute for $n = 0 \rightarrow M - 1$
$r_n(t) = r_{n-1}(t-1) - K_{n-1}^*(t-1)e_{n-1}(t)$
$e_n(t) = x(t) - \hat{x}_n(t)$
$\varepsilon_n = y_n(t) - \hat{y}_n(t)$
$R_n^e(t) = \beta R_n^e(t-1) + [1 - \lambda_n(t)]e_n(t)e_n^T(t)$
$F_n(t) = \beta F_n(t-1) + [1 - \lambda_n(t)]r_n(t-1)e_n^T(t)$
$F_n^y(t) = \beta F_n^y(t-1) + [1 - \lambda_n(t)]r_n(t-1)\varepsilon_n^T(t)$
7. Goto 2.

Figure 4.5: The least squares lattice filter

4.3.1 Resolution of the Diophantine Identity

A problem frequently encountered in linear control system design is resolution of the so-called Diophantine equation of the form (e.g. in q^{-1} operator):

$$A(q^{-1})G(q^{-1}) + q^{-1}B(q^{-1})F(q^{-1}) = P(q^{-1}) \tag{4.9}$$

for unknown polynomials (G, F) given (A, B, P) polynomials. This equation is also commonly called the Bezout equation or the Bezoutian for the case where $P = 1$. The most common area where resolution of this identity is crucial is in an explicit pole-placement control scheme where a model of the form:

$$A(q^{-1})y(t) = B(q^{-1})u(t-1) \tag{4.10}$$

and a controller:

$$Hw(t) = Gu(t) + Fy(t) \tag{4.11}$$

are assumed. Combining equations 4.10 and 4.11 gives:

$$(AG + q^{-1}BF)y(t) = HBw(t-1) \tag{4.12}$$

In order to place the closed-loop poles at the zeros of $P(q^{-1})$, say, eqn.4.9 has to be resolved. Diophantine equations are also used in some model order reduction techniques (Wahlberg 1987). Before embarking on a survey of different methods of solving the equation 4.9 several properties have to be examined.

Property 4.1 *There exist a pair (G, F) which satisfy eqn.4.9 if and only if $P(q^{-1})$, $A(q^{-1})$ and $B(q^{-1})$ share the same greatest common divisor. (MacDuffee 1943)*

Property 4.2 *If A and B are coprime then the Sylvester matrix S*

$$\begin{bmatrix} 1 & 0 & 0 & 0 & 0 & 0 \\ a_1 & 1 & \cdots & b_0 & 0 & 0 \\ \vdots & & 1 & \vdots & b_0 & 0 \\ a_n & & \vdots & \vdots & & b_0 \\ & a_n & & b_m & & \vdots \\ & & a_n & & & b_m \end{bmatrix}$$

is of full rank.

Property 4.3 *If G and F are solutions to the Diophantine equation then:*

$$(G - q^{-1}RB) \quad \text{and} \quad (F + RA)$$

are also solutions for all polynomials R.

The next two sections examine the solution of the Diophantine equation using the above properties.

4.3.1.1 Solution via the Sylvester matrix

The identity 4.9 can be written as a set of simultaneous equations as follows:

$$\begin{bmatrix} 1 & 0 & 0 & 0 & 0 & 0 \\ a_1 & 1 & \cdots & b_0 & 0 & 0 \\ \vdots & & 1 & \vdots & b_0 & 0 \\ a_n & & \vdots & \vdots & & b_0 \\ & a_n & & b_m & & \vdots \\ & & a_n & & & b_m \end{bmatrix} \begin{bmatrix} g_0 \\ \vdots \\ g_{ng} \\ f_0 \\ \vdots \\ f_{nf} \end{bmatrix} = \begin{bmatrix} p_0 \\ \vdots \\ p_{np} \\ 0 \\ \vdots \\ 0 \end{bmatrix}$$

If A and B are coprime then from property 2 we know that the matrix S is of full rank. This in turn implies that the order of G and F polynomials are $deg.B-1$ and $deg.A-1$ respectively. The solution can be performed using any standard method (e.g. Gaussian elimination with pivoting) for the set of equations:

$$S\theta_c = \mathbf{p}$$

where S is the Sylvester matrix, θ_c is the vector of G and F parameters and $\mathbf{p}$ is the vector of P parameters. This method is perfectly adequate and quite simple to program if the matrix S is well conditioned. Unfortunately in adaptive control this can *not* be guaranteed. There are two principal ways where the Sylvester matrix can become ill-conditioned and these are briefly examined below:

1. Presence of common factors

   The Sylvester matrix loses rank when there are common factors present between the A and B polynomials. Although *exact* common factors occur with probability zero, near cancellations are frequently encountered. Three factors can be responsible for this:

   (a) The discrete poles and zeros of a system tend to map to a region close the (1, 0) point in the $\mathcal{Z}$-plane at fast sample-rates (Åström *et al* 1984). For plant with continuous-time zeros fast sampling leads to close cancellation.

   (b) Real plant may only be approximated by high order difference equations. But slow sampling masks high order dynamics. This implies that it is *possible* to over-parameterize real systems during identification if slow sample-rates are used in conjunction with high order models.

   (c) During the tuning transient of self-tuning control it is impossible to guarantee that the estimated parameters are such that there are no common factors even if the final convergence point is *unique*.

2. Small B parameters lead to badly conditioned S matrix. The size of the coefficients of the B polynomial is a function of sample-rate as well as the input/output scaling used (e.g. engineering units, $0 \rightarrow 100\%$, etc).

The discussion above implies that the order of A and B polynomials and the sample-rate of the controller have to be chosen with care if pole-placement control is used without some precautions.

Many solutions have been suggested to remedy this shortcoming of pole-placement control. These include:

1. *Leal and Goodwin (1985)*

   Check the determinant of S. Given a minimum value d it is possible to perturb the estimates using a simple binary search such that a set $\hat{\theta}_c$ is obtained where the determinant is equal to this minimum. The method was shown to be globally convergent. Unfortunately, the schemes for obtaining the determinant of a matrix are not in general very reliable.

2. *Tuffs (1984)*

   Obtain the Moore-Penrose inverse of S' (i.e. $\theta_c' = S'^{\dagger}\mathbf{p}'$), where:

$$S' = \begin{bmatrix} 1 & \cdots & b_0 & 0 & 0 \\ \vdots & 1 & \vdots & b_0 & 0 \\ & \vdots & \vdots & & b_0 \\ a_n & & b_m & & \vdots \\ & a_n & & & b_m \end{bmatrix}$$

$$\theta_c' = [g_1, g_2, \cdots, f_0, \cdots, f_{nf}]^T$$

$$\mathbf{p}' = [p_1 - p_0a_1, \cdots, p_{np} - p_0a_n, 0, \cdots, 0]^T$$

   This can be done using orthogonal transformations (Householder, Givens, Modified Givens) or singular value decomposition (Lawson and Hanson 1974). The drawback here is that there is no longer any guarantee that the closed-loop poles (even asymptotically) are located at or close to the desired locations.

3. *Berger (1986)*

   Split the closed-loop poles into two parts: $T(q^{-1})$ the poles which *must* be placed and $1 + q^{-1}C(q^{-1})$ the poles which may be adjusted within a specified bound to improve the conditioning of the problem. The Diophantine equation 4.9 becomes:

$$A(q^{-1})G(q^{-1}) + q^{-1}B(q^{-1})F(q^{-1}) - q^{-1}T(q^{-1})C(q^{-1}) = T(q^{-1})$$

   The algorithm suggested by Berger (1986) requires the bounds on the coefficients of the C-polynomial, and a preset upper-bound on the coefficients of the F polynomial. Coefficients of C are adjusted to reduce the size of F.

4. *Edmunds (1976), Alix* et al *(1982)*

   The Diophantine eqn.4.9 can be rewritten for any signal $\zeta(t)$ as:

$$G(q^{-1})(A(q^{-1})\zeta(t)) + F(q^{-1})(B(q^{-1}\zeta(t-1)) = P(q^{-1})\zeta(t) \qquad (4.13)$$

   Coefficients of the G and F polynomials can now be *estimated* using the RLS routine described in the earlier section. A very simple approach is to perform a single iteration of this pseudo estimation technique per control sample. ζ is usually chosen as an uncorrelated random sequence.

5. *Middleton and Goodwin (1986)*

   In cases of rapid sampling where the coefficients of the B-polynomial become very small and the poles and zeros of the discrete system cluster around the (1, 0) point in the $\mathcal{Z}$-plane the δ operator can replace the q^{-1} operator. It was demonstrated by the authors that the condition number of the Sylvester matrix improves significantly using such a scheme.

4.3.1.2 The Euclidean Algorithm

A second method of solving the Diophantine eqn.4.9 is to use property 1. The proof of property 1 is constructive (see MacDuffee 1943, Åström and Wittenmark 1984) and makes use of the so-called Euclidean division algorithm to obtain the greatest common divisor (gcd) of the two polynomials A and B. According to Knuth:

> *Euclid's algorithm which is found in Book 7, Propositions 1 & 2 of his* Elements *(circa 300 BC), and which many scholars conjecture was actually Euclid's rendition of an algorithm due to Eudoxus (circa 375 BC), may be called the granddaddy of all algorithms, because it is the oldest nontrivial algorithm which has survived to the present day.*
>
> D. Knuth, Semi-numerical Algorithms pp.249 (1969)

```
Begin
    If deg.A > deg.B Then
        X = A; Y = B;
    Else
        X = B; Y = A;
  Endif
    Begin
        While R ≠ 0 DO
            X/Y = Q + R/Y
            X = Y; Y = R;
    End
End
```

Figure 4.6: The basic Euclid's algorithm

The penultimate value of R in the algorithm of Fig.4.6 is the gcd of the two polynomials A and B. Once the gcd is obtained two basic solutions can be employed:

- The gcd can be factored out of both polynomials and the Diophantine solved using any of the methods discussed.
- Property 1 can be used to arrive at a generalized form of Euclid's algorithm for solving the Diophantine equation (Aho *et al* 1974).

Begin
$X_0 = 1; Y_0 = 0; Y_1 = 1; i = 1; A_0 = B; A_1 = A$
While A_i does not divide A_{i-1}
Begin
$Q = A_{i-1}/A_i$
$A_{i+1} = A_{i-1} - QA_i$
$X_{i+1} = X_{i-1} - QX_i$
$Y_{i+1} = Y_{i-1} - QY_i$
$i = i + 1$
End
End

Figure 4.7: Generalized Euclid's algorithm

The second solution leads to the generalized Euclid's algorithm (see Fig.4.7). This algorithm solves the standard problem $AX + BY = gcd(A, B)$ where A, B, X and Y can be in any operators.

A more efficient and elegant algorithm using a truly recursive formulation was derived by Aho *et al* (1974). In addition, parallel implementations of the Euclidean scheme using systolic arrays have been proposed by Brent and Kung (1982).

Clearly, the most critical stage of the Euclidean algorithm is in obtaining the gcd of the two polynomials and deciding whether the R the remainder at the final stage is zero. Usually a threshold is set to to check the "smallness" of the remainder R. Unfortunately, the size of the final R depends upon the proximity of the cancelling zeros of A and B, their "distance" from the rest of the zeros and the scaling of the polynomials. A "reasonable" choice for the scaling of the polynomials appears to be such that the resulting coefficients are set to numbers between -1 and 1. To the best of the author's knowledge there is no *a-priori* scaling factor which makes the choice of the threshold for R a meaningful one.

4.3.2 Diophantine Equations in Multivariable Systems

Multivariable pole-placement control using dynamic output feedback and matrix fraction description of systems was studied by Wolovich (1974) and later suggested as a viable self-tuning strategy by Wellstead and Prager (1981). Elliot, Wolovich and Das (1984) subsequently extended the approach for more general systems (i.e. systems with different observability and controllability indices). See also Elliot and Wolovich (1984) for a rapprochement of discrete *vs.* continuous designs and model-following *vs.* pole-placement control.

In most designs the multivariable systems are described in matrix fraction forms where:

$$\mathbf{y}(t) = \mathbf{A}_L^{-1}(q^{-1})\mathbf{B}_L(q^{-1})\mathbf{u}(t) \quad (4.14)$$
$$\mathbf{y}(t) = \mathbf{B}_R(q^{-1})\mathbf{A}_R^{-1}(q^{-1})\mathbf{u}(t) \quad (4.15)$$

A and **B** are polynomial matrices and all delays are absorbed in the **B** polynomial matrix. Two basic operations are frequently required with pole-placement control depending on the particular realization:

- Given $\mathbf{A}_L^{-1}\mathbf{B}_L$ find a right coprime matrix fraction description $\mathbf{B}_R\mathbf{A}_R^{-1}$ and *vice versa.*

- Obtain the minimal degree solution to the Diophantine equation

$$\mathbf{A}_L\mathbf{G}_R + \mathbf{B}_L\mathbf{F}_R = \mathbf{P}_L \tag{4.16}$$

$$\mathbf{G}_L\mathbf{A}_R + \mathbf{F}_L\mathbf{B}_R = \mathbf{P}_R \tag{4.17}$$

A method based on the Euclidean algorithm (MacDuffee 1943) is given in Kučera (1979). Here we outline another algorithm due to Wolovich and Antsaklis (1984). A state-space description of eqn.4.14 or eqn.4.15 can be obtained via Wolovich's Structure Theorem (Wolovich 1974, Kailath 1980):

$$\begin{aligned} \mathbf{x}(t+1) &= \mathcal{A}\mathbf{x}(t) + \mathcal{B}\mathbf{u}(t) \\ \mathbf{y}(t) &= \mathcal{C}\mathbf{x}(t) \end{aligned}$$

provided the polynomial matrix $\mathbf{A}_R(\mathbf{A}_L)$ is column (row) proper or reduced (see Kailath 1980). It can be shown that:

$$\begin{aligned}
\mathbf{S}_R^{\nu}\mathbf{B}_R &= \overline{\mathbf{M}}\mathbf{S}_R^{\mu} + \hat{\mathbf{M}}\mathbf{A}_R \\
\overline{\mathbf{M}} &= [\overline{\mathbf{M}}_1, \overline{\mathbf{M}}_2, \cdots, \overline{\mathbf{M}}_p]^T \\
\overline{\mathbf{M}}_j &= [c_j^T, c_j^T\mathcal{A}, \cdots, c_j^T\mathcal{A}^{\nu_j-1}], \text{ where } c_j \text{ is the } j^{th} \text{ row of } \mathcal{C} \\
\hat{\mathbf{M}} &= [\hat{\mathbf{M}}_1, \hat{\mathbf{M}}_2, \cdots, \hat{\mathbf{M}}_p]^T \\
\hat{\mathbf{M}}_j &= \begin{bmatrix} 0 & & \\ c_j\mathcal{B} & 0 & \\ c_j\mathcal{A}\mathcal{B} & c_j\mathcal{B} & \\ \vdots & & \\ c_j\mathcal{A}^{\nu_j-2}\mathcal{B} & & 0 \end{bmatrix} \begin{bmatrix} \mathbf{I} \\ q\mathbf{I} \\ q^2\mathbf{I} \\ \vdots \\ q^{\nu_j-1}\mathbf{I} \end{bmatrix} \\
\mathbf{S}_R^{\mu} &= \text{block diagonal } [1, \cdots, q^{\mu_j-1}]^T \\
\mathbf{S}_R^{\nu} &= \text{block diagonal } [1, \cdots, q^{\nu_j-1}]^T
\end{aligned}$$

where ν_j and μ_j are the observability and controllability indeces. It can then be shown that for any polynomial matrix $\mathbf{P}$ we can have:

$$\mathbf{P} = \overline{\mathbf{P}}\mathbf{S}_R^{\mu} + \hat{\mathbf{P}}\mathbf{A}_R$$

The algorithm then becomes:

$$\begin{aligned}
\overline{\mathbf{G}}_L &= \overline{\mathbf{M}}^{-1}\overline{\mathbf{P}}_R \\
\mathbf{G}_L &= \overline{\mathbf{G}}_L\mathbf{S}_R^{\nu} \\
\mathbf{G}_L\mathbf{B}_R &= \overline{\mathbf{G}}_L\overline{\mathbf{M}}\mathbf{S}_R^{\mu} + \tilde{\mathbf{B}}_R\mathbf{A}_R \\
\mathbf{F}_L &= \hat{\mathbf{M}} - \tilde{\mathbf{B}}_R
\end{aligned}$$

A similar algorithm can be found for the left matrix fraction description of the system by considering the transposed system (see Kailath 1980). Next we consider the algorithm for obtaining the left coprime matrix fraction description of the system given the right matrix fraction description.

$$
\begin{aligned}
\mathbf{P}_L &= \mathbf{D}_R^{\nu}\mathbf{B}_R \\
\mathbf{D}_R^{\nu} &= \text{diag}\,[q^{\nu_j}] \\
\mathbf{P}_L &= \overline{\mathbf{P}}_L\mathbf{S}_R^{\mu} + \hat{\mathbf{P}}_L\mathbf{A}_R \\
\overline{\mathbf{H}}_L &= \overline{\mathbf{M}\mathbf{P}}_L \\
\mathbf{H}_L &= \overline{\mathbf{H}}_L\mathbf{S}_R^{\nu} \\
\mathbf{K}_L &= \hat{\mathbf{P}}_L - \overline{\mathbf{H}}_L\hat{\mathbf{M}} \\
\mathbf{D}_R^{\nu} &= \mathbf{H}_L\mathbf{B}_R + \mathbf{K}_L\mathbf{A}_R \\
\mathbf{A}_L &= \mathbf{D}_R^{\nu} - \mathbf{H}_L \\
\mathbf{B}_L &= \mathbf{K}_L
\end{aligned}
$$

4.3.3 The state-space framework

State-space self-tuners have been subject of some interest in recent years (Lam 1980, Warwick 1982, Samson 1982, Shieh *et al* 1982). Many new theoretical results as well as computational algorithms have been proposed. The majority of these designs are based on the observable canonical representation of the system which is of course closely linked with CARMA and CARIMA models. The system model:

$$A(q^{-1})y(t) = B(q^{-1})u(t-1) + C(q^{-1})\xi(t)$$

can be described as:

$$
\begin{aligned}
\mathbf{x}(t+1) &= \mathbf{A}\mathbf{x}(t) + \mathbf{b}u(t) + \mathbf{e}\xi(t) \\
y(t) &= \mathbf{c}^T\mathbf{x}(t) + \xi(t)
\end{aligned}
\tag{4.18}
$$

First we consider the state-estimation algorithm. Recall that the representation is in the "innovations" form (Åström and Eykhoff 1970) and it is easy to show that the appropriate state-space Kalman filter equations are given by:

$$\hat{\mathbf{x}}(t+1|t) = [\mathbf{A} - \mathbf{e}\mathbf{c}^T]\hat{\mathbf{x}}(t|t-1) + \mathbf{b}u(t) + \mathbf{e}y(t) \tag{4.19}$$

which in transfer-function form becomes:

$$\hat{\mathbf{x}}(t|t-1) = [q\mathbf{I} - \mathbf{A}^*]^{-1}[\mathbf{b}u(t) + \mathbf{e}y(t)] \tag{4.20}$$

where:

$$
\mathbf{A}^* = \begin{bmatrix} -c_1 & 1 & \cdots & 0 \\ -c_2 & 0 & 1 & 0 \\ \vdots & 0 & \ddots & 1 \\ -c_n & 0 & \cdots & 0 \end{bmatrix}
$$

Lam (1980, 1982) suggests an efficient method of obtaining the state-estimates using a transmittance matrix. It can be shown through some tedious but straightforward algebra that:

$$
\begin{aligned}
[q\mathbf{I} - \mathbf{A}^*]^{-1} &= \frac{1}{C(q^{-1})} \begin{bmatrix} q^{-1} & q^{-2} & & \cdots & q^{-n-1} \\ \beta_2 & q^{-1}\alpha_2 & q^{-2}\alpha_2 & \cdots & q^{-n}\alpha_2 \\ q\beta_3 & \beta_3 & q^{-1}\alpha_3 & \cdots & q^{-n+1}\alpha_3 \\ \vdots & & \ddots & & \\ q^{n-2}\beta_n & q^{n-1}\beta_n & \cdots & q^{-1}\alpha_n & q^{-2}\alpha_n \\ 0 & & \cdots & & q^{-1}\alpha_{n+1} \end{bmatrix} \\
&= \frac{1}{C(q^{-1})}\mathbf{M}_q^* \qquad (4.21)
\end{aligned}
$$

where:

$$
\begin{aligned}
\alpha_i &= 1 + c_1 q^{-1} + \cdots + c_{i-1} q^{-i+1} \\
\beta_i &= \alpha_i - C(q^{-1})
\end{aligned}
$$

and subsequently

$$
\begin{aligned}
\mathbf{M}_y\tilde{\mathbf{y}} &= \mathbf{M}_q^*\mathbf{e}y(t) \\
\mathbf{M}_u\tilde{\mathbf{u}} &= \mathbf{M}_q^*\mathbf{b}u(t) \\
\tilde{\mathbf{y}} &= [y(t-1), \cdots, y(t-n)]^T \\
\tilde{\mathbf{u}} &= [u(t-1), \cdots, u(t-n)]^T
\end{aligned}
$$

The shift structure of the $\mathbf{M}_q^*$ matrix can then be used to arrive at a computationally efficient way of computing the states of the system (see Lam 1982 and Clarke *et al* 1985 for `FORTRAN` coding).

Peterka (1986a,b) uses the alternative of Kalman filtering with the **LD** filter update described in the estimation section. He also considers the case where the model is expressed in the δ' operator where $\delta' = q - 1$. It is then relatively straightforward to show that:

$$
\begin{aligned}
\hat{\mathbf{x}}(t|t) &= (\mu\mathbf{I} + \mathbf{C}(t))\hat{\mathbf{x}}(t-1|t-1) - (\bar{\mathbf{a}} - \tilde{\mathbf{c}}(t))y(t) \\
&+ (\bar{\mathbf{b}} - \tilde{\mathbf{c}}(t)b_0)u(t-1) \qquad (4.22) \\
\mathbf{C}^t &= \begin{bmatrix} \tilde{c}_1(t) & 1 & \cdots & 0 \\ \tilde{c}_2(t) & 0 & 1 & 0 \\ \vdots & 0 & \ddots & 1 \\ \tilde{c}_n(t) & 0 & \cdots & 0 \end{bmatrix} \\
\tilde{\mathbf{c}} &= [\tilde{c}_1(t), \tilde{c}_2(t), \cdots, \tilde{c}_n(t)]^T \\
\bar{\mathbf{a}} &= [c_1, a_2, \cdots, a_n]^T \\
\bar{\mathbf{b}} &= [b_1, b_2, \cdots, b_n]^T
\end{aligned}
$$

The update of the $\tilde{c}$ parameters is given by dyadic reduction of the prearray of eqn.4.23 to eqn.4.24.

$$\begin{bmatrix} \mathbf{k}_f & \mathbf{HL}_x(t-1) \end{bmatrix} \begin{bmatrix} 1 & 0 \\ 0 & \mathbf{D}_x(t-1) \end{bmatrix} \begin{bmatrix} \mathbf{k}_f^T \\ \mathbf{HL}_x^T(t-1) \end{bmatrix} \tag{4.23}$$

$$\begin{bmatrix} 1 & 0 \\ \tilde{\mathbf{c}}^T(t) & \mathbf{L}_x(t) \end{bmatrix} \begin{bmatrix} d_y(t) & 0 \\ 0 & \mathbf{D}_x(t) \end{bmatrix} \begin{bmatrix} 1 & \tilde{\mathbf{c}}^T(t) \\ 0 & \mathbf{L}_x^T(t) \end{bmatrix} \tag{4.24}$$

where:

$$\begin{aligned} \mathbf{H} &= \begin{bmatrix} \mathbf{I} \\ 0 \end{bmatrix} + \mu \begin{bmatrix} 0 \\ \mathbf{I} \end{bmatrix} \\ \bar{\mathbf{c}} &= [1, c_1, \cdots, c_n]^T \end{aligned}$$

c_i are the coefficients of the $C(q^{-1})$ polynomial and μ is set to 0 for CARMA models and 1 for δ' models.

Having obtained the state estimates the feedback gains of the controller must be computed. For LQG type designs this takes the form of solving the Riccati Difference Equation (RDE) for finite horizons and the Algebraic Riccati Equation (ARE) for infinite horizons. Recall the state equation is given by eqn.4.18 and for minimizing a quadratic cost of the form:

$$J = \mathcal{E}\left\{\sum_{i=1}^{N} y^2(t+i) + \lambda u^2(t+i-1)|t\right\}$$

we can employ the standard Riccati update:

$$\begin{aligned} \mathbf{Q} &= \mathbf{c}\mathbf{c}^T \\ \mathbf{P}(t+N) &= \mathbf{Q} \\ \textbf{For} \quad i = N \to 1 & \\ \mathbf{k}(t+i-1) &= [\lambda + \mathbf{b}^T\mathbf{P}(t+i)\mathbf{b}]^{-1}\mathbf{b}^T\mathbf{P}(t+i)\mathbf{A} \\ \mathbf{P}(t+i-1) &= \mathbf{Q} + \mathbf{A}^T\mathbf{P}(t+i)\mathbf{A} - \mathbf{A}^T\mathbf{P}(t+i)\mathbf{b}\mathbf{k}(t+i-1) \\ u(t) &= -\mathbf{k}(t)[\hat{\mathbf{x}}(t) + \mathbf{k}_f(y(t) - \mathbf{c}^T\hat{\mathbf{x}}(t)) \end{aligned}$$

The update of the matrix **P**(t+i) as with the parameter estimation and state estimation takes the by now standard form of prearrays and postarray updates of:

$$\begin{bmatrix} 1 & \mathbf{b}^T\mathbf{L}(t+i) \\ \mathbf{c} & \mathbf{A}^T\mathbf{L}(t+i) \end{bmatrix} \begin{bmatrix} \lambda & 0 \\ 0 & \mathbf{D}(t+i) \end{bmatrix} \begin{bmatrix} 1 & \mathbf{c}^T \\ \mathbf{L}^T(t+i)\mathbf{b} & \mathbf{L}^T(t+i)\mathbf{A} \end{bmatrix}$$

to:

$$\begin{bmatrix} 1 & 0 \\ \mathbf{k}' & \mathbf{L}(t+i-1) \end{bmatrix} \begin{bmatrix} \tilde{d} & 0 \\ 0 & \mathbf{D}(t+i-1) \end{bmatrix} \begin{bmatrix} 1 & \mathbf{k}'^T \\ 0 & \mathbf{L}^T(t+i-1) \end{bmatrix}$$

$$\mathbf{k}(t+i-1) = \mathbf{k}'/\tilde{d}$$

There are also "doubling algorithms" and "Chandrasekhar type" updates which in some sense are the control analogues of the fast least squares algorithms discussed earlier. For details see (Kailath 1980, Morf and Kailath 1975). Chandrasekhar type updates were shown to be numerically unstable with respect to round-off and error propagations (Verhagen and Van Dooren 1986).

The solutions to the ARE can be found either by the use of iterative methods to convergence or can be done via the solution of a generalized eigenvalue problem. For details and and the numerical algorithm see Pappas *et al* (1980), Van Dooren (1981).

4.3.4 Finite stage Receding Horizon Techniques

Self-tuning controllers based on receding horizon strategies where a finite-stage quadratic cost is minimized recursively have been applied to real-life processes ranging from laboratory scale to full scale industrial plant with considerable degree of success (see Fig.4.8).

Lambert, E., 1987	MIMO and SISO (GPC), Industrial soap drier
Lambert, M., 1987	SISO multi-tasking (GPC), Compliant robot arm
Al-Assaf, Y., 1987	SISO (GPC), Mill temperature control
De Keyser, R., 1985	SISO (EPSAC), Cutter suction dredger ship
	SISO (EPSAC), Residential heating plant
Karny, M., 1985	SISO (LQ) Thickness of roll metal in a cold rolling mill
	SISO (LQ) Temperature regulation in sintering zone of a rotary furnace
	SISO (LQ) Temperature and pressure regulation in a drum boiler

Figure 4.8: Some real applications of receding horizon control

Most of the methods are based on minimization of a finite stage quadratic cost of the form:

$$J = \mathcal{E}\left\{ \sum_{i=N_1}^{N_2} (y(t+i) - w(t+i))^2 + \sum_{i=1}^{NU} \lambda \Delta u^2(t+i-1) | t \right\}$$

N_1 is the initial costing horizon, N_2 is the final costing horizon, NU the control horizon and λ the control weighting. Here we only consider reliable methods of obtaining the GPC (Generalized Predictive Control) control signal. The relation of GPC and other control strategies is discussed in Mohtadi (1986), Mohtadi and Clarke (1986). In receding horizon GPC a set of future control signal moves is postulated based upon the knowledge of the required future trajectory and the output predictions given that the future control increments are zero (see Clarke *et al* 1987a,b). The set of postulated control signal increments is then given by:

$$\Delta \mathbf{u} = (\mathbf{G}^T\mathbf{G} + \lambda \mathbf{I})^{-1}\mathbf{G}^T(\mathbf{w} - \mathbf{f}) \tag{4.25}$$

where

$$\mathbf{G} = \begin{bmatrix} g_0 & & \\ g_1 & \ddots & \\ \vdots & & g_0 \\ & & \vdots \\ g_{N_2} & & g_{N_2-NU+1} \end{bmatrix}$$

g_is are the coefficients of the sampled step response of the system and λ is the control weighting, NU the control horizon beyond which it is assumed that the control is constant, $\mathbf{f}$ is the vector of output predictions given that there are no changes in the control signal sequence (i.e. open-loop prediction) and finally $\mathbf{w}$ is the vector of future reference trajectory.

The control computation can be split into two basic parts.

- Output prediction: computing $\mathbf{f}$ and $\mathbf{G}^T(\mathbf{w} - \mathbf{f})$
- Inversion of $(\mathbf{G}^T\mathbf{G} + \lambda\mathbf{I})$

The output prediction can be performed by iterating the plant model starting from the present states of the plant $\{y(t), \cdots, y(t-n), u(t-1), \cdots, u(t-m)\}$ assuming that $\Delta u(t+j) = 0$ for $j \geq 1$. $\mathbf{G}^T(\mathbf{w} - \mathbf{f})$ can be found recursively as

$$\sum_{i=1}^{N_2}(w(t+i) - f(t+i))\mathbf{r_i}$$
$$\mathbf{r_i} = [g_{i-1}, g_{i-2}, \cdots, g_0, 0, \cdots, 0]^T, NU \times 1 \text{ vector}$$

The matrix inversion in eqn. 4.25 can be performed by N_2 successive calls to the **LD** filter discussed earlier (note the use of the normal equations and the least squares solution) with the initial values of $\mathbf{L} = \mathbf{I}$ and $\mathbf{D} = \frac{1}{\lambda}\mathbf{I}$. The scalar product of the bottom row of $\mathbf{L}$ with the vector $\mathbf{G}^T(\mathbf{w} - \mathbf{f})$ gives $\Delta u(t)$. This method reduces the coding effort considerably but perhaps is not the most numerically efficient approach. An adavantage of the method is in the fact that the implementation of the multivariable controller is very similar to the SISO case (Mohtadi *et al* 1986). Although the **UD** or **LD** methods are numerically robust the direct method described here can become ill-conditioned with certain class of unstable and nonminmum-phase systems. Higher precision arithmetic or slight variation of the algorithm usually solves the problem. For more details on implementations see Mohtadi (1986).

An indirect approach yields the optimal control synthesis of Peterka (1984). First we need to recall two properties of matrices:

Property 4.4 *Lower-band row-stationary matrices commute*

Property 4.5 *Matrix* $\mathbf{G}$ $(N_2 \times N_2)$ *matrix can be written as* $\mathbf{G} = \mathbf{A}^{-1}\mathbf{B}$ *where*

$$\mathbf{A} = \begin{bmatrix} 1 & & & \\ a_1 & 1 & 0 & \\ a_2 & a_1 & & \\ \vdots & & 1 & \\ a_n & a_{n-1} & \cdots & \\ 0 & \cdots & & 1 \end{bmatrix} \qquad \mathbf{B} = \begin{bmatrix} b_0 & & & \\ b_1 & b_0 & 0 & \\ b_2 & b_1 & & \\ \vdots & & b_0 & \\ b_m & b_{m-1} & \cdots & \\ 0 & \cdots & & b_0 \end{bmatrix}$$

We therfore have

$$\begin{aligned} \Delta\mathbf{u} &= (\mathbf{G}^T\mathbf{G} + \lambda\mathbf{I})^{-1}(\mathbf{G}^T(\mathbf{w} - \mathbf{f}) \\ &= \mathbf{A}(\mathbf{B}^T\mathbf{B} + \lambda\mathbf{A}^T\mathbf{A})^{-1}\mathbf{B}^T(\mathbf{w} - \mathbf{f}) \end{aligned}$$

As we are only interested in the first element of the sequence and $\mathbf{A}$ has unity as its (1,1) element without loss of generality we can write:

$$(\mathbf{B}^T\mathbf{B} + \mathbf{A}^T\mathbf{\Lambda}\mathbf{A})\mathbf{u}' = \mathbf{B}^T(\mathbf{w} - \mathbf{f})$$

with the first element of $\Delta\mathbf{u}$ and $\mathbf{u}'$ the same. This will lead to the following equation:

$$\mathbf{UDU}^T\mathbf{u}' = \mathbf{B}^T(\mathbf{w} - \mathbf{f})$$

If NU and N_1 are also included then we have:

$$\mathbf{UDU}^T = \mathbf{B}^T\mathbf{\Omega}\mathbf{B} + \mathbf{A}^T\mathbf{\Lambda}\mathbf{A}$$

with $\mathbf{\Omega}$ and $\mathbf{\Lambda}$ as diagonal matrices. The **UD** factors can be found directly without formation of the matrix by repeated application of the modified Given's rotation (Gentleman 1973). As the matrix **G** is no longer formed the numerical properties of this update are better than the update of the inverse. The algorithm can stop here but we may require to remove modes of $A(q^{-1})$ from computations of **f**. Then the following decomposition for $\mathbf{B}^T$ may be employed:

$$\mathbf{B}^T = \mathbf{US} + \mathbf{F}^T\mathbf{A}$$

where **S** is a strictly lower-band and **F** is a lower-band matrix. We therefore have:

$$\mathbf{UDU}^T\mathbf{u}' = (\mathbf{US} + \mathbf{FA})(\mathbf{w} - \mathbf{f})$$

But recall that **S** has zeros on the diagonal and the first element of $\mathbf{u}'$ is not affected by **S** and we obtain:

$$\mathbf{UDU}^T\mathbf{u}'' = \mathbf{F}(\mathbf{Aw} - \mathbf{f}')$$

where $\mathbf{f}'$ is the predictions of the output signal before going through the $1/A$ filter. This is in fact the method adopted by Peterka (1984) in the development of the predictor based self-tuner although his derivation is different from the approach adopted here. The algorithm has good numerical properties but again with some unstable and negative going nonminimum-phase systems decomposition of $\mathbf{B}^T$ may become unstable numerically and alternative schemes or double precision arithmetic must be sought.

4.3.5 Spectral Factorization

Much of the polynomial based LQG design strategies, Wiener filtering as well as the design of safety measures to ensure stability of the Extended Least Squares or Recursive Maximum Likelihood estimators require projection of the roots of an appropriate polynomial into the stability region — otherwise known as spectral factorization (Hunt and Grimble Chapter 3 of this book, Peterka 1984, Chang 1961). In short the task of spectral factorization amounts to obtaining the polynomial $C(q)$ with all roots inside the stability region such that:

$$C'(q)C'(q^{-1}) = C(q)C(q^{-1})$$

where $C'(q)$ has roots within and without the stability region.

There are many proposed algorithms for obtaining the spectral factors. Here we only consider three basic methods:

1. Peterka's simple algorithm
2. Bauer-type algorithm
3. Newton-type algorithm

Algorithm 1 is the simplest to program but perhaps the slowest to converge. Algorithm 2 has its basis in extending finite stage LQ methods to infinite stage designs and as such its convergence rate is closely related to the convergence of the appropriate Riccati equation. Algorithm 3 is perhaps the most general and can be shown to have quadratic convergence as the name suggests. Details of each of the algorithms are described below:

4.3.5.1 Peterka's Algorithm (*Bohm* et al *1984*)

Consider the polynomial $C'(q^{-1})$. Multiplying it by its complex conjugate gives:

$$C'(q^{-1}) = c'_0 + c'_1 q^{-1} + \cdots + c'_n q^{-n}$$
$$M(q) = C'(q^{-1})C'(q)$$
$$M(q) = m_n q^{-n} + m_{n-1} q^{-n+1} + \cdots + m_0 + \cdots + m_n q^n$$

Initialise $C = 1$
1.$Q = q^{-n} M/C + R/C$
2.$C = q^{-n} Q(q)$
3.$C = C/c_0$
4.Goto 1

Figure 4.9: Peterka's spectral factorisation algorithm

The algorithm in Fig.4.9 is very efficient requiring only the polynomial division algorithm discussed earlier. A check for convergence could be added by examining the magnitude of the remainder R at each iteration. For practical purposes of self-tuning control a single iteration of the algorithm per control sample is sufficient (Bohm *et al* 1984). In self-tuning the polynomial M is reconstructed every sample but C is retained from the iteration of the previous sample. An important aspect of the algorithm is that at each stage the C-polynomial is stable although it may not be the spectral factor of C'. This is of great importance for the practical applicability of the method.

4.3.5.2 Bauer-type algorithm (*Bauer 1955, Kučera 1979*)

This algorithm is based on **LD** (or **UD**) factorization of a Toeplitz matrix **M** whose entries are the coefficents (m_0 to m_n) of the M polynomial.

$$\mathbf{M} = \begin{bmatrix} m_0 & m_1 & \cdots & m_n \\ m_1 & m_0 & \cdots & m_{n-1} \\ \vdots & & \ddots & \vdots \\ m_n & m_{n-1} & \cdots & m_0 \end{bmatrix}$$

The **LD** factorization of the **M** matrix is obtained. The entries of the **L** matrix are shifted diagonally up and to the left by one and the new entries for the top row of the $\mathbf{L}^T$ matrix are computed. These constitute the spectral factors of the polynomial C'. The rate of convergence of this algorithm is closely linked with the rate of convergence of the related Riccati equation (Anderson *et al* 1974). As before, however, a single iteration of the algorithm can be employed per control sample. The method is easily extendable to the multivariable case (see Kučera 1979 for a square root formulation and Rissanen 1973 for Cholesky factorization of block Toeplitz matrices).

4.3.5.3 Newton-Type algorithm (*Kučera and Vostry 1976, Wilson 1972*)

The final method is due to Wilson (1972) and is based on a Newton gradient algorithm. Briefly the algorithm simply consists of two successive stages:

1. Obtain X_t such that

$$C_t(q)X_t(q^{-1}) + C_t(q^{-1})X_t(q) = 2M$$

2. $C_{t+1} = \frac{1}{2}(C_t + X_t)$

and iterate until convergence. The method is very efficient and can be shown to posses quadratic convergence rates. Obtaining X_t requires solving an $n \times n$ set of simultaneous equations. This can be done very efficiently ($O(n^2)$ operations) if the structure of the appropriate matrix is taken into account. If parallel architectures are used such matrix computations can be carried out in $O(n)$ operations (Kung *et al* 1985). Recently a numerically efficient multivariable version of this algorithm was derived by (Jezek and Kučera 1986).

4.4 Concluding remarks

Some of the algorithms necessary for realization of a self-tuning controller have been discussed. We however glossed over some of the more fundamental issues associated with implementation of digital controllers in general such as:

- Implementation of digital filters: Regression type models can become very sensitive to numerical roundoff and coefficient represenation; δ models (Agarwal and Burrus 1975, Middleton and Goodwin 1986) are superior with respect to roundoff errors but the overall scheme may be more sensitive to noise; cascade of second/first order sections is the most robust form but does *not* allow for feed through terms necessary in control. For a thorough discussion see Hanselmann (1987).

- Effects of roundoff and quantization: With modern hardware 56 bit mantissa in floating point co-processors (e.g. 68881 maths co-processor on the 68020 board) it is inconceivable that roundoff will have an influence on usual problems but even with 12 bit A/D and D/A conversion frequently significant quantization effects are seen (e.g. in robotics applications).

- Choice of the controller structure: As with PID control issues such as forward-path and feedback-path realizations affect properties such as proportional and derivative-type kicks in the control action. Integral desaturation schemes and control limiting are also closely related to the structure adopted.

- Effects of input/output scaling on different algorithms are not fully understood and frequently an appropriate choice of scaling simplifies a numerically difficult problem especially when using signal processors and fixed point arithmetic.
- Choice of the hardware and software is closely linked and perhaps of prime importance. The hardware architecture dictates whether for example parallel processing and pipelining are possible, which in turn raises questions regarding the ability to implement specific algorithms in a parallel form.
- Signal preconditioning and filtering such as Nyquist and logical spike filtering also play very important roles in improving the overall quality of control.

These and many other issues pertaining to practical implementation of digital controllers have been addressed in Åström and Wittenmark (1984), Clarke (1985), Hanselmann (1987) and Wittenmark and Åström (1984).

Not all of the algorithms discussed in this chapter are necessary for developing a useful and numerically reliable self-tuning controller. The most salient and vital algorithm is the **LD** (**UD**) filter whose implementation is both stable and efficient. An issue not addressed here but inseparable from the algorithm design is the associated real-time programming aspects which can be found in Tuffs (Chapter 8 of this book).

4.5 Acknowledgement

The author is supported by the W.W. Spooner Junior Research Fellowship at New College, Oxford whose funding is gratefully acknowledged.

Bibliography

Agarwal, R., C. and C. S. Burrus (1975)
"New recursive digital filter structures having very low sensitivity and roundoff noise ":IEEE Trans. Circuits Syst., Vol. CAS-22, no. 12
Aho, A.V., J.E. Hopcroft and J.D. Ullman (1974)
"The design and analysis of computer algorithms":Addison-Wesley, Reading-Mass.
Al-Assaf, Y. (1987)
"Self-tuning control: Theory and applications":D.Phil Thesis, Dept. of Eng. Science, University of Oxford, in preparation
Alexander, S.T. (1986)
"Fast adaptive filters: A Geometrical approach":IEEE ASSP Magazine, October, Vol.3, No.4
Alix F., J.M. Dion, L. Dugard and I.D. Landau (1982)
"Adaptive control of nonminimum-phase systems: comparison of several algorithms and improvements":Richerche di Automatica, Vol.13, No.1
Anderson, B.D.O., K.L. Hitz and N.D.Diem (1974)
"Recursive algorithms for spectral factorization":IEEE Trans. Circuits and Systems, Vol.CAS-21, No.6, pp.742-750
Åström , K.J. and B. Wittenmark (1984)
"Computer controlled systems":Englewood Cliffs, NJ., Prentice Hall
Åström ,K.J. and Eykhoff,P. (1970)

"System identification, a survey":Report 7006, Lund Institute of Technology
Åström ,K.J. and Wittenmark,B. (1973)
"On self-tuning regulators":Automatica, Vol.9, No.2, pp.185-199
Åström ,K.J., Hagander,P. and Sternby,J. (1984)
"Zeros of sampled systems":Automatica, Vol.20, No.1, pp.31-38
Bard, Y. (1974)
"Nonlinear parameter estimation":Academic press
Bauer, F.L. (1955)
"Ein direktes iterationsverfahren zur hurwitzerlegung eins polynoms":Archiv der electrischen Ubertrangung, Vol.9, pp.285-290
Berger, C.S. (1986)
"A robust pole-placement algorithm for adaptive control":2nd IFAC workshop on adaptive systems in control and signal processing, Lund, Sweden
Bierman, G.J. (1977)
"Factorization methods for discrete system estimation":Academic Press, New York.
Bohm,J., Halouskova,A., Karny,M. and Peterka,V. (1984)
"Simple LQ self-tuning regulators":IFAC 9th World Congress, Budapest, Hungary
Bruijn, P.M., L.J. Bootsma and Verbruggen, H.B. (1980)
"Predictive control using impulse response models":IFAC symposium on digital computer applications to process control, Dusseldorf, Germany
C.L. Lawson and R.J. Hanson (1974)
"Solving least squares problems":Prentice Hall
Chang, S.S.L (1961)
"Synthesis of optimum control systems":McGraw Hill
Cioffi, J.M. and T. Kailath (1984)
"Fast recursive least squares transversal filters for adaptive filtering":IEEE Trans. Acoust. Speech Signal Process, Vol. ASSP-32, pp.304-337
Clarke,D.W. (1982)
"Model following and pole-placement self-tuners":Opt.Control App. and Methods, Vol.3, pp.323-335
Clarke,D.W. (1985)
"PID algorithms and their computer implementation":Trans.Inst.M.C., Vol.6, pp.305-316
Clarke,D.W. and Frost,P.J. (1979)
"Control BASIC for microcomputers":IEE Conference on Trends in On-line Computer Control Systems, Sheffield
Clarke,D.W. and Gawthrop,P.J. (1975)
"Self-tuning controller":Proc.IEE, Vol.122, No.9, pp.929-934
Clarke,D.W. and Zhang,L. (1984)
"Long-range predictive control using weighting-sequence models":OUEL report 1542/84
Clarke,D.W., Cope,S.N. and Gawthrop,P.J. (1975)
"Feasibility study of the application of microprocessors to self-tuning regulators":OUEL report 1137/75
Clarke,D.W., Kanjilal,P.P. and Mohtadi,C. (1985)
"A generalized LQG approach to self-tuning control. Part 1. Aspects of design":Int. J. Control, Vol.41, No.6, pp.1509-1523
Clarke,D.W., Kanjilal,P.P. and Mohtadi,C. (1985)
"A generalized LQG approach to self-tuning control. Part II. Implementation and simulation":Int. J. Control, Vol.41, No.6, pp.1525-1544
Clarke,D.W., Mohtadi,C. and Tuffs,P.S. (1987a)
"Generalized predictive control. Part 1: the basic algorithm":Automatica, Vol.23, No.2, pp.137-148
Clarke,D.W., Mohtadi,C. and Tuffs,P.S. (1987b)
"Generalized predictive control. Part 2: Extensions and interpretations":Automatica, Vol.23, No.2, pp.149-160
De Keyser, R.M.C. and Van Cauwenburghe, A.R. (1985)

"Extended prediction self-adaptive control":7th IFAC symposium on Identification and system parameter estimation, York, U.K.
Edmunds, J.M. (1976)
"Digital adaptive pole-shifting regulators": Ph.D. Thesis, Control Systems Centre, UMIST, U.K.
Elliott,H. and Wolovich,W.A. (1984)
"Parameterization issues in multivariable adaptive control":Automatica, Vol.20, No.5, pp.533-545
Elliott,H., Wolovich,W.A. and Das,M. (1984)
"Arbitrary adaptive pole placement for linear multivariable systems":IEEE Trans.Autom.Control, Vol.AC-29, No.3, pp.221-228
Friedlander, B. (1982)
"Recursive Lattice forms for spectral estimation":IEEE Trans. Acoust. Speech Signal Process, Vol. ASSP-30, pp.920-930
G. C. Goodwin, R. Lozano-Leal, D. Q. Mayne and R. H. Middleton (1986)
"Rapprochment between continuous and discrete model refernce adaptive control":Automatica, Vol.22, No.2, pp.199-209
Gawthrop, P.J. (1987)
"Continuous self-tuning control":Research studies press, Lechworth, U.K.
Gentleman W.M. (1973)
"Least squares computations by Givens transformations without square roots": J. Inst. Maths. Applic., Vol.12, pp.329-336
Goodwin,G.C. and Sin,K.S. (1984)
"Adaptive filtering, prediction and control":Prentice-Hall
Hanselmann, H. (1987)
"Implementation of digital controllers - a survey":Automatica, Vol.23, No.1, pp.7-32
Itakura, F. and S. Saito (1971)
"Digital filtering techniques for speech analysis and synthesis":Conference record, 7th international Congress Acoust. Budapest
Jezek, J. and V. Kučera (1985)
"Efficient algorithm for matrix spectral factorization":Automatica, Vol.21, No.6, pp.663-669
Jones, G. (1987)
"Programming in occam":Prentice Hall
Jota, F.G. (1987)
"The application of self-tuning control technique to a multi-variable process": D.Phil. Thesis, Oxford University Dept. of Eng. Science, U.K.
Jover, J.M. and T. Kailath (1986)
"A parallel architecture for Kalman filter measurement update and parameter estimation":Automatica, Vol.22, No.1,pp.43-59
Kailath, T. (1980)
"Linear Systems":Englewood Cliffs, NJ: Prentice Hall
Karny, M., A. Halouskova, J. Bohm, R. Kulhavy and P. Nedoma (1985)
"Design of linear quadratic adaptive control: theory and algorithms for practice":Supplement to the Journal Kybernetika, Vol.21
Knuth, D. (1969)
"The art of computer programming Vol.II: Seminumerical algorithms":Addison-Wesley, Reading-Mass.
Kučera , V. (1979)
"Discrete Linear Control: The Polynomial Equation Approach":Wiley, Chichester
Kučera , V. and Z. Vostry (1976)
"Expanding spectral density into correlation sequence":IEEE Trans., Vol.AC-21, pp.592-593
Kung, S.Y., H.J. Whitehouse and T. Kailath (1985)
"VLSI and Modern signal processing":Prentice Hall
Kuo,F. and Chen,C.T. (1985)
"Recursive algorithms for coprime fractions and diophantine equations":American Control Conference, Boston

Lam,K.P. (1980)
"Implicit and explicit self-tuning controllers":D.Phil Thesis, Oxford University and OUEL report 1134/80
Lam,K.P. (1982)
"Design of stochastic discrete time linear optimal regulators Part I. Relationship between control laws based on a time series approach and a state space approach":Int.J.Systems Sci., Vol.13, No.9, pp.979-1000
Lambert, E. (1987)
"The industrial application of long range prediction":D.Phil Thesis, Dept. of Eng. Science, University of Oxford, in preparation
Lambert, M. (1987)
"Adaptive control of flexible systems":D.Phil Thesis, Dept. of Eng. Science, University of Oxford, in preparation
Laub, A.J. (1985)
"Numerical linear algebra aspects of control design computers":IEEE Trans., Vol.AC-30, No.2, pp.97-108
Lee. D.T., M. Morf and B. Friedlander (1981)
"Recursive square root ladder estimation algorithms":IEEE Trans. Acoust. Speech Signal Process, Vol. ASSP-29, pp.627-641
Levenberg, K. (1944)
"A method for solution of certain nonlinear problems in least squares":Quart. Appl. Math., Vol.2, pp.164-168
Ljung, S. and L. Ljung (1985)
"Error propagation properties of recursive least squares adaptation algorithms":Quart. Appl. Math., Vol.2, pp.164-168
Ljung, L. and T. Soderstrom (1983)
"Theory and practice of recursive identification":The MIT press, Cambridge, Massachusets
Ljung,L., Morf,F. and Falconer,D. (1978)
"Fast calculation of gain matrices for recursive estimation procedures":Int.J.Control, Vol.27, No.1, pp.1-50
Lozano-Leal,R. and Goodwin,G.C. (1985)
"A globally convergent adaptive pole placement algorithm without a persistency of excitation requirement":IEEE Trans.Autom.Control, Vol.AC-30, No.8, pp.795-798
MacDuffee C.C. (1943)
"Vectors and Matrices"
Marple, S.L., Jr. (1987)
"Digital spectral analysis with applications":Prentice Hall
Marquardt, D.W. (1963)
"An algorithm for least-squares estimation of nonlinear parameters":J.Soc.Indust.Appl.Math., Vol.11, No.2, pp.431-441
Middleton,H.R. and Goodwin,G.C. (1986)
"Improved finite word length characteristics in digital control using delta operators":IEEE Trans., Vol.AC-31, No.11, pp.1015-1021
Mohtadi, C. (1986)
"Studies in advanced self-tuning algorithms":D.Phil Thesis, Dept. of Engineering Science, University of Oxford, U.K.
Mohtadi,C. and Clarke,D.W. (1986)
"Generalised predictive control, LQ, or pole-placement: a unified approach":CDC conference, Athens
Mohtadi,C., Shah,S.L. and Clarke,D.W. (1986)
"Generalized predictive control of multivariable systems":OUEL report 1640/86
Morf, M. and T. Kailath (1975)
"Square-root algorithms for least-squares estimation":IEEE Trans., Vol.AC-20, No.4, pp.1241-1245
Pappas, T., A.J. Laub and N.R. Sandell Jr. (1980)

"On the numerical solution of Discrete time Algebraic Riccati Equation":IEEE. Trans, Vol.AC-25, No.4, pp.631-642
Peterka, V. (1986a)
"Algorithms for LQG self-tuning control based on input-output delta models":2nd IFAC workshop on Adaptive systems in control and signal processing, Lund, Sweden
Peterka, V. (1986b)
"Control of uncertain processes: Applied theory and algorithms":Supplement to the Journal Kybernetika, Vol.22
Peterka,V. (1970)
"Adaptive digital regulation of noisy systems":IFAC Symposium on Identification and Process Parameter Estimation, Prague
Peterka,V. (1984)
"Predictor-based self-tuning control":Automatica, Vol.20, No.1, pp.39-50
R. P. Brent and H.T. Kung (1982)
"Systolic VLSI arrays for polynomial GCD computation":Tech. Rep., Comp. Sci. Dept., Carnegie-Mellon University
Rissanen, J. (1973)
"Algorithms for triangular decomposition of block Hankel and Toeplitz matrices with application to factoring positive matrix polynomials":Mathematics of Computation, vol.27, no.121, pp.147-154
Samson, C. (1982)
"An adaptive LQ controller for nonminimum-phase systems":Int. J. Control, Vol.35, No.1, pp.1-28
Seborg, D.E, T.F. Edgar, and S.L. Shah (1986)
"Adaptive control strategies for process control: a survey":AIChE Journal, Vol.32, No.6, pp.881-914
Shieh, L.S., C.T. Wang and Y.T. Tsay (1983)
"Fast suboptimal state-space self-tuner for linear stochastic multivariable systems":Proc.IEE, Pt.D. Vol.130, No.4, pp.143-154
Tuffs,P.S. (1984)
"Self-tuning control: algorithms and applications":D.Phil Thesis, Dept. Eng. Science, University of Oxford., U.K.
Tuffs,P.S. and Clarke,D.W. (1985)
"Self-tuning control of offset: a unified approach":Proc.IEE, Vol.132, Pt.D, No.3, pp.100-110
Van Dooren P., (1981)
"A generalized eigenvalue approach for solving Riccati equations":SIAM J. Sci. Stat. Comp., Vol.2, No.2
Verhaegen, P. and P. Van Dooren (1986)
"Numerical Aspects of different Kalman filter implementations":IEEE. Trans., Vol.AC-31, No.10, pp.907-918
Wahlberg, B. (1987)
"On the identification and approximation of linear systems":Ph.D. Thesis, Dept. of Electrical Engineering, Linkoping University, Sweden
Warwick, K. (1982)
"Self-tuning controllers via the state-space":Ph.D. Thesis, Dept. Electrical Engineering, Imperial College, U.K.
Wellstead, P.E., D. Prager and P. Zanker (1979)
"Pole assignment self-tuning regulator":Proc.IEE, Vol.126, pp.781-787
Wilson, G.T. (1972)
"The factorization of matricial spectral densities":SIAM J. appl. Math. Vol.23, pp.420
Wittenmark, B. and Åström , K.J. (1984)
"Practical issues in the implementation of adaptive control":Automatica, Vol.20, No.5, pp.595-605
Wolovich, W.A. (1974)
"Linear multivariable systems":Springer Verlag
Wolovich W.A. and P.J. Antsaklis (1984)
"The canonical Diophantine equations with applications":SIAM J. on Control and Opt., Vol.22, No.5, pp.777-787

Chapter 5

Simplified algorithms for self-tuning control

K. Warwick

5.1. INTRODUCTION

When applying a discrete time adaptive control algorithm to a system, it is often the case that a relatively large number of calculations must be made during every sampling interval, this is certainly true for self-tuning controllers. For slower processes; such as chemical, heating, this does not present too much of a problem, however when sample periods of m.secs or less are required; e.g. communications, signal processing, robotics; the inter sampling time available may not be sufficient to allow the desired algorithm to be implemented on the available hardware, simply because all the necessary calculations cannot be made in the available time.

The self-tuning controllers described here are based generally on the philosophy that during every sample period the following sequence of events must be followed: (a) sample the system output signal, (b) estimate the system characteristics, (c) form the new controller, (d) calculate and apply the new control input, then wait for a clock pulse before returning to (a). In this sequence, steps (b) and (c) can be particularly computationally expensive and rather complicated, and this results in problems such as:

(1) Improved, faster hardware is expensive, if the resulting controller is going to be used in a practical sense then the computing power employed must be reasonably inexpensive.

(2) Simplicity of the controller is imperative in order to achieve confidence in the controller and to enable performance monitoring and fault-diagnosis.

Simplifications and approximations to, and simplified versions of, the basic self-tuning control building blocks are considered here in order to not only reduce drastically the amount of computation required in the implementation of a self-tuning control algorithm, but also to reveal adaptive controllers which are merely improved approaches to the widely employed PID digital control alogrithms. In section 3 the discussion centres on approximations to parameter estimation schemes and it is shown how the use of a computationally more efficient estimation algorithm does not necessarily mean loss of convergence or accuracy properties. Model reduction techniques are shown, in section 4, to provide a possible route in the adaptive control of high order systems and simplified controller schemes, namely PID and deadbeat, are discussed in sections 5 and 6. The purpose of this text is to show that it is not always necessary or desirable to retain a complicated controller structure, and to present some of the alternative, simplified techniques which are available.

5.2. SYSTEM DEFINITIONS

A basic assumption is made that the system which it is wished to control can be represented by the discrete-time CARMA equation.

$$Ay(t) = q^{-k}Bu(t) + Ce(t) + d \tag{5.2.1}$$

in which the polynomials A, B and C are defined as:

$$\begin{aligned} A &= 1 + a_1q^{-1} + a_2q^{-2} + \ldots\ldots\ldots + a_{n_a} q^{-n_a} \\ B &= b_1 + b_2q^{-1} + \ldots\ldots\ldots\ldots + b_{n_b} q^{-(n_b-1)} \\ C &= 1 + c_1q^{-1} + c_2q^{-2} + \ldots\ldots\ldots + c_{n_c} q^{-n_c} \end{aligned} \tag{5.2.2}$$

Also $\{y(t)\}$ and $\{u(t)\}$ are the system output and input sequences respectively, obtained at time steps t=0, 1, 2, etc.; and q^{-1} is the backward shift operator such that $q^{-i}\, y(t) = y(t-i)$. $\{e(t)\}$ is regarded as a white noise sequence with zero mean and finite variance and $k\geq 1$ is representative of the integer part of the transport delay, so k time periods will elapse before any event occurring at the system input will have an effect on the system output. From the definition given for k, this implies that $b_1 \neq 0$. Finally the term d is a constant d.c. noise level or offset term, it will be seen that in most of the following sections this term is set to zero, this is solely in order to simplify the explanation process. In each case a non-zero d term can be dealt with quite readily by a simple addition to the procedure used.

In the controller descriptions which follow, it is assumed initially that the coefficients a_i, b_i, c_i, and where appropriate d, are known values, possibly having been previously identified. When the self-tuning controller is actually implemented though, as will be described in the next section, estimates of the parameters within the CARMA model (5.2.1) are recursively obtained, and these estimates are employed in place of the true system values, which are not in fact known at any time instant. The assumption that the system coefficients are known is merely made in order to ease controller explanation, the truth that the coefficients are unknown, but can be estimated, is then introduced later on.

5.3. PARAMETER ESTIMATOR SIMPLIFICATION

Many different identification techniques exist which can be employed to obtain recursive estimates of the parameters within an ARMA or more likely CARMA system model, Ljung and Soderstrom (1983). Important aspects to be considered when deciding on a particular type of estimator to use, are estimator stability properties and convergence rate. However when employed within a real-time adaptive controller, the amount of computation required in order to carry out an estimation update can be a prime consideration.

Within an adaptive controller the computation necessary for the parameter estimation recursion alone can often contribute the majority of the total required, especially if an implicit control objective is being used or if the system model is of high order. Problems can therefore be encountered when attempting to implement an adaptive controller simply because too much time is taken for the calculations to be made, with the blame lying firmly on the shoulders of the parameter estimator. This is especially true where a low cost scheme is required and/or where the available computing power is fairly restricted.

In this section a straightforward simplification is described which can be made on most of the computationally more expensive estimation techniques. Because of its general popularity and reported widespread use in the field of adaptive control, an extended recursive least squares algorithm is considered as the base on which the simplification is made. It must be remembered though that the approximation described can be applied to many other algorithms.

5.3.1 Simplified recursive estimator

If the degree of each of the CARMA model polynomials, n_a, n_b and n_c is standardised as n, then with k=1 and d=0 the system output can be written in terms of parameters and signals as:

$$y(t) = \theta^T \psi(t) + e(t) \tag{5.3.1}$$

where $\theta^T = (a_1, \ldots\ldots, a_n;\ b_1, \ldots\ldots, b_n;\ c_1, \ldots, c_n)$ (5.3.2)

and $\psi^T(t) = (-y(t-1), \ldots, -y(t-n);\ u(t-1), \ldots, u(t-n);\ e(t-1), \ldots\ldots, e(t-n))$ (5.3.3)

It must be noted that by the definition of θ above, the system is assumed to be time invariant. This need not necessarily be so although a time varying system would naturally require the parameters given in (5.3.2) to become functions of time.

At a particular time instant, t, estimates can be made of all the parameters contained in θ, these estimates being denoted by

$$\hat{\theta}^T(t) = (\hat{a}_1(t), \ldots\ldots, \hat{a}_n(t);\ \hat{b}_1(t), \ldots, \hat{b}_n(t);\ \hat{c}_1(t), \ldots, \hat{c}_n(t)) \tag{5.3.4}$$

By employing the parameter estimates and known input/output data, a guess can be made as to the new output value, before it is actually measured, the only unknown value being the noise e(t). The error between the actual output and the predicted output can be found, i.e.

$$\varepsilon(t) = y(t) - \hat{\theta}^T(t-1)x(t) \tag{5.3.5}$$

in which ε(t) is the prediction error and

$$x^T(t) = (-y(t-1), \ldots, -y(t-n);\ u(t-1), \ldots, u(t-n);\ \varepsilon(t-1), \ldots, \varepsilon(t-n)) \tag{5.3.6}$$

At time instant t the right hand side of (5.3.5) is therefore known exactly.

With regard to the modelling error $\varepsilon(t)$, the vector of parameter estimates can be updated by means of the equation:

$$\hat{\theta}(t) = \hat{\theta}(t-1) + K(t)\varepsilon(t) \tag{5.3.7}$$

For a standard extended least squares procedure the sequence of estimates $\{\hat{\theta}(t)\}$ are found by means of the equations

$$K(t) = \left[\beta + x^T(t)P(t-1)x(t)\right]^{-1}P(t-1)x(t) \tag{5.3.8}$$

and $$P(t) = \left[I - K(t-1)x^T(t)\right] P(t-1) /\beta \tag{5.3.9}$$

in which β is a scalar weighting factor, which can be variable with respect to time, see Wellstead and Sanoff (1981) for example, or for simplicity can be time invariant within the range $0.95 < \beta < 0.99$.

There are several minor variations on the theme of least squares algorithms, for instance (5.3.9) is sometimes encountered in the form:

$$P(t) = \left[I - K(t)x^T(t)\right] P(t-1) \tag{5.3.10}$$

with the other necessary equations, i.e. (5.3.5), (5.3.7) and (5.3.8) remaining as before. It is not really the aim of this section to discuss the effects of such modifications, but merely to note their possible occurrence before moving on.

For every recursion the set of equations (5.3.5), (5.3.8), (5.3.9) and (5.3.7) are calculated in order to obtain an update of the extended least squares parameter estimates. Although K(t) is a 3n x 1 column vector, P(t) is a 3n x 3n matrix and hence a matrix multiplicatin is necessary for the P(t) update in (5.3.9). For systems where n is necessarily large, i.e greater than 3 or 4, matrix multiplication of two 3n x 3n matrices can prove to be quite excessive, especially when the sampling time is restricted.

Normally in setting initial conditions for the covariance matrix on estimator start up, all diagonal elements are usually chosen to be very large values with all off-diagonals set to zero, i.e. $P(0) = \phi I$ where $\phi = 1000$ for example. Generally once the parameter estimations have homed in on their true values, the diagonal terms of P(t) reduce considerably and the off-diagonal terms remain relatively small with respect to the diagonals, although they do become non-zero.

By making the assumption that the diagonal terms on their own retain sufficient information, the covarience matrix P(t) can be assumed to be simply a diagonal matrix. The main effect of such a simplification is to drastically reduce the amount of necessary computation per recursion. Consider for example a fairly simple system with n = 3 of dimension 9 x 1. By taking into account the calculation P(t-1)x(t), necessary every iteration of the algorithm, in equation (5.3.8), the update for K(t), the number of operations necessary are as follows:

Standard Least Squares: 81 multiplications + 72 additions

Approximate Least Squares: 9 multiplications.

Obviously from these figures such a simplification to the estimation procedure most definitely reduces the number of calculations. Remember also that for the standard least squares technique a complete matrix multiplication is also required, hence leading to even greater computational benefits from the simplified case. Further, for the simplified version the covariance matrix is regularly symmetrical and all its diagonal elements are squared values, hence they are positive. Stronger stability conditions can therefore be obtained, Farsi, Karam and Warwick (1984).

Modifications can still be made to the basic set of equations and, as described in Farsi, Karam and Warwick (1984), an alternative form is to obtain the update parameter vector from

$$\hat{\theta}(t) = \hat{\theta}(t-1) + K(t-1)\varepsilon(t) \qquad (5.3.11)$$

with the remaining equations necessary, i.e. (5.3.5), (5.3.8) and (5.3.9), being as previously stated.

5.3.2 Simplified estimator implementation

Several studies are reported in Warwick, Farsi and Karam (1985), in which the simplified recursive estimator is incorportated within a pole placement self-tuning controller. In order to summarize the results only one particular example is described here, where pole placement controllers employing firstly a simplified estimator and secondly a standard least squares estimator were applied to the 2nd order system modelled by:

$$y(t) = 1.3y(t-1) - 0.767y(t-2) + u(t-1)$$
$$+2u(t-2) + e(t) - 0.5e(t-1)$$

where e(t) was a white noise signal with variance = 1.11.

For both the simplified and standard cases the desired closed loop pole polynomial was specified as:

$$T = 1 - 0.5q^{-1}$$

and for both types of self-tuner, the controller parameters were found, by means of an off-line calculation to result in a control input

$$u(t) = -\ 0.3032u(t-1) + 0.0032y(t)$$
$$+\ 0.1163y(t-1) \qquad (5.3.12)$$

It is therefore expected that the controller parameters for the simplified case in Fig 5.3.2, should converge to the values given in (5.3.12).

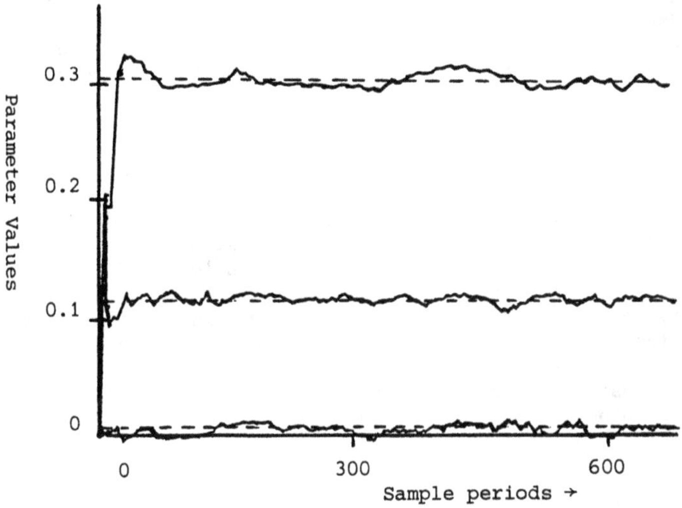

Fig. 5.3.1 Controller parameters for the simplified tuner

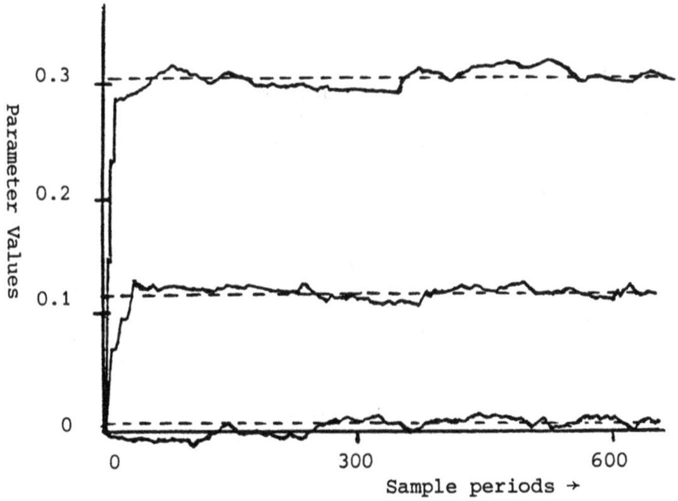

Fig. 5.3.2 Controller parameters for the standard tuner

It can be seen from the figures shown that in terms of parameter convergence and fluctuations there is little or no difference between the plots. Indeed this is also true as far as output signal servo following and regulation is concerned. The point to be remembered though is that the computational requirements for the simplified case are significantly less.

5.3.3 Simplified estimation - conclusions

Computational savings obtained when employing the simplified estimator are enormous if a comparison is made with a standard least squares procedure. Yet convergence properties and stability are similar to those of least squares and hence are much improved from simple techniques such as Stochastic Approximation. Further, the simplified estimator can readily be incorporated within a self-tuning controller and the resulting performance is similar to that obtained when a standard estimator is employed.

An overall conclusion obtained when implementing the simplified estimator is why bother to use the standard more complicated procedure when the simplified case performs just as well but is much easier to implement.

5.4. MODEL REDUCTION METHODS

For real-time controller design an assumption is often made that the system, which it is wished to control, is of low order, usually this means second or first order. For a system which is inherently of high order, the assumption means that the necessary calculations are certainly simplified. Simulation studies, Astrom (1980), have in fact shown that certain high order systems can be adequately controlled by low order representations without any significant deterioration in performance. Indeed when problems do occur, such as poor control or loss of control, an improved control action can in many cases be achieved by altering, usually this means increasing, the sampling interval.

If a low order controller is used directly on a high order system, it is often the case that a working closed loop system is obtained. However in a large number of cases the control action remains relatively poor throughout the range of sampling intervals which are possible. Employing a low order control scheme must necessarily result in a loss of information when the system is of a high order. If the information lost is important in terms of system characteristics, then no matter what adjustments are made to the low order conroller, the resulting performance will remain poor. As far as adaptive controllers are concerned, when a low order system parameter estimation scheme is included, any controller designed around the estimates will be based on, what might be, an incorrect initial picture of the actual system.

In order to avoid information loss due to incorrect system model order assumptions it is a sensible idea to base the controller on calculations made by using a 'correctly' ordered model of the system. For a high order system this would normally mean that a high order controller should result, although implementation of such a scheme means the imposition of a high computational loading. A further, and not unrelated, problem is encountered when a PID type of controller is desired. As will be described in the next section, such a requirement means that a fairly simple system model is employed, although it is preferable for this model to contain information relating to all of the important system characteristics

The method of self-tuning control described in this section employs a system parameter estimator which is of an order necessary to sufficiently characterise the plant, therefore it can be of high order when the plant is of high order. However in order to reduce the necessary computation or when the type of control is limiting, e.g. a PID controller places structural limitations, a model reduction procedure is employed in order to obtain a low order model of the system which nevertheless retains information on the system characteristics. The low order model is then employed for controller design. It must be emphasised that although one particular approach to the model reduction process is described in this section, the overall idea fits in with many reduction procedures, the particular advantage of the method described is its computational simplicity.

5.4.1 Model reduction

Before looking at the overall algorithm, let us consider the technique which is central to this discussion, namely model reduction. In fact there are three basic approaches to the problem of model reduction and control, and these are shown in Fig 5.4.1.

The first of these, which takes the left hand path in Fig 5.4.1., involves obtaining a high order controller from the high order system, the order of controller then being reduced by means of a selected model reduction procedure. The second path, the central limb in Fig 5.4.1, involves a combined reduction/synthesis procedure, whereas the third method, taking the right hand path, means that a low order model of the system is obtained, this model being subsequently employed in the controller synthesis stage.

One reason for using a model reduction procedure within a real-time algorithm is so that the computational requirement every sampling period is less than that necessary if the controller and the system are both of high order, Warwick (1985). The direct costing can in fact be stated as: The computation saved by employing model reduction must be greater than the computation expended in the reduction process. Of the three approaches depicted in Fig 5.4.1., the right hand path, in which the system order is reduced, results in the lowest computational requirement. This can be seen simply if it is considered that whether the left or right hand path is taken a reduction from high to low order must be carried out. For the right hand path however, the controller synthesis involves only a low order description, i.e. it is relatively simple, whereas with the left hand path the controller synthesis involves a high order description, necessarily involving far more computational effort.

In this section it is therefore considered that the order of the system model is reduced, enabling low order controller calculations to be made. Consider the high order system input-output transfer function to be:

$$G(q) = \frac{b_n + b_{n-1}q + \dots\dots + b_1 q^{n-1}}{a_n + a_{n-1}q + \dots\dots + a_1 q^{n-1} + q^n} = \frac{B}{A} \tag{5.4.1}$$

where, for simplicity of explanation, it has been assumed that $n = n_a = n_b$ and $k = 1$, both sides of the delayed version of the transfer function have then been multiplied by q^n, such that

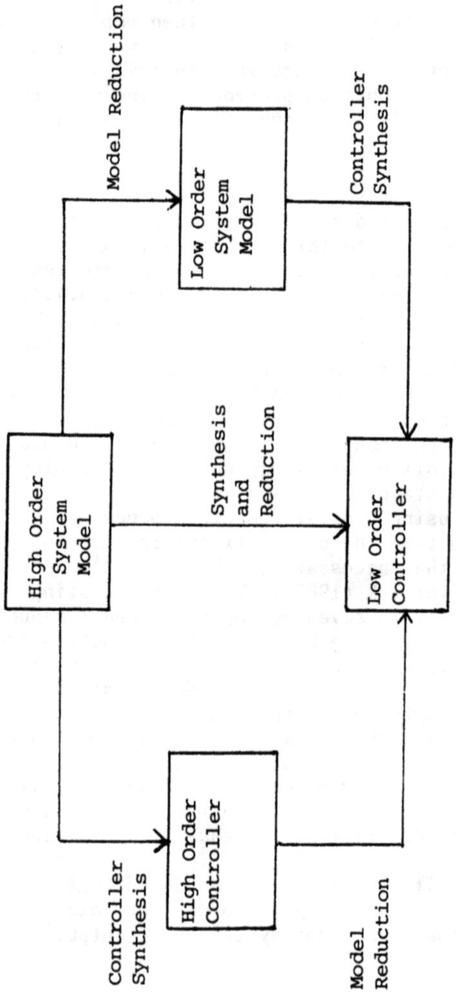

Fig. 5.4.1 Model reduction paths

$$y(t) = G(q)\ u(t) \tag{5.4.2}$$

If the denominator of (5.4.1) is divided into the numerator by taking terms on the right hand side first, i.e. by expanding around $q=\infty$, then we have also that

$$G\ (q) = \sum_{i=1}^{\infty} g_{-i} q^{-i} \tag{5.4.3}$$

where the parameters $\{g_{-i}: i=1\ ,\ldots\}$ are known as the Markov parameters.

The transfer function can also be regarded as a power series expansion about $q=1$, for steady-state conditions, and this requires a substitution $p=q-1$ in order to achieve an expansion about $p=0$ which can be written as

$$G(p) = \sum_{i=0}^{\infty} h_i p^i \tag{5.4.4}$$

where the parameters $\{h_i : i=0,1,\ \ldots\}$ are known as the H-parameters, Warwick (1984).

The low order model, which we wish to find, can also be introduced by writing it in the form of the transfer function

$$R(q) = \frac{D}{E} = \frac{d_z + d_{z-1}q + \ldots + d_1 q^{z-1}}{e_z + e_{z-1}q + \ldots\ e_1 q^{z-1} + q^z} \tag{5.4.5}$$

in which the index $z < n$.

For this reduced order model (5.4.5) its own Markov and H-parameters can be found by taking expansions around $z=\infty$ and $z=1$ respectively.

The model reduction problem can then be stated as one in which the selection of model parameters $\{e_i : i = 1, \ldots, z\}$ and $\{d_i : i = 1, \ldots, z\}$ is made such that the transfer function of the model is a good approximation to that of the system. This process assumes that the system parameters in (5.4.1) are known and that a good approximation can be found such that the important characteristics of the system also appear in the model.

Perhaps the simplest model reduction procedure, and certainly the approach described here, is to match the Markov and H-parameters of the model with those of the system. Only if model and system are of the same order, which cannot be the case if $z<n$, would it be possible for all of the Markov and H-parameters to be matched. Because an exact match is not possible an error will occur between system and model, this can be investigated by means of the error polynomial $W(q)$, where

$$W(q) = B(q)E(q) - A(q)D(q) \tag{5.4.6}$$

in which

$$W(q) = w_m + w_{m-1}q + \cdot \cdot \cdot + w_o q^m \qquad (5.4.7)$$

and $\qquad m = n + z - 1$

Note that the functional (q) is included in A(q) and B(q) for completeness only, these polynomials are A and B respectively.

Then, if w_i is set to zero for i=0 to j, this will result in a match between the first j+1 Markov parameters of the model with those of the system.

Further, the error polynomial is modified by substitution of q=p+1 in order to give:

$$W(p) = \bar{w}_m + \bar{w}_{m-1}p + \cdot \cdot \cdot + \bar{w}_o p^m \qquad (5.4.8)$$

By setting the first $\bar{w}_i$ to zero for i=m to m-λ, this will result in a match between the first λ+1 H-parameters of the model with those of the system.

The overall procedure is arranged as follows:
For a system of given order and a reduced model of a lower specified order, a choice is made of j and λ such that a set of simultaneous equations are obtained. These equations can be solved in order to find the model parameters from any set of system parameters.

5.4.2 Model reduction in real-time control

The overall adaptive controller involves, every sampling period, the following procedure. Once the system output has been sampled, an updated set of system (high order if necessary) parameter estimates are obtained by means of a Recursive Least Squares or other more appropriate technique. The updated estimates are then employed, as though they were the known system parameters, in the predefined set of model reduction equations, hence producing as a result, a low order model. Once this model has been found the new controller parameters can be obtained, for explicit algorithms this will involve further calculations, e.g. pole placement, although for implicit algorithms the low order model parameters can be used directly, see section 5.6.

Once the controller parameters have been obtained, the new control input can be calculated, by use of past input/output information, and subsequently applied. The whole sequence is then repeated, the breakdown of modules necessary being shown in Fig 5.4.2.

In order to carry out an investigation into the performance of an adaptive controller incorporating a real-time model reduction procedure, a simulation exercise was carried out on the system, introduced in Astrom (1980), described by

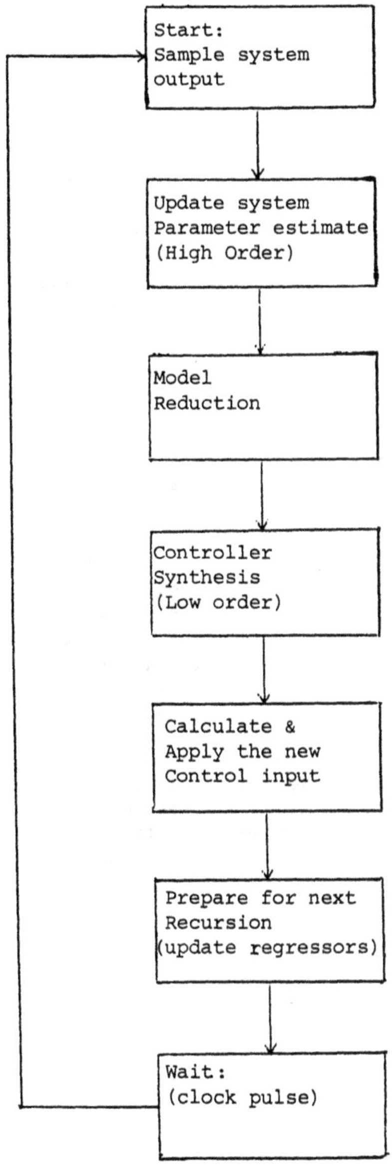

Fig. 5.4.2 Self-tuning control with on-line model reduction

$$y(t) - 0.142y(t-1) - 0.069y(t-2) + 0.0166y(t-3)$$
$$= 0.67u(t-1) + 0.638u(t-2) + 0.171u(t-3)$$
$$+ e(t) + 0.637e(t-1) \tag{5.4.9}$$

in which $\{y(t)\}$ and $\{u(t)\}$ are the sequence of system output and input signals respectively, also $\{e(t)\}$ is a white noise sequence with zero mean and unity variance.

In the exercise the parameters in a third order CARMA process model were recursively estimated using an extended least squares algorithm. For each recursion of the algorithm the updated parameter estimates were then passed to the set of model reduction equations, producing a low (2nd) order approximate model. The control objective was specified as that of pole placement with the closed loop pole polynomial selected as:

$$\text{Pole Polynomial} = 1 - 0.3q^{-1} - 0.18q^{-2}$$

which leads to poles at $q = 0.6, - 0.3$ and 0

The model reduction equations were designed in order to match the first two Markov parameters and the first two H-parameters of the model with those of the system. Hence four equations were required such that the four ($2z$, $z=2$) model parameters could be calculated. However one of the equations results in a model parameter being obtained directly and hence only three equations need to be solved every iteration.

In Fig. 5.4.3 some of the system parameter estimates are shown with respect to time during the simulation run. The low order model parameters are then shown in Fig. 5.4.4 and it can be observed that (i) the low order model parameters converge much more rapidly than the system parameter estimates, and (ii) less fluctuation occurs for each of the model parameters. There is very little point showing system input and output signal values over the simulation run, which merely look like a set of noise values. When compared with a pole placement controller based on a higher order model and applied to the same system, the controller incorporating model reduction exhibits a lighter control action and with the resultant properties common to many 'detuned' algorithms of slightly worse noise regulation but improved controller robustness.

5.4.3 Model reduction methods - conclusions

For an on-line controller as described, which incorporates a model reduction procedure, the computation required for the model reduction process itself is dependent on the order of the reduced model and not on that of the system. This means that computational savings can be great when compared to on-line schemes which use both a high order estimator and controller.

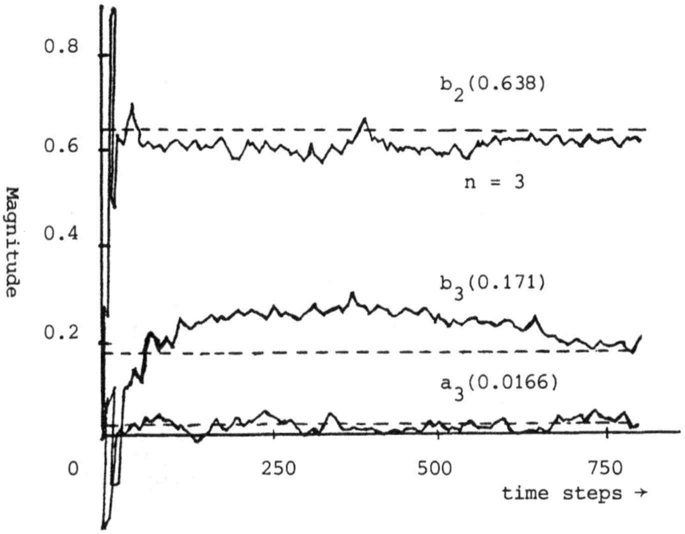

Fig. 5.4.3 Simulation system parameter estimates

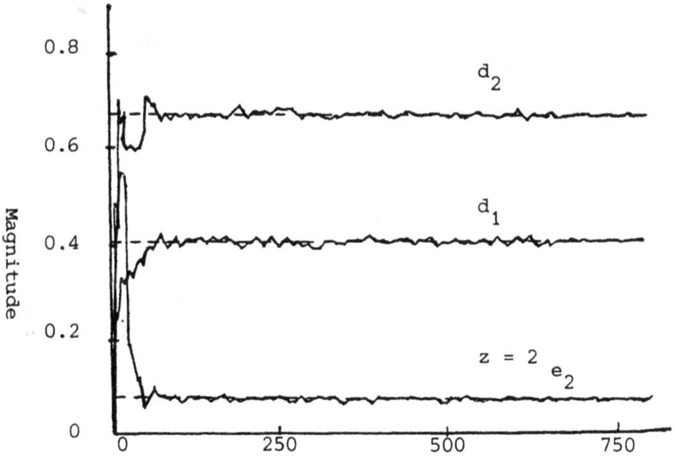

Fig. 5.4.4 Simulation reduced order model parameters

Although a pole placement control objective was employed in section 5.4.2 the model reduction procedure can be incorporated with any objective whether implicit or explicit.

Finally a point must be made as to the reasons behind a model reduction algorithm such as that described in this section. Perhaps it is best to remember that a common method of dealing with high order systems is to employ directly a low order estimator and controller - it seems to work and it is relatively simple. If the resulting control is poor, the sampling interval can be increased until 'reasonable' control is achieved, this necessarily means giving more and more weighting to the low frequency dynamics. When high frequency system components are of sufficient magnitude, increasing the sampling interval can have disastrous effects and, although initially an improved control action might be achieved, often complete loss of control or more likely lack of controller robustness to sudden parameter variations will result. When the model reduction technique is introduced however, both low and high frequency components can be retained in the low order model.

5.5. PID CONTROLLER STRUCTURE

Simplicity of controller design is one of the important features which has resulted in the PID controller form being employed in the vast majority of industrial controllers, this is particularly true in process industries. The widespread use of a PID structure is however also due to the fact that when the controller is well-tuned, satisfactory performance can usually obtained, especially in cases where the plant under control is essentially both Linear and Time Invariant. The tuning procedure is, for off-line manual adjustment, generally not too difficult to carry out as basically only three parameters, i.e. those associated with the Proportional, Integral and Derivative constants, need to be adjusted. However the procedure can be time consuming, particularly if high accuracy is required or if a large number of interactive loops are involved.

Manually tuned PID controllers, whose parameters are fixed, are perfectly capable of dealing satisfactorily with many systems. However plant variations, caused possibly by modifications or ageing, can result in a deterioration of performance and even, in some cases, to loss of control altogether. One solution is to retune the controller periodically, the frequency being dependent on how often or over what time period plant variations occur. This can though be both time consumming and expensive, and may well need to be done all over again very shortly. Further, the plant variations may well take place over a very short space of time, a few seconds for instance, and in these cases periodic manual retuning would not be sensible or even possible.

In this section, three types of adaptive PID controller are discussed. Firstly the classical methods of Ziegler-Nichols (1942) are incorporated into an automated tuning in phase, which is followed by control using fixed PID parameters. The second method is one in which the PID parameters are varied slowly on-line, the intention being to modify the controller parameters, in line with any plant variations, by witnessing response patterns such as the overshoot peaks caused on the plant output when a step input is applied, Higham (1986). The third method is more in keeping with the rest of this text, as a recursive parameter estimator directly feeds new plant parameters to the PID controller, which is therefore based on the very latest estimates. This latter method therefore results in very rapid controller parameter variations if the plant, and hence plant model, parameters vary rapidly.

Whichever approach is employed in order to derive an adaptive PID controller, it must be capable of providing controller flexibility in order to maximize the number of application possibilities. Any necessary adapting or tuning-in phases must be relatively rapid and should not require further tests which disrupt normal plant use. Finally it must be stated that controller reliability and robustness properties are important, especially when the full self-tuning format is applied, any possibility of the controller itself behaving in a peculiar manner, such as becoming unstable, would result in loss of confidence in the control action and/or a higher supervisory control level being used.

5.5.1 PID discrete-time structure

A continuous-time approach to PID self-tuning control can be found in Gawthrop (1986). In this section however the controller will be considered in a discrete-time format in the first instance. Unfortunately this does not mean there is just one correct discrete time PID form, Ortega and Kelly (1984), but rather that many versions are possible as long as proportional, integral and derivative actions are in some way applied. As mentioned in Ortega and Kelly (1984), a common element is the integral action which is applied to the error between the desired and actual ouput values, i.e. $v(t) - y(t)$. Both derivative and proportional actions are then appled in some combination to either the same error or to the actual output signal, $y(t)$, alone. In order to retain generality the form considered here will be one in which the three terms are all applied to the same error signal. In fact this links in with the continuous time PID control, which can be described as:

$$u(s) = K\left[1 + \frac{1}{T_i s} + T_d s\right][v(s) - y(s)] \qquad (5.5.1)$$

in which $u(s)$ is the Laplace Transform of the control input, K is the proportional gain, T_i is the integral (reset) time constant and T_d the derivative time constant. It is perhaps even simpler to consider this equation in the form:

$$u(s) = \left[\bar{K}_p + \frac{\bar{K}_i}{s} + \bar{K}_d s\right][v(s) - y(s)] \qquad (5.5.2)$$

where $\bar{K}_p$, $\bar{K}_i$ and $\bar{K}_d$ are the coefficients associated with proportional, integral and derivative feedback elements respectively, i.e. $K = \bar{K}_p = \bar{K}_i T_i = \bar{K}_d / T_d$.

Several techniques can then be employed in order to find a discrete time version of (5.5.2), the most common is that obtained by means of rectangular integration and Euler derivative approximations, Isermann (1981), whereby one obtains the control input as:

$$u(t) = \left[K_p + K_i(1-q^{-1})^{-1} + K_d(1-q^{-1}) \right] \left[v(t) - y(t) \right] \qquad (5.5.3)$$

where the terms K_p, K_i and K_d are dependent not only on the continuous time gains $\bar{K}_p$, $\bar{K}_i$ and $\bar{K}_d$ but also on the sample period. In fact when the sample period is short, the equation (5.5.3) is a good approximation to (5.5.2), although this is unfortunately not true for larger sample periods. The control action obtained in (5.5.3) also suffers from derivative kick and hence an alternative version which avoids this is also commonly encountered,

$$u(t) = - \left[K_p + K_d(1-q^{-1}) \right] y(t) + K_i(1-q^{-1})^{-1} \left[v(t) - y(t) \right] \qquad (5.5.4)$$

in which it can be seen that the proportional and derivative gains are applied to the actual output signal only. Despite its inherent drawbacks, for completeness, the description of PID controllers given in this section will concentrate on (5.5.3) rather than the modified version of (5.5.4). By rearrangement, equation (5.5.3) can also be written as:

$$(1-q^{-1})u(t) = \left[K_1 + K_2 q^{-1} + K_3 q^{-1} \right] \left[v(t) - y(t) \right] \qquad (5.5.5)$$

where $\quad K_1 = K_i + K_p + K_d$

$$K_2 = -(K_p + 2K_d) \qquad (5.5.6)$$

and $\quad K_3 = K_d$

The effect of the integral action on the control input is then apparent on the left hand side of (5.5.5), where also the relative simplicity of a PID control scheme is exhibited in that only three parameters K_1, K_2 and K_3 need to be selected.

5.5.2 Autotuning of PID controllers

The first methodology considered here is that in which the PID contoller parameters are obtained by means of an initial tuning-in phase, rather in the sense of an identification procedure, Nishikawa and Sannomiya (1981). The controller parameters obtained are then held constant; unless a further tuning-in procedure is demanded some time in the future. So no on-line tuning is incorporated, merely an automation of the initial controller parameter selection. The technique can therefore either be used alone, resulting in time invariant controller parameters, or as a method by which initial controller parameter values can be obtained for an on-line adaptive controller method.

Perhaps the simplest approach is that given by Astrom and Hagglund (1984) in which an automated version of the originally 'manual' Ziegler-Nichols (1942) tuning procedure is described. Essentially, as shown in Fig 5.5.1, limit-cycle oscillations are enforced on the system to be controlled by a relay, in order to obtain both angular frequency and critical gain values, which can be found from measurement of

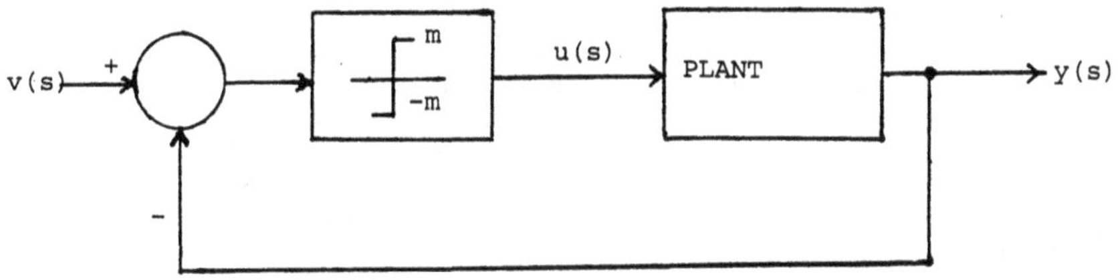

Fig. 5.5.1 Oscillation forcing relay for autotuning

oscillations of the signal [v(s)-y(s)] in terms of amplitude and frequency.

The critical gain is in fact also the relay again, which is given as $4m/\tau v$, where m is the relay saturation level and v is the amplitude of the signal [v(s)-y(s)]. So the level m is adjusted (to make $v \simeq v_{max}/10$) in order to obtain output signal oscillations, and hence it is then possible to find v by direct measurement. Oscillation frequency measurement can prove to be something of a problem in the presence of noise, although a fairly straightforward least-squares procedure can be employed if necessary, the standard method being to simply count the number of times per second the oscillations pass through the mean (zero) value whilst increasing. As well as the relay, an integrator can be included in the loop to make sure that the mean level is in fact zero. It is then fairly straightforward to calculate the PID controller parameters in order to achieve desired phase and gain margins.

5.5.3. PID control using pattern recognition

As a slightly different approach to the selection of PID controller parameter values, the Foxboro 'EXACT' system, Higham (1986), is based on the heuristic approach employed by Control Engineers for many years, rather than on, what can be, a complicated design algorithm.

On start-up of the 'EXACT' controller some initial values must be entered for the proportional, integral and derivative parameters. Although these initial values can be based on any available apriori plant and plant response knowledge, accuracy is not essential at this early stage, really any sensible values will suffice. A step change is then applied to the system and the resultant output response is monitored. A typical response is shown in Fig 5.5.2, where the measurement error = [v(t) - y(t)] and the period of the transient, T, is the time between the maximum value E_1 and the maximum value E_3. The damping provided by the system and any steady-state error can therefore also be recorded. The lead angle of the resultant PID controller is then given by the ratio of integral action to T, whereas the controller lag angle is given by the ratio of the derivative action to T.

Every time a further step change occurs, new controller parameters can be found by monitoring the new transient response. The step magnitude does not itself come into the calculations and hence this is of no consequence. It is imperative though that variations due to noise do not set the algorithm on a false calculation route and hence a noise band is defined, the algorithm not being set in operation until a deviation of magnitude equal to twice the noise band is observed. The controller then operates with its set of latest controller parameters until a step change of sufficient magnitude occurs. Once this has occurred, new controller parameters will be calculated and used in place of the old ones.

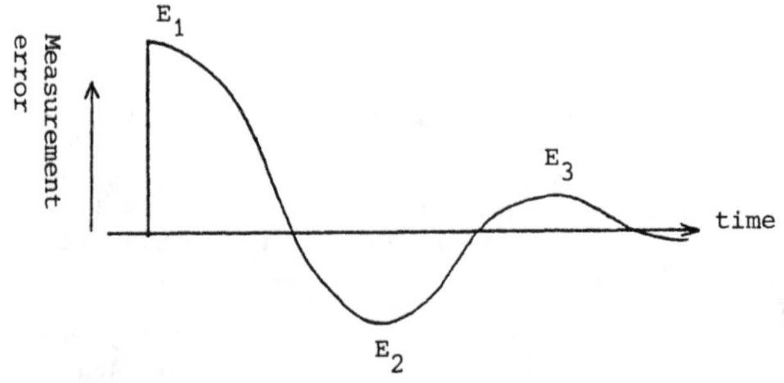

Fig. 5.5.2 Transient error response to a step change

As soon as a sufficient deviation is detected the algorithm requires the magnitude E_1 to be taken, subsequently the algorithm waits for values E_2 and E_3 to be found also. The period of the transient can then be obtained directly from the time taken, the overshoot from the ratio E_2/E_1 and the damping factor from the ratio $(E_3 - E_2)/(E_1 - E_2)$, both of these latter two being absolute values. Once the new controller parameters have been calculated from these values, the algorithm waits for the next transient to occur.

5.5.4 Pole placement PID controller

When considering a self-tuning PID controller in which recursive estimation is carried out in order to obtain the latest updates of the parameters within a model, these estimates can be employed via a desired control objective, subject to that objective being held within a PID framework. Although optimal control objectives can be enforced, or indeed as described in the next section, a deadbeat philosophy, in this section, in keeping with Ortega and Kelly (1984) and Wittenmark (1979), a pole placement approach will be used.

In order to consider a pole placement scheme, firstly let the control feedback equation (5.5.5) be rewritten in the form:

$$R\,u(t) = S\,[v(t) - y(t)] \qquad (5.5.7)$$

where $R = 1 - q^{-1}$ and S is a second order polynomial in the delay operator, q.

When a control input of the type (5.5.7) is applied to a system which can be described in terms of a CARMA model with d = 0 the closed loop equation which results is given by:

$$y(t) = \frac{1}{AR + z^{-k}BS}\,v(t-K) + \frac{CR}{AR + z^{-k}BS}\,e(t) \qquad (5.5.8)$$

The coefficients of the S polynomial, i.e. K_1, K_2 and K_3, must then be selected in order to meet the desired pole placement control objective. If the apriori operator selected pole polynomial is given as the polynomial T, then it is required that the parameters of S be chosen in order to satisfy the equation.

$$AR + z^{-k}BS = T \qquad (5.5.9)$$

in which A, B and k are system characteristics, T is the specified pole polynomial and R is the integral action. The only unknowns in equation (5.5.9) are therefore the coefficients of the S polynomial, i.e. three parameters in total. This means that as only three unknowns exist in (5.5.9), only three equations are required in total, neglecting the fact that unity terms will be equated. The maximum polynomial order on either side of the equation is thus 3, which means that the order of the left hand side of (5.5.9) must be:

$$\max\{n_a + 1 : n_b + k + n_s\} \leq 3 \qquad (5.5.10)$$

where n_a, n_b and n_s are the orders of polynomials A, B and S respectively.

It must be remembered though that n_s has been defined in (5.5.7) to be of order 2, $n_s=2$, and this means that if k=1, which is its minimum value, it follows that we must have $n_b=0$, and also $n_a \leq 2$.

This puts a distinct limit on the complexity of systems which can be dealt with completely by this PID controller in that the deterministic part of the system transfer function can be no more complicated than:

$$y(t) = \frac{b_1}{1 + a_1 q^{-1} + a_2 q^{-2}} u(t-1) \tag{5.5.11}$$

in which a_1 and/or a_2 can be zero, but by definition b_1 is non-zero.

For a self-tuning pole placement PID controller, every sample period the estimated parameters in a CARMA system model, restricted for the deterministic part to the form (5.5.11), must be updated and these new estimates used in (5.5.9) to find controller parameters in the polynomial S. Immediately S has been found the control input (5.5.7) can be employed, this will be of a PID form.

It can be noted that as $S = K_1 + K_2 q^{-1} + K_3 q^{-2}$, so the PID coefficients are given from the equation set (5.5.6) as:

$$\begin{aligned} K_d &= K_3 \\ K_p &= K_2 - 2K_3 \\ K_i &= K_1 + K_2 + K_3 \end{aligned} \tag{5.5.12}$$

5.5.5 PID controller - conclusions

The autotuning type of PID controller is computationally quite simple and has properties which are dependent directly on those of the Ziegler-Nichols tuning procedure itself, that is it does not always work although for the vast majority of cases it works satisfactorily. If the process is slow then so is the tuning-in procedure and hence results can be affected by unexpected disturbances of either high or low frequency.

The Foxboro 'EXACT' tuner is generally robust and is not detramentally affected by disturbances of high magnitude. The system response must follow the sort of pattern expected, see e.g. Fig 5.5.2, otherwise tuning will not take place. The tuning procedure, which is fairly time consuming following a transient can be fooled however and a sinusoidal input for instance is likely to cause the controller coefficients to drift.

PID controllers which are based on a recursively estimated system model, such as the pole placement type of self-tuning PID controller, have properties which are found generally with self-tuning controllers in that although they respond well to system parameter drift, robustness of the controller must always be in question to an extent and only local stability can be ensured. By imposing a PID controller criterion within a self-tuning framework this results in a relatively simple self-tuning algorithm which is computationally not too expensive, especially if the parameter estimates are obtained by an approximate method such as that described earlier.

5.6. DEADBEAT CONTROL ALGORITHMS

In the previous two sections methods have been considered which result in a much simpler controller structure being obtained, either because of, in the first instance, the system being thought of in a simpler way or, in the second case, the control form itself being assumed to have a simple description, namely PID. Controller simplicity can however also be obtained by initially neglecting any desired output value and concentrating purely on output regulatory action in the presence of noise. In this line an optimal type of control action was incorporated in, what has now become widely accepted as, one of the first self-tuning controllers by Astrom and Wittenmark (1973). Simplicity is also obtained by initially neglecting any disturbances affecting the system and concentrating on the servo following properties, rather in the form of a deterministic controller. Once an attempt is made to deal effectively with both the signal and noise at the same time, many technical difficulties arise and the necessary computation is much increased.

Much of the research carried out in the field of self-tuning control makes an implicit assumption either that processing of the noise is by far the most important control objective or that any servo following properties are obtained merely as a by-product of the disturbance reduction procedure. This seems a little disproportionate when it is considered that for many systems, particularly electrical and mechanical, a high signal to noise ratio already exists and the major control requirements are for reference tracking. In this section an emphasis is placed firmly on simple controllers designed with the principal aim of reference signal tracking. It must be noted however that this does not mean that (a) disturbances are ignored, or (b) transient response tailoring is achieved. This latter requirement can in fact be achieved, via a pole placement procedure, although necessarily a certain amount of simplicity is lost. The basic structure of the algorithm described is generally known as a deadbeat controller, Matko and Schumann (1984), although strictly speaking the deadbeat performance is lost once modifications are made to the original case. In this section several variations on the basic deadbeat controller are discussed and their respective properties when incorporated within an adaptive algorithm are considered. All of the control schemes shown have a common factor in that they are both simple to implement and computationally light, Warwick (1986).

5.6.1 Deadbeat Control

A general polynomial feedback controller is introduced as:

$$u(t) = \frac{G}{F}\,[Rv(t) + y(t)] \qquad (5.6.1)$$

in which v(t) is a low frequency reference input signal and R can, in its most general form, be a polynomial although in most cases it will be seen that R is specified as a scalar value. The feedback polynomials G and F are of order n_g and n_f respectively and F is monic for normalisation, i.e. G and F are of the same form as B and A respectively.

In order to obtain a deadbeat control action the feedback polynomials are specified as being.

$$F = S - q^{-k}B$$

$$\text{and} \quad G = A \qquad (5.6.2)$$

Where S is defined, at present, as a general polynomial. It will be shown shortly that the exact choice of S made will determine the type of deadbeat control action employed.

On substitution of the feedback control (5.6.2) into the open-loop system equation, the closed-loop system equation is found to be described by:

$$y(t) = q^{-k}\,\frac{BR}{S}\,v(t) + \frac{F}{SA}\,[Ce(t) + d] \qquad (5.6.3)$$

The choice of polynomial S is therefore not only an important factor in the controller design procedure but also in terms of defining closed-loop system performance. Consider the simple case when S=1. The control input is then found from:

$$u(t) = Bu(t-k) + A[Rv(t) - y(t)] \qquad (5.6.4)$$

Thus, as long as the system polynomials, A and B, the reference input, v(t), and the system output, y(t), are known, along with the delay k, then the new control input value can be calculated subject to R being chosen beforehand. If in this case, R=1/B(1), i.e. R is made equal to the inverse of the sum of the B polynomial coefficients, then the closed-loop equation becomes

$$y(t) = \frac{B}{B(1)}\,v(t-k) + \frac{F}{A}\,[Ce(t) + d] \qquad (5.6.5)$$

The deadbeat property of the controller is now evident in that by considering solely the deterministic transfer function, for any change in reference input v(t), the output will be exactly equal to the reference once $n_b + k$ sample periods have elapsed. The response during the $n_b + k$ sample periods is though entirely dependent on the characteristics of the B polynomial of the system, and hence the transient performance can be extremely poor.

The transfer function relating disturbance to system output shows that when an offset is present, $d \neq 0$, then an offset will almost always occur in the output signal. This is due to the fact that the controller described (5.6.4) does not exhibit an integrating action and hence to get rid of the output offset, steady-state compensation is required as a modification to the basic controller form.

It is possible though, by a simple alteration to S that an integrator can be introduced into the controller. Consider setting $S = B(1)$ and $R = 1$, such that the control input is now

$$u(t) = \frac{B}{B(1)} u(t-k) + \frac{A}{B(1)} [v(t) - y(t)] \tag{5.6.6}$$

where the controller polynomial $F = B(1) - q^{-k}B$ contains an integrating factor $1 - q^{-1}$. The closed-loop equation therefore becomes

$$y(t) = \frac{B}{B(1)} v(t-k) + \frac{FC}{AB(1)} e(t) \tag{5.6.7}$$

which shows that the deterministic transfer function is not altered by the different choice of controller. The disturbance transfer function though is affected particularily in that the offset term d, considered to be time invariant, is removed by the integrating action. A problem with this type of control, also occurring in the previous case, is that the system open-loop denominator appears as the disturbance transfer function denominator. Hence only open-loop stable systems, which in practice means the majority of systems, can be dealt with by this type of control action. As can be seen from (5.6.7) however, when the system contains one denominator root exactly on the z-plane unit circle, i.e. one integrating factor, this is an acceptable system type because the factor will be cancelled in the closed loop by the integrator in F.

5.6.2 Pseudo deadbeat techniques

A general control input scheme (5.6.1), (5.6.2) has been introduced and it was shown how for two particular selections of the controller polynomial S, so a deadbeat action resulted in the closed-loop equations (5.6.5) and (5.6.7). By retaining the basic controller format, various alternative selections can be made for S which although they retain the simplicity and low computational effort of the general technique, do not exhibit deadbeat charactersitics in their closed-loop form.

Consider firstly the simple pole placement controller in which the operator specifies a monic polynomial T which is required to be the closed loop denominator, irrespective of the system characteristics. If the polynomial S = T, then the control input can be written as:

$$Tu(t) = Bu(t-k) + A[Rv(t) - y(t)] \tag{5.6.8}$$

and if R = T(1)/B(1), the closed-loop equation is found to be

$$y(t) = \frac{BT(1)}{BT(1)} v(t-k) + \frac{F}{TA} [Ce(t) + d] \tag{5.6.9}$$

and hence, as $F = T - q^{-k}B$ does not contain an integrating factor so the resultant offset must be removed by an appropriate steady-state compensation modification to the control input. The controller though does achieve a deterministic transfer function with T as its denominator, although this necessarily means that the deadbeat properties are effectively lost. It is perhaps worth noting that as a special case of this pole placement controller, if T is chosen to be unity, i.e. all the closed-loop poles at the origin, the original deadbeat scheme is obtained.

One popular control action, and the final type to be considered in detail here, is that of a PID control scheme. This means that the deadbeat style of controller implementation must be tailored in order to satisfy a PID control objective. There are though several different forms of PID algoritm, dependent on how the three controller coefficients are introduced and subsequently selected, Wittenmark (1979). A general PID controller form is considered here to be:

$$u(t) = \left[K_p + \frac{K_i}{(1-q^{-1})} + K_d(1-q^{-1}) \right] [v(t) - y(t)] \tag{5.6.10}$$

in which K_p, K_i and K_d are coefficients associated with proportional, integral and derivative feedback elements respectively. This equation can, by rearrangement, also be written as

$$u(t) = u(t-1) + \left[(K_p + K_i + K_d) - (K_p + 2K_d)q^{-1} + K_d q^{-2}\right][v(t)-y(t)]$$

and can be linked directly with equation (5.6.6) on condition that the system is of a form such that (a) k=1, (b) B = B(1) = b_1 and (c) $A = 1 + a_1q^{-1} + a_2q^{-2}$ and noting that either or both of the a_i coefficients can be zero. Hence the actual system to be controlled is assumed to have a transport delay of one sample period, a scalar gain as a numerator and to be no more complicated than a second order system. Under these conditions the PID controller (5.6.10) is exactly the same as the deadbeat controller (5.6.6) as long as:

$$K_d = \frac{a_2}{b_1}, \quad K_i = \frac{A(1)}{B(1)}, \quad K_p = \frac{1}{B(1)} - K_p - K_d$$

So for the system described, the control input

$$u(t) = u(t-1) + \frac{A}{B(1)} \; [v(t) - y(t)] \qquad (5.6.11)$$

depicts a controller of the deadbeat type, with a PID structure. Although the system is already limited to a max. of second order system, it is certainly worth pointing out that when the system is actually only of first order, such that $a_2 = 0$, so the differential feedback element will also be zero, thus leaving only a PI controller.

Many other possiblilities arise in terms of controller design via a deadbeat framework, and several of these are considered in depth in Warwick (1986). Further, by making slight modifications to the basic controller polynomial definitions so a much wider selection is available. As a brief example, consider the effect of choosing $F = B(1) - q^{-k}B$ in order to achieve an integrating action, and $G = A$, within the control input

$$u(t) = \frac{G}{F} \left[\frac{A(1)}{A} v(t) - y(t) \right] \qquad (5.6.12)$$

Hence, in this case the reference input is filtered by the system open-loop denominator before the error input is found, although the steady-state effect of the filtering is cancelled due to the A(1) term. This results in a closed-loop expression:

$$y(t) = \frac{BA(1)}{AB(1)} \; v(t-k) + \frac{FC}{AB(1)} \; e(t) \qquad (5.6.13)$$

in which steady-state following is achieved along with offset removal, the open-loop transient response of the system not being affected at all.

5.6.3 Adaptive deadbeat control

If the parameters in the actual system model are unknown, such that they must be recursively estimated with the new estimates being used for the latest control input calculation, some of the questions which must be asked are: (1) Is it possible to produce an adaptive controller by means of a recursive estimator linked with a deadbeat control set up? (2) If so, then what implementation problems occur? (3) Can we get away with using only recursive least squares for the estimation procedure?

An initial assumption is made that a good idea is held as to the structure of the system to be controlled, and hence the problem is reduced to one of parameter estimation and control. More complicated procedures necessary for on-line structure estimation can be incorporated but this is considered to be beyond the scope of this chapter. One point worth remembering is that not all of the control actions described in this section contain an in built integrator, e.g. the pole placement method , and in these cases it is necessary to compensate for the offset d by obtaining an estimate of d, termed $\hat{d}$, and then to introduce a bias into the control input u(t) by use of $\hat{d}$. It is quite possible to carry this compensation out, Warwick (1986), although it must be emphasised that by including $\hat{d}$ in the parameter estimation procedure, often only a poor approximation to d can in practice be obtained and in many cases the system parameter estimates themselves are adversely affected.

As pointed out in Matko and Schumann (1984), controller stability is dependent on (a) $B(1) \neq 0$ and (b) an open-loop stable system. Both of these properties must certainly exist for the algorithms described here although in case (a) for estimated parameters it is required that $\hat{B}(1) \neq 0$, hence an amount of software jacketting is required in order to ensure that this cannot occur. In fact if for the actual system parameters $B(1) = 0$ this means that the system itself has a steady-state gain of zero and although mathematically this is possible, in practice one wonders just what such a system looks like.

If a straightforward recursive least squares algorithm is used to estimate the polynomials A and B, the estimations being denoted by $\hat{A}$ and $\hat{B}$ respectively, then if the noise is coloured, the parameters in $\hat{A}$ and $\hat{B}$ will be biased away from their true values such that on convergence of the least squares estimator $\hat{A} \neq A$ and $\hat{B} \neq B$. The closed loop system equations can then be obtained if it is firstly considered that the actual implemented control input employs the estimated rather than the actual A and B polynomials. So for the control input defined in (5.6.6) we have

$$u(t) = \frac{\hat{B}}{\hat{B}(1)} u(t-k) + \frac{\hat{A}}{\hat{B}(1)} [v(t) - y(t)] \qquad (5.6.14)$$

which results in a closed-loop equation

$$y(t) = \frac{B\hat{A}}{W} v(t-k) + \frac{FC}{W} e(t) \qquad (5.6.15)$$

where $W = \hat{B}(1)A - \hat{B}A + B\hat{A}$ (5.6.16)

This results in two problems. Firstly there can be no guarantee that $\hat{B}A-B\hat{A}$ will tend to zero if $\hat{A}$ and $\hat{B}$ are biased away from A and B respectively, and hence it cannot be guaranteed that W will tend to B(1)A. The second problem is that for steady state following it is required that $B(1)\hat{A}(1) = W(1)$, something which can be seen from (5.6.15). With the biased estimates this point can, unfortunately, not be satisfied either. The conclusion to be drawn is that for this type of deadbeat algorithm, if the noise is coloured, in order to ensure steady-state set point following unbiased $\hat{A}$ and $\hat{B}$ estimates are required, hence an extended estimation algorithm is necessary. This is in fact the case for all of the deadbeat algorithms, Warwick (1986), although it must be remembered that if the noise is white, simply a recursive least squares paramator estimator will suffice.

The self-tuning property of deadbeat controllers when linked with on-line estimators is discussed in Warwick (1986), where it is shown how variable time delays can be dealt with by extending the $\hat{B}$ polynomial. A final point will however be made here concerning the integrating property of algorithms such as that obtained by means of the controller (5.6.6) or in its on-line form (5.6.14). It can be seen from (5.6.14) that the input is applied as:

$$[\hat{B}(1) - q^{-k}\hat{B}]u(t) = \hat{A}[v(t)-y(t)]$$

Hence even if the $\hat{B}$ estimates are biased, the term $(1-q^{-k})$ will still be a common factor on the left hand side and therefore the integrating action will still be applied, thus removing any offset d, even in the presence of biased or poor parameter estimates.

5.6.4 Adaptive deadbeat control - conclusions

The major advantage obtained when employing a deadbeat, rather than any other form of, control strategy is that implementation is simple and results in a relatively low computational requirement. The algorithms are also implicit in that the controller parameters are estimated directly, although the deadbeat algorithms have a further advantage over many other implicit algorithms in that the controller parameters are also the system parameters and hence a good idea is obtained concerning system characteristics and variations.

5.7. CONCLUDING REMARKS

For many digital control applications, computation time per sample period is an important, and in a lot of cases the most important, controller property. If a system to be controlled must, in order to obtain sufficient information, be sampled at a certain rate then all control calculations, along with any other necessary computation, must be carried out within the time period available. With a standard self-tuning control algorithm this means that once the system output has been sampled, an updated model of the system must be formed and subsequently employed to calculate the new controller coefficients. Only after all these calculations have taken place can the new control input value be calculated and applied. Both hardware and software developments ensure that the time taken for a specific amount of computation is reduced, however once a particular set of hardware and software has been chosen, computational improvements can only be obtained by achieving slimlined/simplified algorithms, unless it is deemed appropriate to continually update and modify the existing hardware/software - an expensive business.

Various methods for the simplification of self-tuning controllers have been considered here. Firstly, approximations to full parameter estimation schemes can be used - results from this showed that in practice the simplified schemes work just as well as their complicated counterparts, so it seems reasonable to use the simplified procedure for the majority of systems. Model reduction methods are only really useful on-line when (a) the system is of high order (4th order or above) and/or (b) the system exhibits both high and low frequency dynamics, use of a standard low order self-tuning controller in these cases could result in, at best, a poor control action being applied. Finally for many systems a detuned but fairly robust control objective is much more desirable than an optimal control scheme which can be quite temperamental. Both PID and deadbeat controllers fall into this category and it has been shown how self-tuning controllers can be derived, in many different ways, in order to achieve such goals.

REFERENCES

Astrom, K.J. (1980): 'Self-tuning control of a fixed-bed chemical reactor system', Int. J. Control, Vol. 32, pp 221-256.

Astrom, K.J. and Hagglund, T. (1984): 'Automatic tuning of simple reactors with specifications on phase and amplitude margins', Automatica, Vol. 20 No. 5, pp 645-651.

Astrom, K.J. and Wittenmark, B. (1973): 'On self-tuning regulators', Automatica, Vol. 9, pp 185-189.

Farsi, M., Karam, K.Z. and Warwick, K. (1984): 'Simplified recursive identifier for ARMA processes', Electronics letters, Vol. 20, No. 22, pp. 201-209.

Gawthrop. P.J. (1986): 'Self-tuning PID controllers: algorithms and implementation', IEEE Trans, on Automatic Control, Vol. AC-31, No.3, pp. 201-209.

Higham, E.J. (1986): 'An expert system for process control', Ch.15 in 'Industrial Digital Control Systems', eds. K. Warwick and D. Rees, Peter Peregrinus Ltd.

Isermann, R. (1981): 'Digital Control Systems', Springer-Verlag, New York.

Ljung, L. and Soderstrom, T. (1983): 'Theory and practice of recursive identification', MIT Press.

Matko, D. and Schumann, R. (1984): 'Self-tuning deadbeat controllers', Int.J.Control, Vol.40, No. 2, pp. 393-402.

Nishikawa, Y. and Sannomiya, N. (1981): 'A method for auto-tuning of PID- control parameters', Proc. IFAC congress, Kyoto, Japan.

Ortega, R. and Kelly, R. (1984): 'PID self-tuners: some theoretical and practical aspects', IEEE Trans. on Industrial Electronics, Vol. IE-31, No.4, pp. 332-338.

Warwick, K. (1984):'A new approach to reduced order modelling', Proc. IEE, Pt.D, Vol 131, pp. 74-78.

Warwick, K.(1985): 'Reduced order self-tuning control', Proc. 7th IFAC Symposium on Identification and System Parameter Estimation, York, U.K., pp. 1305-1309.

Warwick, K. (1986):'Adaptive deadbeat control of stochastic systems', Int.J.Control, Vol 44, No.3, pp 651-663.

Warwick, K., Farsi, M. and Karam, K.Z. (1985): 'A simplified pole placement self-tuning controller', Proc. Int. Conference Control 85, Cambridge, pp. 1-6.

Wellstead, P.E. and Sanoff, S.P. (1981): 'Extended self-tuning algorithms', Int.J. Control, Vol.34, pp. 433-455.

Wittenmark, B. (1979): 'Self-tuning PID Controllers based on pole placement', Lund Inst. of Technology, Report No. TFRT-7179.

Ziegler, J.G. and Nichols, N.B. (1942): 'Optimum settings for automatic controllers', Trans. ASME, vol.64, pp. 759-768.

Chapter 6

A unified approach to adaptive control

G. C. Goodwin, R. H. Middleton and M. Salgado

6.1 INTRODUCTION

This chapter presents an integrated approach to the design of adaptive control systems. Emphasis will be given to the unification of analogue and sampled data adaptive controllers. This unification allows continuous time intuitions to be used in the discrete time case. It also gives confidence in the discrete time algorithms when implemented with rapid sampling.

We adopt the view that a practical adaptive control algorithm should be considered as a combination of a robust parameter estimation scheme together with robust control synthesis. A key issue in the design of robust parameter estimators and robust control laws is the effect of unmodelled errors.

6.2 The Delta Operator

The unification of continuous and discrete systems is facilitated by the use of the Delta operator, δ, rather than the usual shift operator, q. The Delta operator is defined as

$$\delta = \frac{q-1}{T} \tag{6.1}$$

where T is the sample period.

Clearly, from (2.1), as $T \to 0$, the Delta operator reduces to the derivative operator D, in continuous time. Thus, for practical sampling rates which should be chosen about 10 times the system bandwidth, results expressed in terms of δ are closely related to the corresponding continuous results.

An unexpected bonus arising from the use of the Delta-operator is that it generally has superior numerical properties compared with the shift operator (Middleton and Goodwin, 1986). These numerical advantages are a consequence of the shift in the origin to the point $1+j0$ in the complex z-plane.

Continuous and discrete systems analyses can be treated simultaneously by use of the following notation:

TABLE 6.1 Unified Notation

Description	Unified Notation	Continuous Form	Discrete Form
Operator	ρ	$D = \frac{d}{dt}$	$\delta = \frac{q-1}{T}$
Integral	$\mathcal{S}$	$\int$	$T\sum$
Exponential	$\mathcal{E}(A,t)$	e^{At}	$(I+AT)^{t/T}$
Square integrable functions	$\mathcal{L}_2$	L_2	l_2
Stability Region	$\lambda\epsilon\mathcal{C}$	$C\equiv$left half plane	$C_\delta=\{\lambda: \|1+\lambda T\|<1\}$

6.3 THE SYSTEM MODEL

Consider a plant described by the following model:

$$A_a y = B_a u + C_a \upsilon \tag{6.2}$$

where the degrees of the polynomial operators, A_a, B_a and C_a satisfy $\partial_{A_a} > \partial_{B_a}$, $\partial_{A_a} \geq \partial_{C_a}$; and υ is a bounded term which may include noise and deterministic disturbances. We rewrite (6.2) by introducing a nominal transfer function, $H_0 = \frac{B}{A}$, as

$$A\ y = B\ u + \eta + d \tag{6.3}$$

where u, y denote the plant input and output respectively, η denotes an 'unmodelled' component and d denotes a purely deterministic noise term. The polynomial operators A, B are coprime and of degree ∂_A, ∂_B respectively where $\partial_B < \partial_A$ and A is monic.

Models of the of the form (6.3) have been used extensively in the literature e.g. in Praly, 1983; Ioannou and Tsakalis, 1986; Kreisselmeier, 1986. The term η includes bounded noise and unmodelled components in the system response. For example, if the plant is modelled in transfer function form as

$$y = [H_0\ (1+\bar{H}) + \bar{\bar{H}}]u \tag{6.4}$$

where $H_0 = \frac{B}{A}$, $\bar{H} = \frac{\bar{B}}{\bar{A}}$ and $\bar{\bar{H}} = \frac{\bar{\bar{B}}}{\bar{A}}$, then

$$\eta = \left[\frac{B\bar{B}}{\bar{A}} + \frac{A\bar{\bar{B}}}{\bar{A}} \right] u \tag{6.5}$$

The deterministic component of the noise, d, is assumed to satisfy a model of the form:

$$Sd = 0 \tag{6.6}$$

where S is a known monic polynomial of degree ∂_S having non-repeated zeros on the stability boundary. For example, if d is a sinewave of angular frequency ω_0, then

$$S = \rho^2 + 2T\left[\frac{1-\cos\omega_0 T}{T^2}\right]\rho + 2\left[\frac{1-\cos\omega_0 T}{T^2}\right]$$

6.4. UNIFIED PARAMETER ESTIMATION

We introduce a bandpass filter to eliminate the deterministic disturbance and also to mitigate the effect of unmodelled dynamics on the estimator. (see Johnson et.al., 1984; Goodwin, 1985; Anderson et. al., 1986). We define the following filtered variables:

$$\bar{y} = \frac{GS}{JQ} y \quad ; \quad \bar{u} = \frac{GS}{JQ} u \tag{6.7}$$

where G, J, Q are monic "Hurwitz" polynomials, $\partial_G = \partial_J$ and $\partial_Q = \partial_S$.

Using (6.7) and (6.6), the model, (6.3), can be rewritten as

$$A\bar{y} = B\bar{u} + \bar{\eta} \tag{6.8}$$

where

$$\bar{\eta} = \frac{GS}{JQ} \eta \tag{6.9}$$

Equation (6.8) will in general be unsuitable for parameter estimation since the "equation error" term $\bar{\eta}$ involves "differentiation" of the unmodelled error which may include noise. We therefore introduce an additional filter, $\frac{1}{E}$, where E is monic, "Hurwitz" and $\partial_E = \partial_A$. We then define:

$$y_f = \frac{1}{E}\bar{y} \; ; \; u_f = \frac{1}{E}\bar{u} \; ; \; \eta_f = \frac{1}{E}\bar{\eta} \tag{6.10}$$

Substituting (6.10) into (6.8) gives

$$\bar{y} = (E - A)\, y_f + Bu_f + \eta_f \tag{6.11}$$

This equation can be expressed in regression form as

$$\bar{y} = \phi^T \theta + \eta_f \tag{6.12}$$

where $n \overset{\Delta}{=} \partial_A$, $m \overset{\Delta}{=} \partial_B$, $s \overset{\Delta}{=} \partial_S$, $g \overset{\Delta}{=} \partial_G$ and

$$\phi^T = \left[y_f, \ . \ . \ . ,\rho^{n-1} y_f, \ u_f \ . \ . \ \rho^m u_f \right] \tag{6.13}$$

$$\theta^T = \left[e_0 - a_0, \ldots \ e_{n-1} - a_{n-1}, \ b_0, \ldots b_m \right] \tag{6.14}$$

$$E(\delta) \overset{\Delta}{=} \rho^n + e_{n-1} \rho^{n-1} + \ldots + e_0 \tag{6.15}$$

$$A(\delta) \overset{\Delta}{=} \rho^n + a_{n-1} D^{n-1} + \ldots + a_0 \tag{6.16}$$

$$B(\delta) \overset{\Delta}{=} b_m \rho^m + b_{m-1} \rho^{m-1} + \ldots + b_0 \tag{6.17}$$

The filtering operation outlined above is crucial to the success of adaptive controllers since it focuses the parameter estimator on the relevant frequency band thereby significantly reducing the deliterious effects of unmodelled dynamics and disturbances. There are two aspects, namely high pass filtering to eliminate d.c. offsets, load disturbances etc. and low pass filtering to eliminate irrelevant high frequency components incuding noise and system response. The nett effect is a bandpass filter. The rule of thumb governing the design of the filter is the upper frequency should be about twice the desired system bandwidth and the lower frequency should be about one tenth the desired bandwidth. Small unmodelled terms remain and it is important to ensure that these do not "throw the parameter estimator off-course". This can be achieved by reducing the estimator gain when one suspects that the response is dominated by the unmodelled components. In the case where we are treating unmodelled dynamics then the filtered unmodelled error, $\left\{ |\eta_f(t)| \right\}$, can be overbounded by a function $\left\{ \sigma(t) \right\}$ which is an exponentially weighted function of 'past' values of $|\bar{u}(t)|$ and $|y(t)|$, (Middleton and Goodwin, 1987).

We then propose the following parameter estimator:

$$\rho\hat{\theta} = bP\phi\epsilon \tag{6.18}$$

$$\rho P = -bP\phi\phi^T P + R \quad ; \quad P(0) = P(0)^T > 0 \tag{6.19}$$

where

$$\epsilon = \bar{y} - \phi^T\hat{\theta} \tag{6.20}$$

$$R = R^T \geq 0 \tag{6.21}$$

and where b represents a time varying gain. The basic idea in the algorithm is to reduce b when the unmodelled dynamics are large.

One possible choice for b is the following deadzone algorithm:

Select $\beta > 1$, and define:

$$\omega(t) \triangleq \begin{cases} 0 & \text{if } |e(t)| \leq \beta(\sigma(t)) \\ f(\beta\sigma(t), e(t))e(t) & \text{otherwise} \end{cases} \tag{6.22}$$

where

$$f(g,e) \triangleq \begin{cases} e-g & \text{if } e>g \\ 0 & \text{if } |e| \leq g \\ e+g & \text{if } e<-g \end{cases} \tag{6.23}$$

we then use

$$b = \frac{\alpha_1\omega}{1+\phi^T P\phi} \quad ; \quad \alpha_1 < \frac{1}{T}\left[\frac{\beta^2-1}{\beta^2}\right] \tag{6.24}$$

The term $R(t)$ in (6.19) has been introduced to ensure the algorithm gain doesn't 'turn off'. Any choice of $R(t) = R(t)^T \geq 0$ that ensures $P(t)$ and $P(t)^{-1}$ are bounded may be used, indeed it can be shown that many of the usual least squares modifications, including covariance resetting, variable forgetting factors, gradient algorithms and constant trace algorithms, are described by a suitable choice of R.

It is also possible to add additional refinements to the algorithm. For example, it is possible to constrain the parameter estimates to lie in prespecified convex regions, and/or to add parameter searches to satisfy additional constraints.

The convergence properties of the least squares parameter estimator are summarized in the following.

<u>Lemmma 6.1</u>

Consider the parameter estimator (6.18), (6.19), (6.24)

applied to any system of the form (6.3) the following properties hold irrespective of the control law:

i) $$\tilde{f} \triangleq \frac{f(\beta\sigma(t), \epsilon(t))}{1 + \phi(t)^T \phi(t)} \in \mathcal{L}_2 \tag{6.25}$$

ii) $$\rho\hat{\theta} \in \mathcal{L}_2 \tag{6.26}$$

iii) $$\text{For all } t, \|\hat{\theta}(t) - \theta\| \leq \nu(P_0)\|\hat{\theta}(0) - \theta\| \tag{6.27}$$

where $\nu(P_0)$ denotes the condition number of the matrix P_0.

Proof:

See Middleton and Goodwin (1987) or Goodwin et.al. (1986).

∇∇∇

6.5 CONTROL SYSTEM DESIGN

When the nominal model is known, there exists many possible choices for the control system design procedure including frequency domain methods, model reference control (for minimum phase plants), pole assignment, linear quadratic optimal control etc. To illustrate the technique we will use pole assignment together with the Internal Model Principle so that the classical three term control law will be a special case of our design procedure.

The input u is generated using

$$LGSu = Pv - Py \tag{6.28}$$

or equivalently

$$Lu_f = -Py_f' + \Gamma v_f' \; ; \; y_f' \triangleq \frac{1}{FQ}y \; ; \; v_f' = \frac{1}{FQ}v \tag{6.29}$$

where v is the reference input and L, P are polynomials in ρ determined from the folloing pole-assignment equation:

$$ALGS + BP = A^* \tag{6.30}$$

where A^* is a monic Hurwitz polynomial of degree $\partial_{A^*} = 2\partial_A + \partial_G + \partial_S$. The degree of L, Γ and P are respectively $\partial_L = \partial_A$ and $\partial_P = \partial_A + \partial_G + \partial_S - 1$. When substituted into (6.29) this yields a strictly proper control law which can be implemented as

$$\bar{u} = (E-L)u_f - Py_f' + \Gamma v_f' \tag{6.31}$$

$$u = \frac{JQ}{GS}\bar{u} \qquad (6.32)$$

The term Γ in (6.28) allows for reference feed-forward (sometimes known as set-point forcing). To retain the properties of the Internal Model Principle it is important to ensure that S divides $\Gamma - P$. In the case of simple integral control, $S=\rho$, in which case we require $\Gamma(0) = P(0)$. This has the simple interpretation of requiring the same d.c. gain in the feedforward and feedback paths. Typically the remaining degrees of freedom in Γ are used to cancel some of the closed loop poles.

The internal model polynomial S, will invariably include integral action. This is essential to eliminate load disturbances, input offsets etc. One should only attempt to cancel sinewave type disturbances if they are well within the achievable closed loop bandwidth. If an internal model polynomial is used then it is desirable to place stable zeros in A^* near the zeros of S, this being the classical equivalent of having an integral bandwidth well below the system bandwidth.

6.6 UNIFIED ADAPTIVE CONTROL ALGORITHM

The essential idea to obtain an adaptive control law is to combine the least squares algorithm of section 6.4 with the certainty equivalence form of the feedback control law (6.29) i.e. given $\hat{\theta}(t)$ we form $\hat{A}$, $\hat{B}$ and $\hat{A}^*$ and solve

$$\hat{A}\hat{L}GS + \hat{B}\hat{P} = \hat{A}^* \qquad (6.33)$$

for $\hat{L}$ and $\hat{P}$. Then implement the control law as

$$\hat{L}u_f = -\hat{P}y'_f + \hat{\Gamma}v'_f \qquad (6.34)$$

which may be rewritten as

$$\bar{u} = (E-\hat{L})u_f - \hat{P}y'_f + \hat{\Gamma}v'_f \qquad (6.35)$$

$$u = \frac{JQ}{GS}\bar{u} \qquad (6.36)$$

In equation (6.33) we have allowed the desired closed loop polynomial to depend upon the parameter estimates (Hence the notation $\hat{A}^*$). This issue is discussed in detail in Middleton and Goodwin (1986b)

For example, the stable well damped part of $\hat{B}$ should be cancelled, i.e. included in $\hat{A}^*$ (see Astrom and Wittenmark, 1984). The unstable part of $\hat{B}$ places an upper bound on the achievable bandwidth; a sensible choice being to reflect the unstable zeros through the stability boundary. The part of $\hat{A}$ which is not stable and well damped places a lower limit on the bandwidth. When these various requirements conflict, one is faced with a very difficult control problem and one ought to investigate alternative architectures including additional measurements if possible.

The structure of the complete adaptive control law is shown in Figure 6.1.

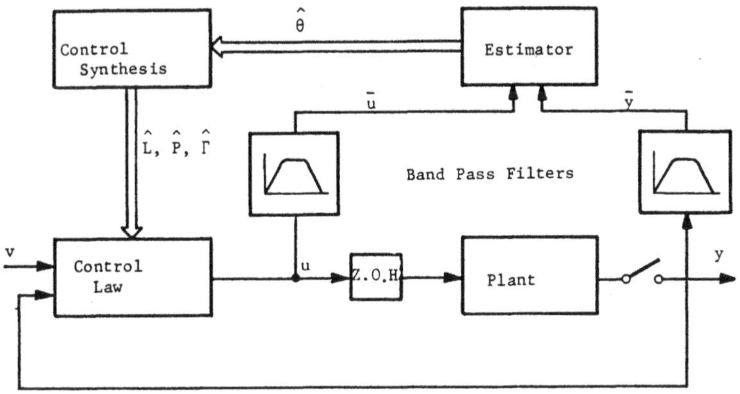

Fig.6.1 Adaptive Control Law Structure

The above adaptive control law can be shown (Middleton et al, 1987) and Middleton and Goodwin, 1987a) to have a number of interesting properties for both time invariant and time varying systems.

For the case of time invariant systems, provided the unmodelled dynamics are sufficiently small, then the system is globally stable in the sense that u, y remain bounded. Moreover, if v (the reference input) is purely deterministic and satisfies $Sv = 0$, then it follows that asymptotically perfect tracking occurs, i.e.

$$\lim |y-v| = 0 \tag{6.37}$$

For the case of time varying systems, provided the parameter variations, i.e. $\rho\theta$, are small in the mean then boundedness of the response can be assured. Note that small in the mean includes the case of slowly drifting parameters as well as the case of parameters with infrequent jumps (in which case $\rho\theta$ includes impulses).

6.7 SOME PRACTICAL ISSUES

In this section we make brief remarks about some practical issues. If the system has a pure time delay then this can be fitted into the above framework by using rational approximations for the delay. By considering the phase shifts of relevance in practice, it can be shown that, over the bandwidth of interest to control system design, a delay approximation having, at most, two poles and two zeros suffices. The errors introduced by use of the rational approximation occur at frequencies above the achievable closed loop bandwidth and can be treated as unmodelled dynamics in the usual way. For further discussion see de Souza et al. (1987).

Another practical question concerns model reference adaptive control of continuous time systems having relative degree greater than 1. It is well known (Astrom et al. 1984) that when sampled rapidly these systems give rise to non-minimum phase zeros in the discrete time model. This would appear to rule out discrete time model reference control of this class of systems at fast sampling rates. However, this is paradoxical since continuous time model reference control is clearly possible. The resolution of this paradox is facilitated by using the Delta operator.

Let the continuous time system be

$$A'(D)y = B'(D)u \tag{6.38}$$

where

$$A'(D) = D^n + a'_{n-1}D^{n-1} + \ldots a'_0 \tag{6.39}$$

$$B'(D) = b'_m D^m + b'_{m-1}\, D^{m-1} + \ldots + b'_0 \tag{6.40}$$

When sampled with period T, the corresponding δ-model is

$$A(\delta)y = B(\delta)u \tag{6.39}$$

where

$$A(\delta) = \delta^n + a_{n-1}\,\delta^{n-1} + \ldots a_0 \tag{6.40}$$

$$B(\delta) = b_{n-1}\delta^{n-1} + \ldots + b_0 \tag{6.41}$$

In the limit as $T \to 0$, we have the following results:

$$\lim_{T\to 0} a_i = a'_i \quad ; \quad i=0, \ldots, n-1 \tag{6.42}$$

$$\lim_{T\to 0} b_i = \begin{Bmatrix} b'_i & ; & i=0, \ldots, m \\ 0 & ; & i=m+1, \ldots, n-1 \end{Bmatrix} \tag{6.43}$$

The extra zeros which arise due to sampling (including

the non minimum phase zeros) are thus seen to move to the far left δ-plane as $T \to 0$. This suggests that the discrete time model can (and should) be considered as

$$A(\delta)y = \bar{B}(\delta)u + \eta \tag{6.44}$$

where $A(\delta)$ is as in (6.40) and $\bar{B}(\delta)$ is given by

$$\bar{B}(\delta) = b_m\delta^m + \ldots + b_0 \tag{6.45}$$

The error term η can then be treated in the normal way as small high frequency unmodelled dynamics (see Goodwin et al. 1986 for further discussion).

There has also been some discussion in the literature on the relative advantages of robust linear control versus adaptive control (Astrom et al., 1986). One view of robust control is that one detunes the control system in such a way that its performance in the worst possible senario, is still acceptable. This common strategy is a satisfactory answer to many practical control system design problems. However, this approach to design means that the control system is a compromise solution between performance and robustness to variations. Adaptive control offers an alternative in these cases since the control system design can then be matched to the plant characteristics as they vary.

A good way to approach the design of an adaptive controller seems to be to start with a fixed robust solution to the control problem and then to gradually increase the performance demands where these increased demands are taken up by the adaptive mechanisms.

Another practical question concerns the provision of integral action. In section 6.6 this was done as part of the overall design strategy. However, in some practical cases, the integrator bandwidth is required to be much slower than that of the plant. This separation in bandwidths implies that one can do the control system design without considering the integrator and then to retrofit the integrator with a suitably small gain. (See Feuer and Goodwin, 1987 for further discussion).

Finally, it should be remarked that the parameter estimation module is, itself, a complex dynamical system and care must be taken to ensure that this system has appropriate dynamic properties. Apart from the aspects discussed in section 6.4, other features are significant. For example, to achieve good response to time varying parameters, requires modifications to be made to the up-date law. An example of the kind of considerations involved shall be briefly presented below.

Qualitatively, the desirable features in a least squares based algorithm are as follows:

- exponential forgetting of past data
- resetting to an unprejudiced treatment of data in absence of excitation.
- an upper bound for P, to prevent bursting phenomena.

- an upper bound for P^{-1}, to avoid the algorithm "going to sleep" in certain directions, and also to avoid numerical ill-conditioning in the P update.

The above requirements can be met by making suitable choices in the algorithm (6.18), (6.19). For example, we may choose b as in (6.24) and

$$R = \gamma_1 I + \gamma_2 P - \gamma_3 P^2 \tag{6.46}$$

where $\gamma_1, \gamma_2, \gamma_3$ are positive instants. The term $\gamma_1 I$ basically ensures that P has a lower bound i.e. P^{-1} has an upper bound. The term $\gamma_2 P$ allows for exponential forgetting and resetting. The term $-\gamma_3 P^2$ ensures that P has an upper bound. These properties can be formally established (Salgado et.al, (1987)) provided reasonable constraints are placed in the choices of the various constants, α_1, γ_1, etc. Under these conditions, the maximum eigenvalue of, $\bar{\upsilon}$, P and the minimum eigenvalue, $\bar{\sigma}$, of P satisfy respectively:

$$\bar{\upsilon} \simeq \frac{\gamma_2}{\gamma_3} + \frac{\gamma_1}{\gamma_2} \tag{6.47}$$

$$\bar{\sigma} \simeq \frac{\gamma_1}{\alpha_1 - \gamma_2} \tag{6.48}$$

It can also be shown that under the conditions referred to above, R in (6.46) is positive semi definite and hence the properties of the parameter estimator given in Lemma 6.1 are retained in this case.

The above algorithm has been called the Exponential Forgetting and Resetting Algorithm (EFRA). Details of the analysis of the algorithm can be found in Salgado et. al., (1987).

6.8 SIMULATION STUDIES

We will illustrate the theory by reference to one of the examples described in Astrom et. al (1986). The plant is

$$G(s) = \frac{K}{(Ts+1)^2} \tag{6.49}$$

where $K \in [1,4]$. $T \in [0.5,1]$.

We make the following choices : sampling period 0.1 seconds ; simplified discrete model based on δ operator as in equation (6.44) ; low pass filter in parameter estimator of $\frac{1}{(\delta+4)^2}$; parameter estimator based on EFRA algorithm (with constants chosen so that $0.2I \leq P \leq 2I$; pole assignment design where A^* is made a function of $\hat{B}$ and $\hat{A}$ as mentioned in section 6.6 so that in this case, the closed loop system approximates $\frac{8.6}{\delta^2+5\delta+8.6}$.

Figure 6.2 shows a simulation where the gain, K, changes from 1 to 4 after 35 seconds. Figure 6.3 shows a simulation where the time constant, T, changes from 1.0 to 0.5 after 35 seconds. Note that these results have a significantly faster transient performance than achieved by the robust controller in Astrom et al. (1986). The trade-off is that a slightly larger transient occurs when the parameters change, but this is quickly corrected by the adaptive mechanism.

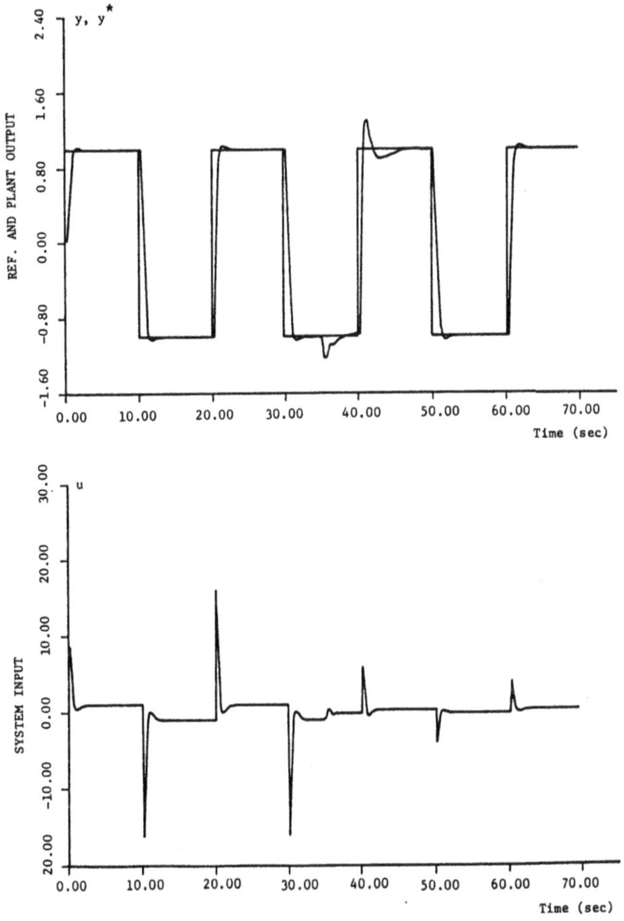

Fig.6.2 Adaptive response to gain change

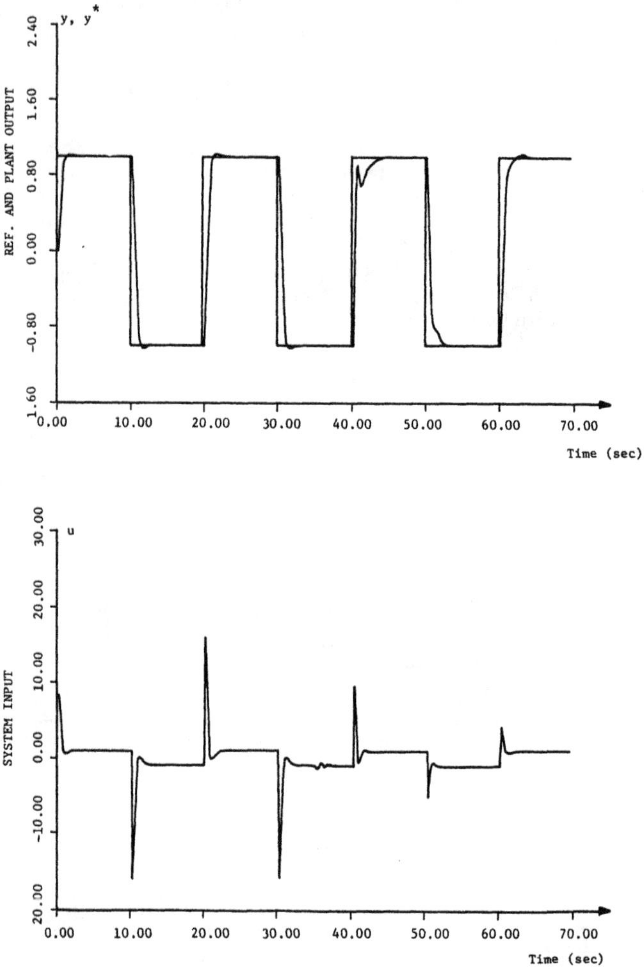

Fig.6.3. Adaptive response to change in time constant

REFERENCES

1. Anderson, B.D.O., Bitmead, R.R., Johnson C.R. Jr., Kokotovic, P.V., Kosut, R.L., Mareels, I.M.Y., Praly L., and Riedle, B.D., 1986, "Stability of adaptive systems : passivity and averaging analysis", MIT Press.

2. Astrom, K.J., and Wittenmark, B., 1984, "Computer controlled Systems : theory and design", Prentice Hall, Englewood Cliffs.

3. Astrom, K.J., Hagander, P., and Sternby J., 1984, "Zeros of sampled systems", Automatica, Vol.20, 1, pp.31-39.

4. Astrom, K.J., Newman, L., and Gutman, P.O., 1986, "A comparison between robust and adaptive control of uncertain systems" IFAC Workshop of Adaptive Control, Lund.

5. de Souza, C.E., Goodwin, G.C., Mayne, D.Q., and Palaniswami, 1987, "An adaptive control algorithm for systems having unknown time delay", to appear, Automatica.

6. Feuer, A., and Goodwin G.C., 1987, "Integral action in robust adaptive control", Tech. Report, Faculty of Electrical Engineering, Technion, Israel Institute of Technology, Haifa, Israel.

7. Goodwin, G.C., 1985, "Some observations on robust estimation control", plennary address, 7th IFAC Symposium on Identification York.

8. Goodwin, G.C., Lozano Leal, R., Mayne, D.Q., and Middleton R.H., 1986, "Rapprochement between continuous and discrete model reference adaptive adaptation, control", Automatica, Vol.22, No.2, pp.199-208.

9. Ioannou, P.A., and Tsakalis, K., 1986, "Robust discrete-time adaptive control", IEEE, AC-31, N11, pp.1033-1043.

10. Johnson, C.R., Jr., Anderson B.D.O., and Bitmead, R.R., 1984, "Improving the robustness of adaptive model-following via information vector filtering", Tech Report, ANU.

11. Kreisselmeier, G., 1986, "A robust indirect adaptive control approach", Int. J. Control, 43, 1, pp.161-175.

12. Middleton, R.H., Goodwin, G.C., Hill, D.J. and Mayne, D.Q., 1987, "Design issues in adaptive control", to appear, IEEE Transactions on Automatic Control.

13. Middleton, R.H., and Goodwin, G.C., 1987, "Adaptive control of time varying linear systems", to appear, IEEE Transactions on Automatic Control.

14. Middleton, R.H., and Goodwin, G.C., 1987b, Digital Estimation and Control : A Unified Approach, to appear.

15. Praly, L., 1983, "Robustness of indirect adaptive control based on pole placement design", IFAC Workshop on Adaptive Control, San Francisco, June.

16. Salgado, M., Goodwin, G.C. and Middleton R., 1987, "A modified least squares algorithm incorporating exponential resetting and forgetting", to appear, Int. Journal of Control.

Chapter 7

Implementation of continuous-time controllers

P. J. Gawthrop

7.1. INTRODUCTION

Most systems of interest to the control engineer exist in a continuous-time setting - they are described by differential equations. In contrast, most controllers which are sophisticated enough to have a self-tuning capability are implemented using digital microprocessor technology and as such exist in a discrete-time setting - they are described by difference equations. It follows that controllers must often be designed by starting off with a continuous-time system and ending up with a discrete-time controller.

We contrast two approaches to such design: discrete-time design and continuous-time design. The former - conventional - approach starts with making a discrete-time model of the (continuous-time) system together with a discrete-time version of the cost-function polynomials (see e.g[1] for a discussion of this). The design then proceeds in a purely discrete-time setting. In contrast, the latter approach as considered in this chapter starts with a purely continuous-time design of the adaptive controller in terms of the continuous-time system; the resultant analogue controller is then converted into a discrete-time design.

Both design methods start with a continuous-time system and end with a discrete-time controller; but the design and continuous-discrete transformation steps are transposed between the two methods.

Some advantages of the continuous-time, as opposed to the discrete-time, approach are as follows:

- The (continuous-time) design method is matched to the actual (continuous-time) system to be controlled. Thus system characteristics such as relative degree and zero location can be directly addressed.

- Artefacts of sampling such as sampled minimum phase systems having zeros outside the unit disc[2,3] are avoided.

- The controller coefficients arising from the self-tuning controller correspond to continuous-time (Laplace domain) transfer functions. Most control engineers find these easier to interpret than coefficients of discrete-time (z-domain) transfer functions. An example of this is that the self-tuning PI (proportional plus integral) controller described elsewhere[4] directly estimates the integral time-constant of the controller.

- The controller sample interval is chosen <u>after</u> the design stage, not before.

The design of the continuous-time self-tuning controllers discussed here is discussed in some detail in a recent book[5]; the purpose of this chapter is to consider the second stage of the continuous-time design process: the implementation of a continuous-time self-tuning controller using a digital algorithm.

As discussed in a recent book, [5] one possible approach to continuous-time self-tuning control is via the concept of an <u>emulator</u>. It is beyond the scope of this chapter to discuss this concept in detail, but a brief outline is given to orientate the reader. A particular algorithm, detuned model-reference control, is chosen to illustrate the main implementation aspects. Further details are to be found in a forthcoming book[6].

Within the limited space available, attention has been focused on the implementation issues directly connected with the conversion of a continuous-time algorithm to a discrete-time algorithm. The other issues connected with the continuous-time design itself appear elsewhere. [5]

7.2. OUTLINE OF CONTINUOUS-TIME SELF-TUNING

This section provides a brief outline of one particular self-tuning control algorithm as a prototype continuous-time algorithm for conversion to a discrete-time equivalent. The discussion has two parts:

1 Controller design for known system parameters.

2 Controller design for unknown system parameters via self-tuning.

The former is considered in this section, the latter is deferred to the next section.

The class of single-input single-output systems considered in this chapter can be described in Laplace transform form as:

$$\bar{y}(s) = \frac{B(s)}{A(s)}\bar{u}(s) + e(s) \tag{7.2.1}$$

where

Symbol	Quantity
$\bar{y}(s)$	System output
$\bar{u}(s)$	Control signal
$e(s)$	Disturbance

The single-input single-output feedback controllers considered in this chapter can be written as a classical feedback controller but with an <u>emulator</u> in the feedback path. This control law can be written in two equivalent forms:

$$\bar{u}(s) = \frac{1}{Q(s)}[R(s)v(s) - \bar{\phi}^{*}(s)] \qquad (7.2.2)$$

and

$$\bar{\phi}^{*}(s) + Q(s)\bar{u}(s) - R\bar{w}(s) = 0 \qquad (7.2.3)$$

where

Symbol	Quantity
$\bar{u}(s)$	Control signal
$\bar{\phi}^{*}(s)$	emulator output
$v(s)$	setpoint
$Q(s)$	control weighting
$R(s)$	setpoint filter

$1/Q(s)$ and $R(s)$ are proper transfer functions. $\bar{\phi}^{*}(s)$ is the <u>emulator</u> <u>output</u> given by:

$$\bar{\phi}^{*}(s) = \frac{G(s)}{C(s)}\bar{u}_z(s) + \frac{F(s)}{C(s)}\bar{y}(s) \qquad (7.2.4)$$

The filtered signal $\bar{u}_z(s)$ is given by

$$\bar{u}_z(s) \triangleq \frac{1}{P(\varepsilon s)}\bar{u}(s); \tag{7.2.5}$$

The emulator polynomials F and G are the (unique) solutions of:

$$\frac{P(s)C(s)}{A(s)P(\varepsilon s)} = \frac{E(s)}{P(\varepsilon s)} + \frac{F(s)}{A(s)}; \quad G(s) = E(s)B(s) \tag{7.2.6}$$

and ε is a small number (eg 0.1).

As discussed elsewhere, the motivation for this feedback controller can be illustrated by considering the corresponding closed-loop system. Defining the transfer function L(s) as

$$L(s) \triangleq \frac{1}{Q(s)} \frac{P(s)B(s)}{P(\varepsilon s)A(s)} \tag{7.2.7}$$

then in the absence of disturbances, the closed-loop system output is given by:

$$\bar{y}(s) = \frac{L(s)}{1+L(s)} \frac{P(\varepsilon s)}{P(s)} R(s)\bar{v}(s) \tag{7.2.8}$$

In the limiting case as $\varepsilon \to 0$ and $Q(s) \to 0$ then $L(s) \to 0$ and

$$\bar{y}(s) = \frac{R(s)}{P(s)}\bar{v}(s) \tag{7.2.9}$$

Thus the closed loop system output follows that of a reference model R(s)/Ps with respect to the setpoint $\bar{v}(s)$. The need for a non-zero Q(s) to give robustness is discussed elsewhere[7].

The emulator equation may be rewritten in <u>linear-in-the-parameters</u> form as

$$\phi^{\star}(t) = \underline{X}^T(t)\underline{\theta} \tag{7.2.10}$$

where the data vector $\underline{X}(t)$ and the parameter vector $\underline{\theta}$ are given, in Laplace transform terms by

$$\bar{\underline{X}}(s) \triangleq \begin{vmatrix} \bar{\underline{X}}_u(s) \\ \bar{\underline{X}}_y(s) \end{vmatrix}; \quad \underline{\theta} = \begin{vmatrix} \underline{\theta}_u \\ \underline{\theta}_y \end{vmatrix} \tag{7.2.11}$$

Where

$$\bar{\underline{X}}_u(s) = \frac{1}{C(s)} \begin{vmatrix} s^n \\ s^{n-1} \\ . \\ 1 \end{vmatrix} \bar{u}_z(s); \quad \bar{\underline{X}}_y(s) = \frac{1}{C(s)} \begin{vmatrix} s^n \\ s^{n-1} \\ . \\ 1 \end{vmatrix} \bar{y}_z(s) \tag{7.2.12}$$

and the components of $\underline{\theta}$ are given by

$$\underline{\theta}_u = \begin{vmatrix} g_0 \\ g_1 \\ . \\ g_n \end{vmatrix}; \quad \underline{\theta}_y = \begin{vmatrix} f_0 \\ f_1 \\ . \\ f_n \end{vmatrix} \tag{7.2.13}$$

The vectors $\bar{\underline{X}}_u(s)$ and $\bar{\underline{X}}_y(s)$ are the Laplace transforms of vectors in controllable form (see section 1.6). The time-domain versions may therefore be computed from the corresponding differential equations.

This particular form provides a convenient means for implementing an emulator. In particular, the data vector $\underline{X}(t)$ is clearly distinguished from the parameter vector $\underline{\theta}$.

7.3. ISSUES IN OBTAINING A DIGITAL ALGORITHM

The linear-in-the parameters equation

$$\phi^{\star}(t) = \underline{X}^T(t)\underline{\theta} + e^{\star}(s) \tag{7.3.1}$$

is non-dynamic; it is just an algebraic relation. It may thus be sampled at any time t_m to give

$$\phi^{\star}_m = \underline{X}_m^T \underline{\theta} + e_m \tag{7.3.2}$$

where

$$\phi^{\star}_m \triangleq \phi^{\star}(t_m);\ \underline{X}_m^T \triangleq \underline{X}^T(t_m);\ e_m = e^{\star}(t_m) \tag{7.3.3}$$

Note that this relation holds whether or not the samples at the times t_m are equispaced, or indeed in the correct order. For the purposes of this chapter, however, uniform samples with interval T will be considered:

$$t_m = mT \tag{7.3.4}$$

Thus there are two main issues involved in obtaining a discrete implementation of an adaptive emulator:

1 Generating accurate approximations to the data vector $\underline{X}(t)$ at each time $t=mT$.

2 Generating the parameter vector estimate $\hat{\underline{\theta}}(t)$ in terms of these samples.

These two issues are considered in the following sections.

7.4. THE STATE-VARIABLE FILTER

The differential equation for a strictly proper system of the form

$$y'(s) = \frac{B'(s)}{A'(s)}u'(s) =$$

where

$$A'(s) = a'_0 s^n + a'_1 s^{n-1} + \ldots + a'_n \tag{7.4.1}$$

$$B'(s) = b'_0 s^n + b'_1 s^{n-1} + \ldots + b'_n \tag{7.4.2}$$

may be rewritten in <u>controllable</u> <u>state-space</u> <u>form</u> as:

$$\frac{d}{dt}\underline{X}^c = \underline{AX}^c + \underline{U}u' \tag{7.4.3}$$

$$y'(t) = \underline{B}^T\underline{X}^c(t) \tag{7.4.4}$$

where the <u>companion</u> <u>matrix</u> $\underline{A}$ is given by

$$\underline{A} = \begin{vmatrix} -a'_1 & -a'_2 & -a'_3 & . & -a'_n \\ 1 & 0 & 0 & 0 & 0 \\ 0 & 1 & 0 & 0 & 0 \\ . & . & . & . & . \\ 0 & 0 & 0 & 1 & 0 \end{vmatrix} \tag{7.4.5}$$

$$\underline{B}^T = [b_1', b_2', .., b_n'] \tag{7.4.6}$$

$$\underline{U}^T = [1,0,0,...,0] \tag{7.4.7}$$

If the subsystem is not strictly proper ($b_0 \neq 0$) the system has a direct feedthrough term. For the purposes of this chapter, we handle this in a rather unconventional way by using an <u>extended</u> <u>state</u> <u>vector</u>. The single nth order differential equation 7.4.3 is recast as n first order differential equations and an algebraic equation

$$\frac{d}{dt}x^c_i = x^c_{i-1} \qquad i=1..n \tag{7.4.8}$$

$$a_0 x^c_0 = u - a_1 x^c_1 - a_2 x^c_2 - \ldots - a_n x^c_n \qquad (7.4.9)$$

The <u>extended</u> <u>state</u> <u>vector</u> is then defined as

$$\underline{X}^c = [x^c_0, x^c_1, \ldots, x^c_n]^T \qquad (7.4.10)$$

(x^c_1 - x^c_n forms the state; x^c_0 is the extension)

Taking Laplace transforms (with zero initial conditions) of equations 6 and 7 gives

$$s\bar{x}^c_i = \bar{x}^c_{i-1} \qquad (7.4.11)$$

$$a_0 \bar{x}^c_0 = \bar{u}(s) - \sum_{i=1}^{n} s^{-i} \bar{x}^c_0 \qquad (7.4.12)$$

and so

$$\bar{x}^c_0 = \frac{s^n}{A'(s)} \qquad (7.4.13)$$

It follows that (with zero initial conditions) the Laplace transform of the extended state vector is

$$\bar{\underline{X}}^c(s) = \frac{1}{A'(s)} \begin{vmatrix} s^n \\ s^{n-1} \\ \cdot \\ 1 \end{vmatrix} \bar{u}(s) \qquad (7.4.14)$$

In this formulation, the states are all derivatives of x^c_n. For this reason, x^c_n is sometimes called the <u>partial</u> <u>state</u> ξ of the system[8]. With zero initial conditions, the partial state ξ can be written in terms of the system input and output as

$$\xi = \bar{x}^c_n = \frac{1}{A'(s)}\bar{u}(s) = \frac{1}{B'(s)}\bar{y}(s) \tag{7.4.15}$$

If the input to each filter was known for all time, then the differential equation could be exactly implemented, and the resulting state exactly computed at each sample instant. However, in a digital implementation, the filter input is only known at discrete instants of time; it is the lack of knowledge of the signal between the samples which introduces an approximation error. At this stage, then, the following approximation is introduced:

Approximation

The filter input is a straight line between each adjacent pair of samples:

$$u'(mT + \tau) = u'_0 + u'_1\tau \quad 0 < \tau < T \tag{7.4.16}$$

where

$$u'_0 = u'(mT); \; u'_1 = \frac{u'(mT+T) - u'(mT)}{T} \tag{7.4.17}$$

Given this approximation, the differential equation (1) can be exactly solved between sample instants to give:

$$\underline{X}(mT+T) = \int_0^\tau e^{\underline{A}(T-\tau)} u'_0 + u'_1\tau \, d\tau \tag{7.4.18}$$

$$= e^{\underline{A}T}\underline{X}(mT)$$

$$+ \underline{A}^{-1}[e^{\underline{A}T} - 1]\underline{U}u_0$$

$$+ \underline{A}^{-2}[e^{\underline{A}T} - \underline{A}T - 1]\underline{U}u_1$$

$$= [1 + (\underline{A}T) + \frac{(\underline{A}T)^2}{2!} + \ldots]\underline{X}(mT)$$

$$+ T[1 + \frac{(AT)}{2!} + \frac{(AT)^2}{3!} + \ldots]\underline{U}u_0$$

$$+ T^2[\frac{1}{2!} + \frac{(AT)}{3!} + \frac{(AT)^2}{4!} + \ldots]\underline{U}u_1$$

This formulation gives rise to

<u>Algorithm 1</u>.

A. Precompute:

$$\underline{\alpha} = e^{\underline{A}T} \tag{7.4.19}$$

$$\underline{\beta}_0 = \underline{A}^{-1}[e^{\underline{A}T} - 1]\underline{U} \tag{7.4.20}$$

$$\underline{\beta}_1 = \underline{A}^{-2}[e^{\underline{A}T} - AT - 1]\underline{U} \tag{7.4.21}$$

B. At each time t=mT update each component of $\underline{X}$ as:

$$\underline{X}_u := \underline{\alpha X}_u + \underline{\beta}_0 u_0 + \underline{\beta}_1 u_1 \tag{7.4.22}$$

$$\underline{X}_y := \underline{\alpha X}_y + \underline{\beta}_0 y_0 + \underline{\beta}_1 y_1 \tag{7.4.23}$$

where:

$$u_0 = u(mT-T); \; u_1 = \frac{u(mT) - u(mT-T)}{T} \tag{7.4.24}$$

and

$$y_0 = y(mT-T); \; y_1 = \frac{y(mT) - y(mT-T)}{T} \tag{7.4.25}$$

However, inspection of the equations reveals that a recursive solution is also possible. In particular, defining

$$\Delta_1 = T\underline{U}u_0 + X(mT-T) \tag{7.4.26}$$

$$\Delta_2 = \frac{T}{2}[\underline{A}\Delta_1 + T\underline{U}u_1] \tag{7.4.27}$$

$$\Delta_k = \frac{T}{k}\,\underline{A}\Delta_{k-1} \quad k>2 \tag{7.4.28}$$

it follows that

$$X(mT) = \sum_{k=1}^{\infty} \Delta_k \tag{7.4.29}$$

This gives rise to

<u>Algorithm 2</u>

At each time interval perform the following recursion:

$\Delta := T\underline{U}u_0 + \underline{X};$
$\underline{X} := \Delta;$

FOR k := 2 TO max DO
  BEGIN
  $\Delta := \frac{T}{k}\underline{A}\Delta;$
  IF k=2 THEN $\Delta := \Delta + \frac{T^2}{2}\underline{U}u_1;$
  $\underline{X} := \underline{X} + \Delta;$
  END;

For practical purposes, ∞ has been replaced by the integer max in this algorithm; typically, max 5. Note that the matrix multiplication is considerably simplified by the companion form of $\underline{A}$.

Both emulator transfer functions (G(s)/C(s)) and (F(s)/C(s)) can be implemented in this fashion.

7.5. DISCRETE ESTIMATION OF CONTINUOUS-TIME PARAMETERS

Digital implementation of the continuous-time estimator implies a sample rate similar to that of the corresponding digital controller. In this section, it is shown that discrete-time estimation of continuous-time parameters is possible[9,10] without introducing any sampling error. This allows the estimation sample rate to be divorced from the controller sample rate.

As discussed in section 7.3, the linear-in-the-parameters equation

$$\phi^*(t) = \underline{X}^T(t)\underline{\theta} \tag{7.5.1}$$

can be sampled at any set of times t_m indexed by m to give the sampled equation

$$\phi^*_m = \underline{X}_m^T \underline{\theta} + e_m \tag{7.5.2}$$

This sampled model is of the correct form for parameter estimation using, for example, recursive least-squares. The discrete-time least-squares algorithm appropriate to the discrete-time linear in the parameters model is well known and will not be derived here. See any of the textbooks[11,12,13,14,15] for details.

The two important point of implementation to be emphasised here are:

1 Discrete-time estimation can be used to estimate $\underline{\theta}$ which contains continuous-time parameters.

2 The samples of $\phi^*(t)$ and $\underline{X}$ can be taken at any set of sample instants - which need not be uniformly spaced.

7.6. CONCLUSION

With reference to a particular continuous-time self-tuning controller, it has been found possible to obtain a discrete-time version. The approximation due to discretization is localised in the state-variable filter; thus a good implementation of the state-variable filter will lead to accurate parameter estimates. Such accurate estimates are particularly important when either the parameters themselves are of interest, or when the controller parameters are to be sent to another algorithm such as a PID controller.

It has been shown how a discrete-time parameter estimator may be used to estimate continuous-time parameters.

Apart from the consequent loss of information, this estimator may run at a longer sample interval than the state-variable filter without any loss in accuracy. Indeed, the estimator can run as a background task asynchronously with the state-variable filter; this is particularly useful when the estimator task has to compete with some other task (such as keyboard driver) within a microprocessor-based controller.

A more compete discussion of the discrete-time implementation of continuous-time self-tuning controllers will appear in a forthcoming book[6],

which will also contain complete software listings of the relevant algorithms.

References

1. Gawthrop, P.J., "Some interpretations of the self-tuning controller," Proceedings IEE, vol. 124, no. 10, pp. 889-894, 1977.

2. Astrom, K.J., Hagander, P., and Sternby, J., "Zeros of sampled systems," Automatica, vol. 20, no. 1, pp. 31-38, 1980.

3. Wellstead, P.E., Zanker, P., and Edmunds, J.M., "Pole assignment self-tuning regulator," Proceedings IEE, vol. 126, pp. 781-787., 1979.

4. Gawthrop, P.J., "Self-tuning PID controllers: Algorithms and implementation," IEEE Transactions on Automatic Control., vol. AC-31, no. 3, pp. 201-209, 1986.

5. Gawthrop, P.J., Continuous-time Self-tuning Control. Vol 1: Design, Research Studies Press, Mechanical Engineering Research Studies, Engineering control series., Lechworth, England., 1987.

6. Gawthrop, P.J., Continuous-time Self-tuning Control. Vol 2: Implementation, Research Studies Press, Mechanical Engineering Research Studies, Engineering control series., Lechworth, England., 1988 (In preparation).

7. Gawthrop, P.J., "Robust continuous-time self-tuning control of single-input single-output systems," International Journal of Adaptive Control and Signal Processing, 1987 (to appear).

8. Kailath, T., Linear Systems, Prentice-Hall, 1980.

9. Young, P.C., "Process parameter estimation and self-adaptive control," in Theory of self-adaptive systems, ed. P.H. Hammond, Plenum Press, New York, 1966.

10. Young, P.C., "Parameter estimation for continuous-time models - A survey," Automatica, vol. 17, no. 1, pp. 23-39, 1981.

11. Clarke, D.W., "Introduction to self-tuning controllers," in Self-tuning and adaptive control: Theory and applications., ed. Harris, C.J. and Billings, S.A., pp. 36-71, Peter Peregrinus, 1981.

12. Ljung, L. and Soderstrom, T., Parameter identification, MIT Press., London, 1983.

13. Eykhoff, P., System identification, Wiley, 1974.

14. Astrom, K.J. and Wittenmark, B., Computer controlled

systems, Prentice-Hall, 1984.

15. Goodwin, G.C. and Sin, K.S., Adaptive filtering prediction and control, Prentice-Hall, Englewood Cliffs, New Jersey, USA, 1984.

Chapter 8

Software aspects of self-tuning control

P. S. Tuffs

8.1 INTRODUCTION

Self-tuning control has been an active field of research for the past fifteen years. During this time many problems have occurred with the method, ranging from theoretical issues such as stability and robustness, to applications needs such as integral action and set-point following.

Past research into self-tuning control can be broken down into roughly three areas. Theoretical studies have concentrated on issues such as the robustness and stability of controllers in the presence of false assumptions about the process model. Numerical studies have addressed the need for stable parameter estimation updates and prevention of blow-up when using forgetting factors. Applications studies have revealed the need for incremental data in parameter estimation, integral action in the controller, and methods for dealing with actuator saturation.

Many of the initial problems in self-tuning control have been overcome. Stability and robustness have been improved by the use of long-range predictive controllers (De-Keyser and Van-Cauwenberghe (1981), Ydstie (1982), Clarke et al (1984)). Numerically stable covariance updates have been developed together with variable forgetting factors that prevent blow-up (Peterka (1975), Biermann (1977), Ydstie (1982), Kulhavy and Karny (1984)). Integral action and actuator desaturation methods have been added (Box and Jenkins (1976), Hodgson (1982), Tuffs and Clarke (1985), Campo and Morari (1986), Lawson and Hanson (1974)). The problems of rapidly sampled systems have been solved by use of hybrid controllers (Gawthrop (1980), Goodwin (1985)).

Despite the advances made in research, self-tuning controllers have not been placed in widespread use and currently show no signs of replacing conventional three term regulators in the process control industry. There are many possible explanations for this. Self-tuning controllers are inherently more complex than conventional fixed-parameter algorithms, since a parameter estimator is running at the same time as the control calculation is being performed.

The number of parameters needed to commission a self-tuner is on the order of ten or twelve, many more than the three term regulator, and the parameters are less simple to understand than a gain, reset time and derivative time constant.

Complexity alone is not sufficient to explain a lack of widespread implementation, since the control calculation is often a very small part of a highly complex control system. The success of the Foxborough EXACT self-tuning controller (Bristol, 1985) indicates that there is a need for this technology within the industry, so why has there not been a rapid deployment of the technique in other process control computers?

One answer may be that a self-tuning controller is a software entity, yet little research has been published on the software engineering of self-tuning controllers. In order to embed a self-tuning controller into an existing process control system the mathematical algorithms must be translated into a computer program, interconnected with the rest of the system, and interfaced to the process Operator and commissioning Engineer.

The EXACT system solves these problems by "hiding" the complexity in a single enclosure, where the connections are made via analog signals. This solution is not appropriate if the existing control scheme consists of a process computer programmed, say, in FORTRAN, and the final system must operate in an integrated fashion.

This chapter considers some of the software aspects of self-tuning control, and attempts to use some of the principles of software engineering in devising possible solutions. Solutions are presented in both the conventional numerical processing language FORTRAN, and the more modern and powerful language Ada.[1] The real-time aspects of the algorithms are also considered, since to be useful within a control computer the software must be small, fast, simple to use, and must be reliable, understandable and maintainable.

8.2 THE STRUCTURE OF A SELF-TUNING CONTROLLER

A self-tuning controllers consists of a parameter estimation algorithm coupled to a discrete-time control law. In a general purpose self-tuning scheme, such as the Generalized Predictive Controller of Clarke et al. (1984) the control law is obtained by setting performance requirements for the closed loop system in terms of a generalized cost function, equation 8.1:

---

1. Ada is a registered trademark of the U.S. Government (Ada Joint Program Office).

$$J = \sum_{i=N_1}^{N_2} [P(q^{-1})y(t+i) - w(t+i)]^2 + \sum_{i=1}^{N_U} [\lambda(1 - q^{-1})u(t+i-1)]^2 \qquad (8.1)$$

The process model for the system to be controlled is given in equation 8.2:

$$y(t) = \frac{B(q^{-1})}{A(q^{-1})} u(t-k) + \frac{e(t)}{A(q^{-1})(1 - q^{-1})} \qquad (8.2)$$

The usual notation has been used for the measured variable $y(t)$, the control signal $u(t)$, the integrated disturbance model $e(t)$ and the set-point $w(t)$. The polynomials $A(q^{-1})$, $B(q^{-1})$ contain the process dynamics, the pure delay in the system is k-1 samples, the inverse-reference model polynomial $P(q^{-1})$ specifies the desired set-point following response, and a cost λ is included on increments of the control signal $\Delta u(t)$. Δ is the difference operation $1-q^{-1}$. Three prediction horizons are used in the cost function to achieve different control objectives. For a discussion of these terms see Clarke et al. (1984).

The parameter estimation scheme can be summarized as the following one-step ahead predictor of the measured variable, derived from the process model equation 8.2:

$$y(t) = y(t-1) + q(1 - \hat{A})\Delta y(t-1) + \hat{B}\Delta u(t-k) + \varepsilon(t) \qquad (8.3)$$

The estimated quantities are indicated by a hat superscript.

In essence, all the information needed to implement the interface to a self-tuning controller is contained in equations 8.1 and 8.3. These equations, however do not capture the full requirements for a control algorithm, which must be able to deal with things such as not being in command of the process during manual operation, saturation of the control actuators, bumpless transfer from manual to automatic mode and smooth start-up when the control program is initiated. Also, the equations do not reflect the implementation details of the algorithm. These issues are best handled at two distinct levels: the high-level structure and interface requirements (top-down) and the low-level implementation mechanisms (bottom-up).

8.2.1 Top-down Structure

At a high level of abstraction (Booch, 1983) a self-tuning controller can be formulated as a black-box, which receives the process measurement and set-point and produces a control signal for the process (Fig. 8.1). It is possible to produce software that implements this abstraction, but such a solution is undesirable for a number of reasons.

Firstly, there is no interface for setting the performance requirements for the system, so a default set of characteristics must be assumed by the estimator and the control law. This does not model the differing requirements of control systems that may range from sluggish temperature control loops to high performance flexible systems. It does not even match the equations 8.1 and 8.3 which call for parametric estimation and a cost function specification. Second, there is no information about the internal state of the algorithm, which makes monitoring of its behaviour difficult if not impossible. Just as a physical black-box is not a realistic structure for a self-tuning controller, a software black-box is not a viable solution.

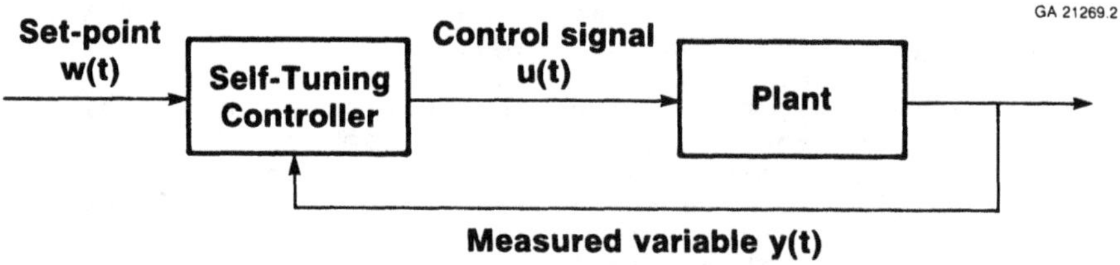

Fortran interface:

Subroutine STC (W,U,Y)

Ada interface:

Procedure STC (Set_Point, Measured_Variable: in Process_Variable;

Control_Signal: out Actuator_Variable);

Fig. 8.1 The "Black-Box" Model of a Self-Tuning Controller

At the other end of the scale is a solution in which all internal states of the self-tuning controller are visible to the programmer, but which has the same macro-structure as the black-box. The estimator, the performance requirements and the control calculation are all viewed as being part of the same module. No attempt is made to isolate the distinct elements of the algorithm, and data flow paths are not defined or controlled. This solution is only marginally better than including the physical text of the subprogram into the control system itself, since all the data vectors required for the low-level implementation must be created and managed from the calling environment. However, the solution does allow tailoring of the control algorithm to meet differing performance needs.

Such a solution might appear as shown in Fig. 8.2. Since all internal states are visible, the procedural interface would be extremely complex, even with the ability to represent collections of related variables in record structures in an Ada solution. Even though the algorithm could be tailored to meet the needs of different systems, and the internal performance could be monitored, the software would be unwieldy and difficult to use.

A compromise must be reached between simplicity of use and access to internal structures. Such a compromise can be derived by considering the problem in an object-oriented framework.

8.2.2 Object-Oriented Design

A self-tuning controller is a type of control algorithm and has a particular set of operations that apply to it. In software engineering terms it is an abstract data type, and it is amenable to packaging in a modular form.

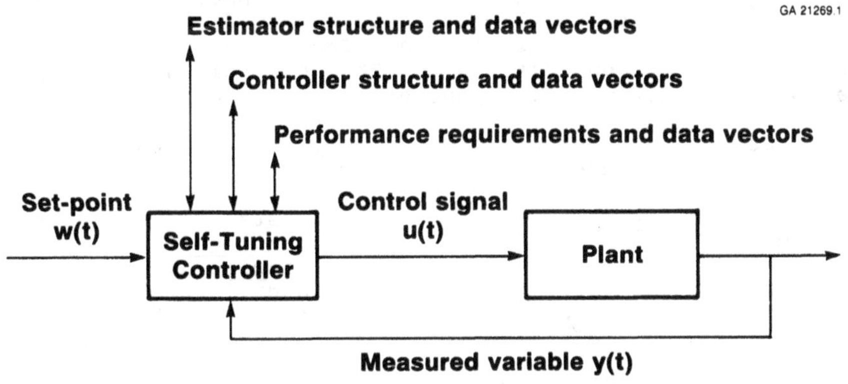

Fortran interface:

Subroutine STC (W,U,Y,N1,N2,NU,RLAMDA,PQ,...)

Fig. 8.2 The "Transparent-Box" Model of a Self-Tuning Controller

An object is some entity in the description of a problem that can be represented as a variable or constant in a software solution. In many cases the objects in a problem are physical items, such as a sensor or an actuator, in other cases objects are more abstract such as the prediction horizon of a controller.

An object-oriented approach to designing software is discussed in Booch (1983). In brief, once the problem has been formulated (section 8.2) an informal strategy is developed as a textual description of the solution. The nouns in the description become the objects (variables) in the software, and the verbs and verb-phrases yield the procedures and functions that allow access to the objects. The only operations permitted on the objects of a particular type are those defined by the verbs.

The following paragraph serves to illustrate this idea for a self-tuning controller.

"A self-tuning controller consists of a recursive estimation update, a set of performance requirements, and a control calculation. The process variables are filtered to provide data for the estimator and controller. The performance requirements are set in terms of the prediction horizons, the reference model speed and the control cost. The estimator can be set to an initial state, the covariance matrix can be boosted by a multiplicative factor, and the forgetting factor can be modified, or turned on and off. The parameter estimates from the estimator can be accepted or rejected for the controller. The controller is either in or out of command of the process, but must always calculate a control 'bid' based on its saved state and the process variables. The control signal can be clipped by an actuator before it acts on the process. Desaturation is performed after the control calculation. The process variables must be 'balanced' on startup. The control and estimation algorithms must work independently of each other."

Using the method proposed by Booch, the nouns and adjectives are underlined.

"A <u>self-tuning controller</u> consists of a recursive estimation update, a set of <u>performance requirements</u>, and a control calculation. The <u>process variables</u> are filtered to provide data for the <u>estimator</u> and <u>controller</u>. The <u>performance requirements</u> are set in terms of the <u>prediction horizons</u>, the <u>reference model speed</u> and the <u>control cost</u>. The <u>estimator</u> can be set to an <u>initial state</u>, the <u>covariance matrix</u> can be boosted by a multiplicative factor, and the <u>forgetting factor</u> can be modified, or turned on and off. The <u>parameter estimates</u> from the <u>estimator</u> can be accepted or rejected for the <u>controller</u>. The <u>controller</u> is either in or out of command of the <u>process</u>, but must always calculate a <u>control 'bid'</u> based on its saved <u>state</u> and the <u>process variables</u>. The <u>control signal</u> can be clipped by an <u>actuator</u> before it acts on the <u>process</u>. Desaturation is performed after the <u>control calculation</u>. The <u>process variables</u> must be 'balanced' on startup. The <u>control</u> and <u>estimation algorithms</u> must work independently of <u>each other</u>."

Not all implementation details are contained in this description, but that is as it should be since this specification merely defines the software interface presented to a user of the algorithm. From this description the following objects relevant to the solution are identified as:

```
SELF_TUNING_CONTROLLER

     PERFORMANCE_REQUIREMENTS
       PREDICTION_HORIZONS
       REFERENCE_MODEL_SPEED
       CONTROL_COST
```

```
ESTIMATOR
  COVARIANCE_MATRIX
  FORGETTING_FACTOR
  PARAMETER_ESTIMATES

CONTROLLER
  CONTROL_BID
  CONTROL_SIGNAL
  CONTROL_STATE

PROCESS_VARIABLES
 FILTERED_VARIABLES
```

Indentation is used to imply relationships between the objects. Some nouns such as "actuator" and "process" are ignored since they are part of the environment and not the controller.

The operations on these objects and classes of objects are obtained from the paragraph by examining the verbs and verb-phrases:

"A self-tuning controller consists of a recursive estimation update, a set of performance requirements, and a control calculation. The process variables are filtered to provide data for the estimator and controller. The performance requirements are set in terms of the prediction horizons, the reference model speed and the control cost. The estimator can be set to an initial state, the covariance matrix can be boosted by a multiplicative factor, and the forgetting factor can be modified, or turned on and off. The parameter estimates from the estimator can be accepted or rejected for the controller. The controller is either in or out of command of the process, but must always calculate a control 'bid' based on its saved state and the process variables. The control signal can be clipped by an actuator before it acts on the process. Desaturation is performed after the control calculation. The process variables must be 'balanced' on startup. The control and estimation algorithms must work independently of each other."

```
SELF_TUNING_CONTROLLER

   PERFORMANCE_REQUIREMENTS
     SET
     EXAMINE

   ESTIMATOR
     SET_INITIAL
     EXAMINE
     MODIFY_FORGETTING
     UPDATE_PARAMETERS
     BOOST_COVARIANCE

   CONTROLLER
     SET_INITIAL
     CALCULATE_CONTROL_BID
```

```
    DESATURATE
    ACCEPT_PARAMETERS

PROCESS_VARIABLES
    SET_INITIAL
    BALANCE
    FILTER
```

The names of the operations have been derived from the verbs in the description. Some extra operations have been added to reflect the ideas of information hiding, for example the operation EXAMINE was added as the inverse of the SET operation for data-types such as the PERFORMANCE-REQUIREMENTS. Also, since the strategy is informal, some of the operations such as TURN_FORGETTING and MODIFY have been implicitly combined, since they act on the same object (FORGETTING_FACTOR) in a similar way.

Note that statements such as "The control and estimation algorithms must work independently of each other" will impact the data flow between objects in the solution during implementation, and may lead to the selection of independent tasks as the method of implementation. This is discussed further in section 8.4.

The structure of a self-tuning controller has been defined, and is seen to consist of four distinct elements: performance requirements, a parameter estimator, a controller and a set of process variables. This is consistent with the mathematical representation given in section 8.2.

8.2.3 Bottom-up Structure

Top-down design is applicable when designing the interface and macro-structure of a piece of software. During implementation of the operations, however, it is necessary to define a set of common tools from which the higher-level operations can be constructed, so that the final product will be coherent and manageable.

In a self-tuning controller these tools must perform operations such as FILTERING of variables and manipulation of VECTORS. One way to identify the needed tool-set is to repeat the object-oriented design for each segment of the solution, this time concentrating on implementation details rather than functionality. Another contributor to the tool-set design is the mathematics of the algorithm. For example a scalar product operation is often found in a self-tuning controller (Clarke, 1981) and this can be implemented as a procedure which operates on two vectors to produce a scalar result.

The following categories of tools have been found useful in implementing a wide variety of FORTRAN-based adaptive control algorithms (Tuffs, 1985).

FILTER	Filtering operations on time-series data, moving average, auto-regressive, Runge-Kutta pseudo-continuous, spike
VECTOR	Scalar product, add, subtract, scalar multiple, move
POLYNOMIAL	Add, subtract, multiply, divide, Diophantine equation, load from vector, save to vector, stability projection (Kucera, 1979)
MATRIX	Add, subtract, multiply, divide (inverse), U-D factorization algorithm (Bierman, 1977)

This is not an exhaustive list, but it does indicate the types of operations that are required for self-tuning control. Other tools of a more general control systems nature such as dead-bands, limits and PI algorithms (Ebert et al. 1981) would be needed in a full implementation.

It is important to get the tools right before starting the coding of the algorithms so that the overall solution has a consistent feel and structure.

8.3 FORTRAN and Ada Implementations

Now that the objects and operations have been defined the specification must be translated into a programming language. The ease with which this can be performed is greatly influenced by the nature of the language chosen for the implementation. For example, FORTRAN does not support record structures or private types (data types where the only operations allowed are those provided by the programmer). Objects such as PERFORMANCE_REQUIREMENTS and ESTIMATOR must be coded as general purpose ARRAY structures, and there are no guarantees that a user will restrict operations on these variables to SET_INITIAL etc. This does not mean that the object-oriented approach has to be discarded. However, the final solution will not be as attractive or as manageable as one implemented in a language like Ada which is able to represent the needed data types, and to group related entities in packages.

Two solutions are given to illustrate the coding of a self-tuning controller designed in this way, one in FORTRAN and one in Ada. Since the implementation details are not being considered, the deliverable at this stage is a set of procedures and data structures, and a skeleton driver program which indicates the way that the algorithm will be used.

8.3.1 A FORTRAN Solution

The following code segment illustrates one way of implementing the performance requirements section of the controller. The operations are SET and EXAMINE, and the requirements call for the PERFORMANCE_REQUIREMENTS to be set in terms of the PREDICTION_HORIZONS, REFERENCE_MODEL_SPEED and CONTROL_COST.

In FORTRAN, assume that the performance requirements are stored in an ARRAY PRQ, that the prediction horizons are INTEGER's N1, N2, NU, that the inverse reference model is a first order moving average filter with a time-constant PTC (in samples of the controller), and that the control cost is a REAL variable RLAM. The body of the subroutine has been included to clarify the operations performed by the SET and EXAMINE procedures. This reveals that the subroutines translate the performance requirements to and from an internal form for subsequent use in the control calculation. The subroutine SETPRQ (SET performance requirements) might be coded as:

```
      SUBROUTINE SETPRQ(N1,N2,NU,PTC,RLAM,PRQ)
      REAL PRQ(*)
      PRQ(1) = FLOAT(N1)
      PRQ(2) = FLOAT(N2)
      PRQ(3) = FLOAT(NU)
      PRQ(4) = 1.0
      PRQ(5) = -EXP(1.0/PTC)
      PRQ(6) = RLAM
      RETURN
      END
```

and the inverse routine EXAPRQ (EXAmine performance requirements) as:

```
      SUBROUTINE EXAPRQ(PRQ,N1,N2,NU,PTC,RLAM)
      REAL PRQ(*)
      N1 = IFIX(PRQ(1))
      N2 = IFIX(PRQ(2))
      NU = IFIX(PRQ(2))
      PTC = 1/ALOG(-PRQ(5))
      RLAM = PRQ(6)
      RETURN
      END
```

The object PERFORMANCE_REQUIREMENTS has been implemented as a REAL array whose elements are the prediction horizons, the polynomial $P(q^{-1})$ coefficients, and the cost of increments of the control signal. In a driver program the SET procedure might be used as follows:

```
      PROGRAM STC
      DIMENSION PRQ(6)
      ...
      CALL SETPRQ(1,10,1,2.0,0.1,PRQ)
      ...
      END
```

There are a number of problems with this implementation. All internal operations on the performance requirements variable PRQ have to be written in terms of the array elements PRQ(i), making the implementation of the controller rather messy. Also, if at a later date the storage locations of the variables in the array change, for example if the degree of the reference model is increased to 2, then all of the dependent routines such as CALCULATE_CONTROL_BID will have to be modified to reflect the change.

There is no checking that the parameters are even of the correct type in the calls to SETPRQ and EXAPRQ, in the 'typeless' FORTRAN manner. Thus, an unfamiliar user of the procedure may make a mistake (the prediction horizons are never zero or negative) in using the procedure that would not be detected unless an IF-THEN construct was added to the code.

The rest of the implementation of a self-tuning controller in FORTRAN reflects these difficulties, and the code would be unwieldy and unreadable as a result. A possible set of FORTRAN subprogram definitions and documentation is given in appendix A, and a skeleton driver program is given below.

Since the storage requirements of the EST, VAR and CON arrays are not yet known, an arbitrarily large number has been inserted so that compilation can occur. This would have to be remedied in an actual implementation.

```
C DESCRIPTION -
C   Skeleton driver program for a FORTRAN based self-tuning
C   controller. The process variables, including the
C   set-point are input using subroutine ADC (Analog to
C   Digital Converter). The actuator is driven using
C   DAC (Digital to Analog Converter) and the control signal
C   limits are set at +/- 100 Units. Sample rate is set by
C   the calls to MARK and WAIT, which have to be implemented
C   on the computer used for control in terms of the system
C   services. The intention is that WAIT puts the
C   controller to sleep for H seconds after the call to MARK
C   was made, and issues a system level message if overruns
C   occur.
C
      PROGRAM STC
      DIMENSION PRQ(6), EST(100), VAR(100), CON(100)
      DATA UMAX, UMIN /100.0, -100.0/
      DATA H/0.1/
      CALL SETPRQ(1, 10, 1, 2.0, 0.1, PRQ)
      CALL SETEST(2, 2, 1, 1000.0, 0.9, EST)
      CALL SETCON(PRQ, EST, CON)
      CALL SETVAR(PRQ, EST, VAR)
C
C Obtain initial state of system and set control signal to
C zero. Balance the process variables.
C
      CALL ADC(W,Y)
      U = 0
      CALL DAC(U)
      CALL BALVAR(W, U, Y, VAR)
```

```
C
C Real time control loop.
C
10        CONTINUE
          CALL MARK
          CALL ADC(W,Y)
          CALL FILVAR(W, U, Y, VAR)
          CALL UPDEST(VAR, EST)
          CALL CALCON(PRQ, EST, VAR, CON, UBID)
          IF (UBID .GT. UMAX) THEN
            U = UMAX
          ELSE IF (UBID .LT. UMIN) THEN
            U = UMIN
          ELSE
            U = UBID
          ENDIF
          CALL DAC(U)
          CALL DESCON(U,CON)
          CALL WAIT(H)
          GOTO 10
        END
```

8.3.2 An Ada Solution

The package structure in Ada allows the four elements of a self-tuning controller to be grouped together in a meaningful way. In keeping with the concept of an abstract data-type the implementation details of the data structures needed for the self-tuning controller can be made 'private', so that only the operations declared in the packages can be applied to objects (variables) declared in a driver program. The compiler will enforce this logical abstraction and check that the use of the packages is consistent with their design.

Based on the design in section 8.2.2 the structure of these Ada packages will be similar to that shown below. Ada key words appear in bold face type, and ellipsis indicate that details have been omitted.

```
package SELF_TUNING_CONTROLLER is
  package PERFORMANCE is
    ...
    type REQUIREMENTS is limited private;
    procedure SET(...);
    procedure EXAMINE(...);
    ...
  end PERFORMANCE;
  package ESTIMATION is
    ...
    type ESTIMATES is limited private;
    procedure SET_INITIAL(...);
    procedure EXAMINE(...);
    procedure MODIFY_FORGETTING(...);
    procedure UPDATE_PARAMETERS(...);
    procedure BOOST_COVARIANCE(...);
    ...
```

```
  end ESTIMATION;
  package CONTROLS is
    ...
    type CONTROL is limited private;
    procedure SET_INITIAL(...);
    procedure CALCULATE_CONTROL_BID(...);
    procedure DESATURATE(...);
    procedure ACCEPT_PARAMETERS(...);
    ...
  end CONTROLS;
  package PROCESS is
    ...
    type VARIABLES is limited private;
    procedure SET_INITIAL(...);
    procedure BALANCE(...);
    procedure FILTER(...);
    ...
  end PROCESS;
end SELF_TUNING_CONTROLLER;
```

Fig. 8.3 gives a graphical interpretation of these packages, using the notation of Booch (1983). Packages are represented as boxes, data-types exported from the packages appear in rounded windows, operations on the data-types appear in square windows, and the body of the package appears as a 'blob' to reflect the hidden knowledge about its internal mechanisms.

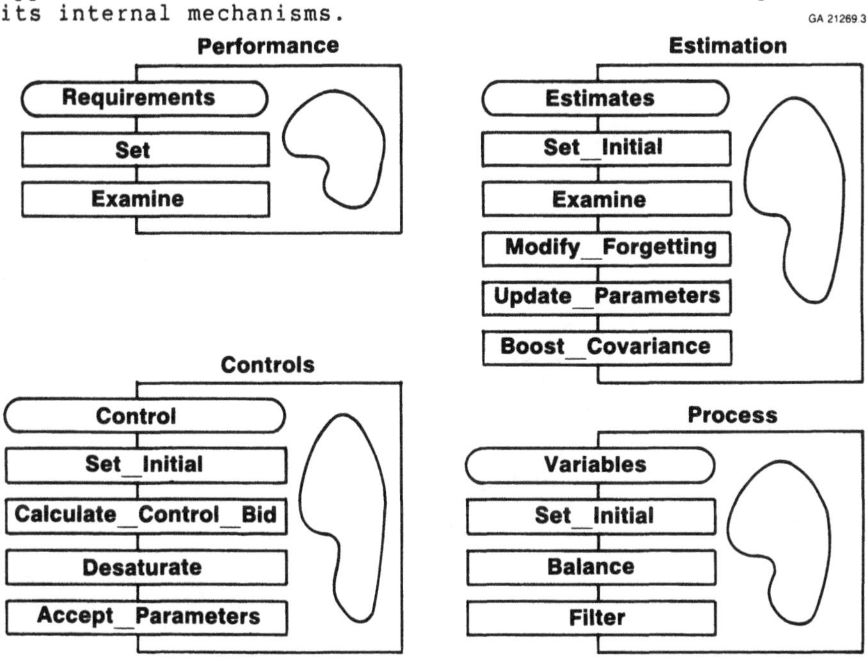

Fig. 8.3 "Booch" Diagram of a Self-Tuning Controller

The following Ada specification shows how an interface to the PERFORMANCE requirements package might be implemented.

```
package SELF_TUNING_CONTROLLER is
  package PERFORMANCE is
    type COST is new FLOAT range 0.0 .. FLOAT'LAST;
    type TIME_CONSTANT is new FLOAT range 0.0 .. FLOAT'LAST;
type HORIZONS is
      record
        N1 : POSITIVE := 1;
        N2 : POSITIVE := 10;
     NU : POSITIVE := 1;
      end record;
    type REQUIREMENTS is limited private;
    procedure SET(DESIRED      : out REQUIREMENTS;
                  TO_HORIZON : in HORIZONS := (1, 10, 1);
                  WITH_SPEED : in TIME_CONSTANT := 1.0;
                  WITH_COST  : in COST := 0.0);
    procedure EXAMINE(CURRENT     : in REQUIREMENTS;
                      HORIZON_IS : out HORIZONS;
                      SPEED_IS   : out TIME_CONSTANT;
                      COST_IS    : out COST);
  private
    type POLYNOMIAL is array(NATURAL range <>) of FLOAT;
    type REQUIREMENTS is
      record
        STORED_HORIZONS      : HORIZONS;
        REFERENCE_MODEL      : POLYNOMIAL(0 .. 1);
        STORED_CONTROL_COST : COST;
      end record;
  end PERFORMANCE;
end SELF_TUNING_CONTROLLER;
```

As part of the implementation some new floating point data types have been defined, to reflect the different nature of numbers such as the control COST's and TIME_CONSTANT's. These are available to the user of the package and allow local variables to be defined in the implementation. The range constraint explicitly identifies variables of these types as being non-negative, so the compiler and run-time environment can check that the parameters conform with the intentions of the implementor. Default values have been given for all the input parameters in SET. These may be omitted from the parameters list of a call to SET, thus simplifying its use.

The naming of the formal parameters in the SET and EXAMINE procedures becomes clear when the driver code is written. This code is quite readable even in the absence of comments. The highest level objects identified during the design phase appear as variables in this program.

```
with SELF_TUNING_CONTROLLER;
procedure STC is
use SELF_TUNING_CONTROLLER.PERFORMANCE;

  PERFORMANCE_REQUIREMENTS : REQUIREMENTS;
  PREDICTION_HORIZONS      : HORIZONS;
  CONTROL_COST             : COST := 0.1;
  REFERENCE_MODEL_SPEED    : TIME_CONSTANT := 2.0;

begin

  SET(DESIRED    => PERFORMANCE_REQUIREMENTS,
      TO_HORIZON => PREDICTION_HORIZONS,
      WITH_SPEED => REFERENCE_MODEL_SPEED,
      WITH_COST  => CONTROL_COST);

  EXAMINE(CURRENT    => PERFORMANCE_REQUIREMENTS,
          HORIZON_IS => PREDICTION_HORIZONS,
          SPEED_IS   => REFERENCE_MODEL_SPEED,
          COST_IS    => CONTROL_COST);
end STC;
```

This code can be compiled before the procedure bodies have been implemented, so that the design can be checked for completeness and consistency. Note that the PREDICTION-HORIZONS are implicitly initialized to (1, 10, 1) from the type declaration of HORIZONS.

It is a simple and obvious step to replicate the FORTRAN self-tuning controller of section 8.3.1 with the Ada packages defined above.

8.3.3 Testing the Software

One of the most important parts of software development is testing that the solution meets the design requirements. Self-tuning controllers are extremely difficult to test since there are many ways that they may fail. Some failure modes are inherently part of the algorithm while others may be coding or implementation errors that are difficult to detect.

One sound method for testing the implementation of a self-tuning controller is to simulate a variety of process models and pick performance criteria that yield solutions which are easy to verify. For example, (at least) four classes of plant can be defined:

STABLE, MINIMUM-PHASE
STABLE, NON-MINIMUM-PHASE
UNSTABLE, MINIMUM-PHASE
UNSTABLE, NON-MINIMUM-PHASE

and four classes of control objectives can be tested:

MINIMUM-VARIANCE	the tightest control possible
MINIMUM-RIPPLE	equivalent to minimum variance for plant without zeroes
LOW-GAIN	the closed loop process is identical to the open loop process, for set-point changes
OPEN-LOOP	zero-mean white noise at the output of the process should give control structures without feedback

Each control objective has specific behaviour on the different types of plant. For instance minimum variance controllers should yield an unstable mode in the control signal for non-minimum-phase processes.

Another important class of process arises when the estimator model contains more parameters than the physical process (Clarke et al. 1984). A parameter estimator on such a simulation should produce canceling factors in the $B(q^{-1})$ and $A(q^{-1})$ polynomial estimates, whose locations cannot be predicted: they may be stable or unstable.

These tests serve to exercise the numerical properties of a control algorithm, but that is only one aspect of a the code. For example, the storage space allocated for the data vectors must be large enough to deal with the complexity of the process model. This is critical in FORTRAN where storage is allocated statically at compile time, but is less of a problem in Ada where storage can be allocated as needed. Off-by-one errors are a common mistake in the implementation of a self-tuner, and many of these can be detected by simulating degenerate process models such as $y(t) = u(t-1)$.

The importance of testing in digital simulation mode cannot be overemphasized, since once a controller has to interface though an A/D or D/A converter and run in a multi-tasking environment many of the simple tests (such as a minimum-variance controller having a 1-sample set-point transition) disappear in the face of fractional time-delays and quantization effects.

Other issues in software verification and testing are beyond the scope of this chapter, but much work has been done in this area including Structured Walk-throughs Yourdon (1985), and The Art of Software Testing, Myers (1979) all of which is applicable to the testing of software for control systems implementation.

8.4 REAL-TIME ASPECTS

The design phase of the self-tuning controller called for the "... control and estimation algorithms (to work) independently of each other." This implies a degree of parallelism which cannot be expressed in a FORTRAN solution, but which leads naturally to the selection of independent tasks in Ada. It also implies that the controller must not depend on data-vectors being filled by the estimator, hence the introduction of the PROCESS_VARIABLES entity in section 8.2.2.

8.4.1 Cyclic vs. Data Driven Executives

There are a number of different models that can be used to express real-time control schemes. The simplest, and the one most often used in current control computers, is the Cyclic Processing Model (Softech, 1986) which has the main program looping at a pre-determined rate. The FORTRAN driver program in section 8.3.1 is an example of this technique.

For this method to work correctly each element of the cycle must take less time than the repetition period. If one element overruns then all the other elements will be delayed in their execution. In general, parameter estimation requires more computational resources than calculating a control signal. Also, the parameter estimator need not run in real-time, since it is only required to provide a new set of estimates when the controller 'accepts' them.

A multi-tasking implementation of a self-tuner is shown in Fig. 8.4. Tasks are represented as parallelograms whose implementation details are hidden inside "blobs" (Booch, 1983). Interface procedures to other tasks are shown as windows into the parallelogram, and data flow between tasks is represented by the connecting lines. The head of the arrow points to the server task.

The CONTROLLER and ESTIMATOR are formulated as independent tasks which communicate as needed by calling the entries in each other's interfaces. The CONTROLLER demands data from the PROCESS_VARIABLES task only when it is needed, so the CONTROLLER must set its own sample-rate. The overall system is monitored by a MANAGER task which may have a user in the loop deciding when to accept or decline the new parameter estimates.

This scheme is known as a Data Driven executive, and it has the advantage that the elements are structured independently. For example, failure of the parameter estimator need not affect the control calculations. The concept of Data Driven executives is an area of current research.

8.4.2 Other Aspects

The question most often asked about a new piece of control software is how much space does it require and how fast can it run? In 1975 when the first self-tuning controllers were being implemented on custom-built micro-processors (Clarke et al. 1975) these were important issues, but the technology has since expanded at a rate which has made them almost redundant. Attention still has to be paid to the size and speed of the algorithms, since there will always be an application that needs the code to run twice as fast as it currently does. However, Booch notes that concentrating on understanding the problem can have far more impact than bit-twiddling on a poor solution; this is certainly true for self-tuning control.

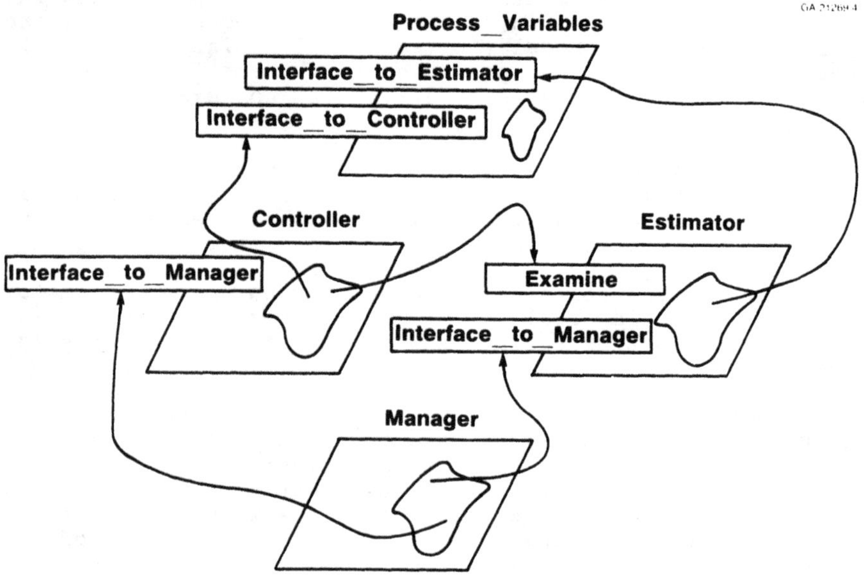

Fig. 8.4 Multiple Task Structure of Self-Tuning Controller

For example, the recursive formulation for the least-squares parameter estimator has more impact on the speed of execution than any fast floating-point tricks would have on a Gauss elimination procedure for matrix inversion. Likewise identifying that a symmetric matrix can be stored in a triangular factorized form (Bierman, 1977) saves far more space than reducing the number of significant digits used to represent a floating point number.

Fortunately, the implementors of self-tuning control have long been concerned with areas of efficiency so most of the algorithms now in the literature can be implemented in high-speed real-time embedded systems.

8.5 CONCLUSIONS

Many of the problems standing in the way of implementing self-tuning control have been resolved with the introduction of long-range predictive controllers and numerically stable algorithms. Practical improvements such as integral action and anti-windup mechanisms have been added as the result of many industrial trials, and the technology seems to be ready for full scale use.

The one area that has received relatively little attention is the software engineering of the self-tuning concept. This is just as important as the other engineering aspects, since without properly designed and tested software, together with verification methods and implementation support tools, the algorithms will not reach a majority of the embedded control systems in the aerospace and commercial process industries.

This chapter has considered some of the ways that software engineering can be applied to the design of a self-tuning controller. The need for this engineering will not depart with the advent of automatic code generation schemes such as Auto-Code/Matrix-X (ISI, 1986) which can generate real-time control code directly from block diagram models of the system. Such tools will simplify the interconnection of algorithms and structuring of control systems in much the same way that a full screen text editor facilitates the production of code and documentation.

There are still many problems that must be addressed in the implementation of self-tuning control. Further research is needed in the areas of real-time paradigms, multi-tasking controllers, knowledge-based performance monitors and interactive support systems. Most of these lie in the realm of software engineering.

8.6 REFERENCES

Bierman, G. J., (1977). Factorization methods for discrete system estimation. Academic Press.

Booch, G., (1983). Software engineering with Ada. Benjamin/Cummings Publishing Company Inc.

Box, G. E. P., and Jenkins, G. M., (1976). Time series analysis: forecasting and control. Holden-Day series in time series analysis and digital processing.

Bristol, E. H., (1985). The Exact pattern recognition adaptive controller, a commercial success, in English. 4th Yale Workshop on Adaptive Control.

Campo, P. J., Morari, M., (1986). Infinity-norm formulation of model predictive control problems. ACC Seattle, Washinton, USA.

Clarke, D. W., (1981). Implementation of adaptive controllers. In 'Self-tuning and adaptive control' (ed. Harris and Billings) Peter Peregrinus.

Clarke, D. W., Cope, S. N. and Gawthrop, P. J., (1975). Feasibility study of the application of microprocessors to self-tuning regulators. OUEL report 1137/75.

Clarke, D. W., Mohtadi, C. and Tuffs, P. S., (1984). Generalized predictive control. Part 1: the basic algorithm. OUEL report 1555/84.

Clarke, D. W., Mohtadi, C. and Tuffs, P. S., (1984). Generalized predictive control. Part 2: Extensions and interpretations. OUEL report 1557/84.

De Keyser, R. M. C. and Van Cauwenberghe, A. R., (1981). Self-tuning predictive control. Journal A, Vol.22, No.4, pp. 167-174.

Ebert, R. J., Kosakowski, J. J. and Snavely, S. L., (1981). Alcoa RDDC Functions and Subroutines. Aluminum Company of America.

Goodwin, G. C., (1985). Continuous and discrete adaptive control. Technical report EE518, NSW.

Gawthrop, P. J, (1980). Hybrid self-tuning control. Proc. IEE, Vol. 127, Pt.D, No. 5, pp. 229-236.

Hodgson, A. J. F., (1982). Problems of integrity in applications of adaptive controllers. OUEL report 1436/82.

Integrated System Inc. (ISI), (1986). Automatic code generation for real-time control systems. 2500 Mission College Blvd., Santa Clara, CA 95054-1215 USA.

Kucera, V., (1979). Discrete linear control. John Wiley & Sons.

Kulhavy, R. and Karny, M., (1984). Tracking of slowly varying parameters by directional forgetting. IFAC 9th World Congress, Budapest, Hungary.

Lawson, C. L. and Hanson, R. J., (1974). Solving least-squares problems. Prentice-Hall.

Lim, K. W., (1982). Robustness of self-tuning controllers. OUEL report 1422/82.

Myers, G. J, (1979). The art of software testing. John Wiley & Sons, Inc, New York, NY.

Peterka, V., (1975). A square-root filter for real-time multivariable regression. Kybernetika, Vol. 11, No. 1, pp. 53-67.

Softech Inc., (1986). Designing real-time systems in Ada. Softech Inc, 460 Totten Pond Road.

Tuffs, P. S., (1985). FAUST: introduction and overview. OUEL report 1568/85.

Tuffs, P. S. and Clarke, D. W., (1985). Self-tuning control of offset: a unified approach. Proc. IEE, Vol. 132, Pt.D, No. 3, pp. 100-110.

Ydstie, B. E., (1982). Robust adaptive control of chemical processes. Thesis, Department of Chemical Engineering and Chemical Technology, Imperial College.

Ydstie, B. E., (1984). Extended horizon adaptive control. IFAC 9th World Congress, Budapest, Hungary.

Yourdon, E., (1985). Structured walk throughs. Yourdon Press, New York, NY, USA.

APPENDIX A: A FORTRAN IMPLEMENTATION OF A SELF-TUNING CONTROLLER

```
C DESCRIPTION -
C   This package is a sketch of the FORTRAN implementation  of a
C   self-tuning controller designed from an object oriented view.  This
C   implementation is incomplete in the sense that not all of the
C   operations  identified in the design are included here.  However,
C   this package  does indicate the general structure of a FORTRAN based
C   self-tuner.
C
C NOTES -
C   1)  The term "private" indicates that the structure of a
C   variable need only be known inside the procedure and not outside the
C   procedure in order to use the procedure.  In practice a private type
C   could be a real array, with an appropriate dimension.
C
C   2)  Missing code needed for implementation is indicated by
C   elipses (...).
C
C   3)  This code has been compiled on a VAX/VMS FORTRAN-77 compiler
C   but the bodies of the code have not yet been implemented.
C
C PERFORMANCE_REQUIREMENTS -
C   SET (SETPRQ) -
C     Set performance requirements in terms of the prediction horizons,
C     the reference model speed, and the control cost
C
C     INPUT PARAMETERS -
C       N1    Integer   Prediction horizon start for error signal
C       N2    Integer   Prediction horizon end for error signal
C       NU    Integer   Horizon on control signal
C       PTC   Real      Reference model time constant in samples
C       RLAM  Real      Cost on control signal increments
C
C     OUTPUT PARAMETERS -
C       PRQ   Private   Contains static information about performance
C                       requirements of controller.
C
        SUBROUTINE SETPRQ(N1, N2, NU, PTC, RLAM, PRQ)
        REAL PRQ(*)
C       ...
        RETURN
        END
C
C ESTIMATOR -
C   SET_INITIAL (SETEST) -
C     Set initial parameter estimator structure, covariance elements and
C     forgetting factor.
C
C     INPUT PARAMETERS -
C       NA    Integer   Number of A coefficients to estimate
C       NB    Integer   Number of B coefficients to estimate
C       K     Integer   Lower bound on delay of process
C       DCOV  Real      Initial covariance value (Identity matrix)
C       FORG  Real      Forgetting factor between 0.0 and 1.0
C
C     OUTPUT PARAMETERS -
C       EST   Private   Contains information about the structure of
C                       estimator and workspace for calculation of
C                       parameter estimates.
C
        SUBROUTINE SETEST(NA, NB, K, DCOV, FORG, EST)
        REAL EST(*)
C       ...
        RETURN
        END
C
```

```
C   UPDATE_PARAMETERS (UPDEST) -
C     Update parameter estimates from process variables.
C
C     INPUT PARAMETERS -
C       VAR   Private   Dynamic state of the process.
C
C     INPUT OUTPUT PARAMETERS -
C       EST   Private   Updated estimates of the process model.
C
        SUBROUTINE UPDEST(VAR, EST)
        REAL VAR(*), EST(*)
C       ...
        RETURN
        END
C
C CONTROLLER -
C   SET_INITIAL (SETCON) -
C     Set initial structure of controller workspace from performance
C     requirements and parameter estimator.
C
C     INPUT PARAMETERS -
C       PRQ   Private   Performance requirements.
C       EST   Private   Parameter estimator structure.
C
C     OUTPUT PARAMETERS -
C       CON   Private   Initialized controller workspace.
C
        SUBROUTINE SETCON(PRQ,EST,CON)
        REAL PRQ(*), EST(*), CON(*)
C       ...
        RETURN
        END
C
C   CALCULATE_CONTROL_BID (CALCON) -
C     Calculate control signal from filtered signals, performance
C     requirements and parameter estimates.
C
C     INPUT PARAMETERS -
C       PRQ   Private   Performance requirements
C       EST   Private   Estimates of process model
C       VAR   Private   Filtered variables
C
C     INPUT OUTPUT PARAMETERS -
C       CON   Private   Workspace for calculation of control signal.
C
C     OUTPUT PARAMETERS -
C       UBID  Real      Control signal bid from controller
C
        SUBROUTINE CALCON(PRQ, EST, VAR, CON, UBID)
        REAL PRQ(*), EST(*), VAR(*)
C       ...
        RETURN
        END
C
C   DESATURATE (DESCON) -
C     Desaturate controller for actuator limits.
C
C     INPUT PARAMETERS -
C       UACT  Real      Actual control signal applied to process
C
C     INPUT OUTPUT PARAMETERS -
C       CON   Private   Updated to reflect saturation in the process.
C
        SUBROUTINE DESCON(UACT, CON)
        REAL CON(*)
C       ...
        RETURN
        END
C
```

```
C PROCESS_VARIABLES -
C   SET_INITIAL (SETVAR) -
C     Set the structure for the saved variables from the performance
C     requirements and estimator structure.
C
C     INPUT PARAMETERS -
C       PRQ   Private   Performance requirements
C       EST   Private   Estimator structure
C
C     OUTPUT PARAMETERS -
C       VAR   Private   Sets the structure for the dynamic plant
C                       information, updated in FILVAR.  This
C                       dynamic information is used in both estimation
C                       and control.
C
      SUBROUTINE SETVAR(PRQ, EST, VAR)
      REAL PRQ(*), EST(*), VAR(*)
C     ...
      RETURN
      END
C
C   BALANCE (BALVAR)
C     Balance the initial state of the saved variables to reflect the
C     state of the process.
C
C     INPUT PARAMETERS
C       W     Real      Set point of process
C       U     Real      Control signal of process
C       Y     Real      Measured variable of process
C
C     OUTPUT PARAMETERS
C       VAR   Real      Dynamic state of the process.
C
      SUBROUTINE BALVAR(W,U,Y,VAR)
      REAL VAR(*)
C     ...
      RETURN
      END
C
C   FILTER (FILVAR) -
C     Filter process variables for control and estimation algorithms.
C     Updates the dynamic information stored in VAR.
C
C     INPUT PARAMETERS
C       W     Real      Set point of process
C       U     Real      Previous control signal sent to process
C       Y     Real      Measured variable of process
C
C     INPUT OUTPUT PARAMETERS
C       VAR   Private   Updated dynamic information about the process.
C
      SUBROUTINE FILVAR(W, U, Y, VAR)
      REAL VAR(*)
C     ...
      RETURN
      END
```

Chapter 9

Multistep predictive control and adaptation

B. E. Ydstie, A. H. Kemna and L. K. Liu

9.1. INTRODUCTION.

Most chemical processes have complex nonlinear dynamics and interactions between the different operating variables. Currently these processes are almost exclusively controlled using linear time-invariant theory in the form of decentralized PI regulators and occasionally derivative and feed forward controllers are used. The tuning of these controllers is usually based on a qualitative judgement of response patterns and, once an acceptable tuning is found, this setting is usually left unchanged for the life of the process or until the quality of the closed loop response has deteriorated significantly. Although the majority of chemical processes are designed to be self stabilizing and can be controlled via the use of simple heuristic approaches like the one described above, it is not at all clear that this is the best alternative. It is likely that energy, raw material and operating costs can be reduced significantly if more advanced control and identification theory is used.

One of the more exciting recent developments in the use of advanced control methods in chemical process systems has been the application of multistep predictive control methods as well as adaptive estimation techniques. Much of the early interest in the application of adaptive control to chemical process systems can be traced back to the articles by Åstrøm and Wittenmark [1] and Clarke and Gawthrop [2]. Both of these articles communicated the adaptive control theory in a manner that made it readily available to practicing control engineers with little training and background in advanced control theory. Many applications of these ideas to chemical process control problems were reported in the literature (See e.g. Seborg, Edgar and Shah [3]). However, the general conclusion of much of this work was that the adaptive controllers often did not have the necessary robustness properties to be applied unsupervised in a chemical plant context. Consequently, this theory has not yet had a significant impact in the chemical process industries. Further research has been aimed towards developing more robust approaches to the control and estimation problems. Much of this research effort has concentrated on improving the *recursive least squares* procedure for parameter estimation and the development of *finite horizon linear quadratic controllers* based on heuristic arguments.

The recursive least squares algorithm is based on a quadratic performance index and it performs a Newton iteration at each step. This algorithm converges about as fast as can possibly be expected and it has good convergence and stability properites in the linear time invariant case. There are a few practical problems associated with this approach that need to be addressed before it can be applied to adaptive control problems, however. Firstly, the algorithm needs to be modified so that it does not loose its sensitivity. This property is often achieved via a reset or a forgetting factor. Another problem with the recursive identification approach concerns the slow parameter drift. During quiescent periods there is little or no data available for process identification; the parameters may drift and produce instabilities. This problem is often avoided by including a deadzone which stops the identification when the process is not excited.

Several different approaches to the finite horizon LQC problem have been taken. In practice the main idea turns out be this: Choose a prediction horizon which is at least as long as the combined effect of the deadtimes and the inverse response and compute the sequence of future control inputs that minimizes the deviation between predicted process outputs and and a given desired trajectory over the horizon. The idea behind this approach is strikingly similar to the finite horizon linear quadratic controller. The difference lies in the fact that the trajectory is recomputed at each time instant in a *receding horizon* fashion. The receding horizon approach to the finite horizon LQC problem is for example discussed by Kwon and Pearson [4], Cutler and Ramaker [5] and Richalet [6]. The former is a theoretically motivated paper, the two latter papers are motivated by the practical applications of the finite horizon receding horizon idea to industrial control problems.

The multi-step receding horizon approach has recently been applied in the context of adaptive control by a number of authors. (See for example de Keyser and van Cauwenberghe [7,8], Lee and Lee [9], Tung [10], Casalino et al. [11], Goodwin et al. [12], Ydstie and Liu [13], Ydstie et al. [14], and Clarke et al. [15] to name a few). The idea appears to have been developed independently by a number of researchers. This is evidenced by the fact that the different algorithms seem to vary in minor ways. The basic principle of extending the prediction horizon and using the receding horizon concept appears to unify these different approaches somewhat and, similar, if not identical, performance is experienced.

In this contribution an identification algorithm based on a new formula for the variable forgetting factor is presented. The proposed algorithm includes a leakage term that prevents the problem of parameter drift. The use of the leakage is different from that suggested by Tsakalis and Ioannou [16]. We follow a new tack in developing the receding horizon strategy. Instead of developing a parametric feedback controller we formulize the control problem as a *Linear Program* (LP). This approach is normally used by Shell Chemical Co. in the QDMC. The approach has been developed further by Campo and Morari [17]. The purpose of the LP formulation is to make the multistep algorithm more directly compatible with the operating and design constraints. The prize one pays is that the controller cannot be formulated in the standard way as a transfer function.

9.2. LEAST SQUARES ESTIMATION AND FORGETTING FACTORS.

The theory is developed for the following linear time invariant model of a stable plant:

$$y(t) = L(q^{-1})u(t) + d(t) \tag{9.1}$$

where $L(q^{-1})$ is an m×m strictly proper transfer matrix in the backward shift operator. The m vector y(t) consists of output signals that are measured, the m-vector u(t) consists of manipulated input signals and d(t) is an m-vector of unmeasured disturbance signals supposed to capture the errors made in the modeling. In chemical process control applications it is often realistic to model the disturbances as random walks. For the purposes of parameter estimation and adaptive control it is convenient then to define a new disturbance variable using an incremental model

$$v_i(t) = y_i(t) - y_i(t-1) - \phi_i(t-1)^T\theta_i\,, \quad \text{for } i=1,2,...m \tag{9.2}$$

where ϕ is a vector past process input and output increments so that

$$\phi(t)^T = (1-q^{-1})[y(t),y(t-1),...y(t-n_y),u(t),u(t-1),...,u(t-n_u)] \tag{9.3}$$

and θ is a vector of parameters. The variable v now becomes an estimate of the driving term in the random walk. The regression formulation (9.2) implies that we can write Eq. 9.1 as m semidecoupled Multiple-Input Single-Output (MISO) models

$$y_i(t) = y_i(t-1) + \phi_i(t-1)^T\theta_i + v_i(t), \quad \text{for } i=1,2,...m \tag{9.4}$$

and we can apply a linear identification technique to obtain estimates of θ and v. Sometimes it is possible to use additional process information and include nonlinear terms in the regression vector without changing the linear properties of the parameter estimation [18].

EXPONENTIALLY WEIGHTED LEAST SQUARES ALGORITHM:

Given $\{P(0)>0, \theta(0), \phi(-1), \lambda(t) \text{ for } t>0\}$

Step 1: $P(t)=[P(t-1)-P(t-1)\phi(t-1)\phi(t-1)^TP(t-1)/n(t)]/\lambda(t)$ (9.5)

where $n(t)=\lambda(t)+\phi(t-1)^TP(t-1)\phi(t-1)$ (9.6)

Step 2: $\theta(t) = \theta(t-1) + P(t)\phi(t-1)e(t)$ (9.7)

where $e(t) = y(t) - y(t-1) - \phi(t-1)^T\theta(t-1)$ (9.8)

Step3: Set t=t+1 and go to step 1

The algorithm should be coded so that the matrix P is updated in factored manner. In the multivariable case m-copies of this algorithm are executed at each step.

We assume that we have the following conditions fulfilled.

ASSUMPTIONSET 1: For i=1,2,...m there exists m finite vectors θ_i so that $|v_i| \in l_\infty$, and positive constants M and c so that ϕ is persistently exciting in the sense that

$$\sum_{t-M}^{t} \phi_i(i)\phi_i(i)^T \geq cI > 0 \text{ for all } t > M$$

We now have

RESULT 9.1: Suppose that Assumption 1 is satisfied and that the forgetting factors are computed so that $0<\lambda_{min}\leq\lambda(t)\leq 1$ for all t, then lim sup $|\theta(t)| \leq k_1$, lim sup $|\theta(t) - \theta(t-1)| \leq k_2$, and

$$\lim\sup \frac{1}{t}\sum_{i=1}^{t} \frac{e(i)^2}{n(i)} \leq q < \infty \tag{9.9}$$

where k_1, k_2, and q are finite positive constants.

PROOF: See [26]

If the persistent excitation condition is not fulfilled then the parameters may drift as shown below.

Fortescue et al.[19] suggested updating the forgetting factors to keep a measure of the information content of the estimator constant. This algorithm has been extended to the Kalman filter [20]. The formula they suggested has been applied to a few industrial processes and it was shown that the algorithm has desirable convergence characteristics in the deterministic case. Numerous studies have indicated that the proposed method may be sensitive to choice of the user definable parameter N_0, however. Some of these problems may be overcome by modifying the update slightly and choose the forgetting factors to satisfy the relationship:

$$\lambda(t)=\frac{\mathrm{TraceP}(t-1)}{\mathrm{TraceP}(t-1)+e(t)^2/[(r+\phi(t-1)P(t-1)\phi(t-1))]} \qquad (9.10)$$

where r is an estimate of the measurement variance. In practice it suffices to set r = 1. Update (9.11) prevents the trace of P(t) form growing to large values since we have $\lim\lambda(t) = 1$ as trace P(t) tends to infinity. It will also prevent the trace P(t) to vanish since $\lim \lambda(t) = 0$ as trace P(t) tends to zero. In fact, in our experiments and simulations we have found that this algorithm is quite insensitive with respect to choice of r. The experiment shown in Figure 9.1 below demonstrates this effect. The long term steady state value of the trace of the covariance matrix is here plotted against choice of r . The experimental set-up is explained in section 9.5.

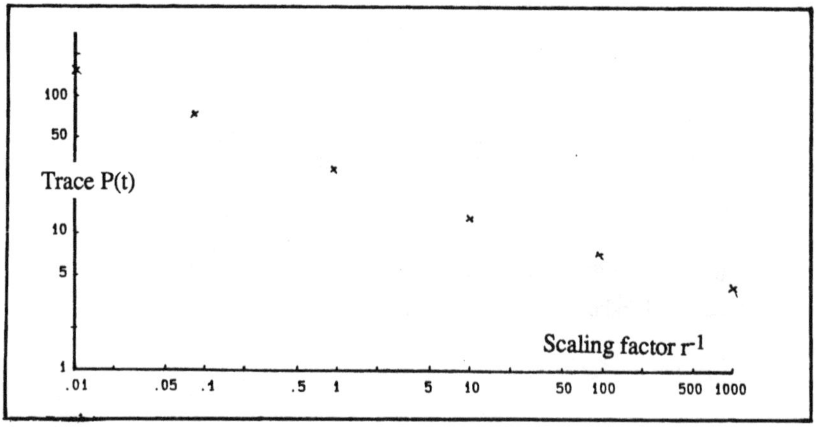

Figure 9.1: Sensitivity with respect to choice of r.

Cordero and Mayne [21] showed that the original Fortescue algorithm converges if an additional boundedness condition on the covariance matrix is introduced. Ydstie and Sargent [22] suggested a modified update and demonstrated convergence without the additional boundedness assumption. It is straightforward, using the analysis given here to show that $\lim\lambda(t)=1$ when the forgetting factor is updated via Eq. 9.11. All of results in this paper then follow as before.

Recursive parameter estimators exhibit slow parameter drift when there is not sufficient process excitation. This can be seen from the simulation example illustrated in Figure 9.2. Here we use the least squares algorithm with a fixed forgetting factor and deadbeat control. The system is described by

$$y(t) = 0.5y(t-1) + u(t-1) + 0.05, \quad u(t) = - \theta(t)y(t)$$

There is a small output disturbance and this causes the estimate $\theta(t)$ to drift when there is little process exitation. After a period of transient drift the closed loop system becomes unstable. This instablity in turn leads to process exitation and good conditions for the parameter estimator. The estimate is then returned to a region where the process is asymtotically stable but not persistently excited. The parameter then drifts until a new burst occurs. This apparently periodic behaviour changes its characteristics very slowly. It is possible to show that the fixed point point of this map corresponds to an area preserving elliptic attractor. The dynamic behaviour associated with such fixed points is usually quite complex and sensitive to the choice of the initial conditions. A further discussion is given by Ydstie and Golden [23].

There exist methods that can be used to prevent some of the problems associated with slow parameter drift. These methods usually involve either a deadzone or leakage term. The following modification combines the effect of both:

$$\theta_i(t) = \theta_i(t-1) + K_i(t)e_f(t) - \sigma(\theta_i(t-1) - \theta_i^*(t))$$

where $\theta_i^*(t) = \theta_i^*(t-1)$ if $K_i(t)e_f(t) \leq \varepsilon$ and $\theta_i^*(t)=\theta_i(t-1)$ elsewise

ε is here a deadzone variable, e_f is a filtered version of the prediction error, σ is the leakage term and K is the adaptation gain. Index i denotes the ith parameter estimate.

9.3 MULTI-STEP PREDICTIVE CONTROL AND CONSTRAINTS.

Multi-step predictive control involves the solution of an optimization problem on-line to obtain the sequence of future control inputs that minimizes the deviations between the predicted future outputs and a given sequence of setpoints. In practice there are also constraints imposed on the control problem. The constraints may involve control weights as well as physical limitations on the input and output magnitudes. These are relatively straightforward to deal with in the SISO case. In the multivariable case they become difficult to deal with in a traditional control setting. Thus, we propose the following optimization approach.

$$\min_{u(t+i),i=0,1,\ldots,M-1} \sum_{T=1}^{N} |p(t+T) - y^*(t+T)|w(T) \tag{9.11}$$

Subject to: $\underline{u} \leq u(t+i) \leq \bar{u}$, $\underline{y} \leq p(t+T) \leq \bar{y}$, $i=0,1,..,M-1$, $T= 1,2,...N$

where $|x|$ is the maximum norm of a vector, i.e. $|x| = \max|x_i|$,
$w(T)$ is a sequence of weights,
$p(t+T)$ is the prediction of $y(t)$ T steps into the future,
N is the prediction horizon, and
M is the control horizon with $u(t+M+i)=u(t+M-1)$ for $i=1,2,3.....$

The formulation above is easily changed so that the magnitude of change in the control action is constrained as well. We then simply include an additional constraint, the "move suppression factor"

$$c_j \geq |u_j(t+i)-u_j(t+i+1)| \quad \text{for } i=0,1,\ldots,M-1 \tag{9.12}$$

This has a similar effect as weighting the input changes in the LQ formulation. We now combine the optimization problem above with the adaptive estimator described in the previous section to form the multistep adaptive predictive policy. Feedback is produced by updating the control sequence at each sampling instant and by providing an estimate of the disturbances acting on the process. Additional correction follows from the continuous updating of the parameter estimates.

From the least squares estimator we get a time varying predictor model of the system in the left fraction form

$$y(t+1) = [I - A(q^{-1})]y(t+1) + B(q^{-1})u(t) + d(t). \tag{9.13}$$

Where A is a diagonal polynomial matrix with monic terms and B is a full polynomial matrix. A and B are defined so that $A^{-1}B = L$. A T-step ahead predictor for this system can be developed by following this procedure: Define two polynomial matrices $F(q^{-1})$ and $G(q^{-}$

1) with appropriate orders satisfying the generalized Diophantine identity

$$I = F(q^{-1})A(q^{-1})(1-q^{-1}) + T(q)^{-T}G(q^{-1}). \qquad (9.14)$$

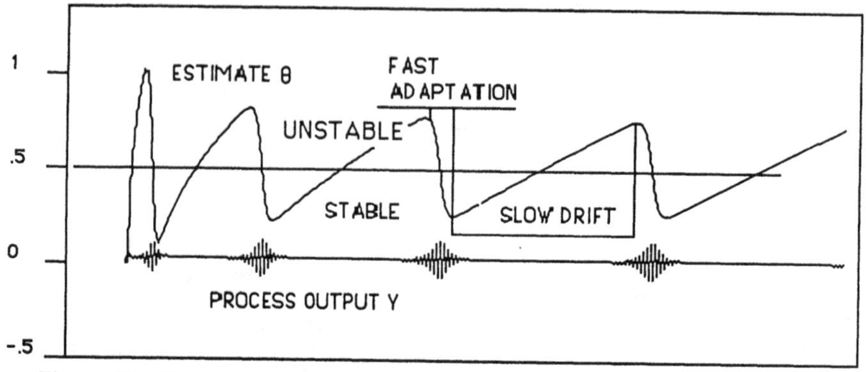

Figure 9.2: Slow parameter drift due to model mismatch. $\lambda = 0.99$.

$T(q)$ is an m×m predictor matrix defined in the forward shift operator q so that

$$T(q) = \mathrm{diag}(q,q,....,q) \qquad (9.15)$$

By pre-multiplying Eq. 9.13 with $F(q^{-1})A(q^{-1})$ and using the Diophantine identity we obtain the T-step ahead predictor.

$$p(t+T) = H_0(q)\Delta u(t) + H_1(q^{-1})\Delta u(t-1) + G(q^{-1})y(t) + D(q^{-1})\upsilon(t), \text{ for } T=1,2,...,N$$

where (9.16)

$$D(q^{-1}) = T(q)^T F(q^{-1}),\ H_0(q) + H_1(q^{-1}) = q^T F(q^{-1})B(q^{-1}), \text{ and}$$
$$\upsilon(t) = e(t)/n(t)$$

where $\Delta = 1-q^{-1}$ is the incremental operator and the coefficients of $H_0(q)$ are simply related to the first T-1 impulse response coefficients of the estimated model and p(t+T) denotes the T syep ahead predicted value for the process output. We have here followed established practice and included an estimate of the effect of disturbances through the variable $\upsilon(t)$ which is an estimate of v(t) in equation 9.2. For T = 1 the expression above reduces to a one-step ahead prediction. For finite N we can use the predictor equation (9.16) and write:

$$P(t) = H\Delta U(t) + Z(t) \qquad (9.17)$$

prediction — current and future controls — past outputs, inputs and disturbances

where $P(t)^T = (p(t+T)^T, p(t+T-1)^T, ..., p(t+1)^T)$
$U(t)^T = (u(t+T-1)^T, u(t+T-2)^T, ..., u(t)^T)$,
$Z(t)^T = (z_T(t)^T, z_{T-1}(t)^T, ..., z_{T+1}(t)^T)$, and
$z_T(t) = H_1(q^{-1})\Delta u(t-1) + G_T(q^{-1})y(t) + D_T(q^{-1})\upsilon(t)$

(Index T indicates the Diophantine equation associated with the T-step ahead prediction)

This formulation gives a mathematical program for the control sequence $U(t)^T = (u(t),u(t+1),...u(t+M-1),u(t+M-1),...u(t+M-1))$ indicating that only M controls are computed and the rest of the future controls are left unchanged. It has been suggested that good values are $N \in [5\text{-}20]$ and $M \in [1\text{-}5]$. This choice depends somewhat on the samplingtime. In practice N has to be large enough to capture the essential dynamics of the process. M, the control horizon, is a tuning parameter. For open loop stable processes that are overdamped it usually suffices to set M=1.

Although the objective function above is nonlinear, we can use the constraints and recast the problem in the standard *Linear Programming* format

$$\text{Minimize } a^T U, \text{ subject to } AU \leq b \text{ where } \min u_i \geq 0 \text{ for } i=1,2,..,M. \qquad (9.18)$$

The program involves few variables (M×m) and a larger number of inequality constraints. The large number of constraints indicate that significant savings in computer time can be realized by solving the dual problem

$$\text{Minimize } -b^T x, \text{ subject to } AU \leq a \text{ where } \max u_i \leq 0 \text{ for } i=1,2,..,M. \qquad (9.19)$$

This program is then solved during each sampling interval.

The LP approach to multistep control is motivated by the fact that in many practical process control applications it is appropriate to formulize the control problem using time domain considerations. Moreover, constraints constitute an important part of most chemical process control problems and these can be included in a transparent manner in the LP formulation. Efficient numerical routines (Simplex or the more recent Karmarkars methods for example) exist that can be used to solve for u(t) at each sampling instant. A number of factors contribute to make this a possible problem to solve on-line. The special structure of the problem can be used to accelerate the convergence rate of the algorithm; only small changes are allowed at each sampling instant so that the previous solution provides a good starting point for the control calculation, and lastly; this starting point is in the feasible set provided the constraints do not change. Stablity is not an issue here since the process is assumed to be open loop stable.

We now outline an approach which leads to an explicit solution for the control law calculation. Several routes are possible to take when there are no inequality constraints. Ydstie and Liu (1984) for example set M=1 and w(T)=0 for T≤N-1. The control law becomes

$$R(q^{-1})\Delta u_f(t) = [S(q^{-1}) + T(q)]r_f(t) - S(q^{-1})\upsilon_f(t) \qquad (9.20)$$

where the subscripts f indicate that the variables are filtered. Here we have used the parallel compensation technique shown in Fig. 9.3. This approach is used with the IMC discussed in Ref [24] and it is implicit in the DMC approach [5]. The polynomial matrices R and S are computed from the least squares estimates via the Diophantine equations. We get:

$$R(q^{-1}) = H_0(1) + H_1(q^{-1})$$
$$S(q^{-1}) = G(q^{-1})$$

Thus the control equation becomes

$$\Delta u(t) = H_0(1)^{-1}\{[S(q^{-1}) + T(q)]r_f(t) - S(q^{-1})\upsilon_f(t) - H_1(q^{-1})\Delta u(t)\} \qquad (9.21)$$

$H_0(1)$ can be guaranteed to be nonsingular and $R(q^{-1})$ can similarly be guaranteed to be stable provided that the process itself is exponentially stable [26]. The latter condition may not always be satisfied in the adaptive case and a stability check may have to be included. If R is unstable in a step then it can be stabilized by the following method. By elementary row operations bring R to triangular form. I.e. set

$$R'=RU \tag{9.22}$$

where U is unimodular and R is a triangular polynomial matrix. This can for example be done by following part of the algorithm which reduces R to the Smith form. If there are zeros of R' outside a domain ϑ with acceptable relative damping then these are reflected orthogonally onto the boundary of ϑ to form a new polynomial matrix R''. Finally, the R is obtained by setting

$$R = R''DK. \tag{9.23}$$

where K is a matrix of constants that ensures that the steady state gains do not change. R is now a stable polynomial matrix.

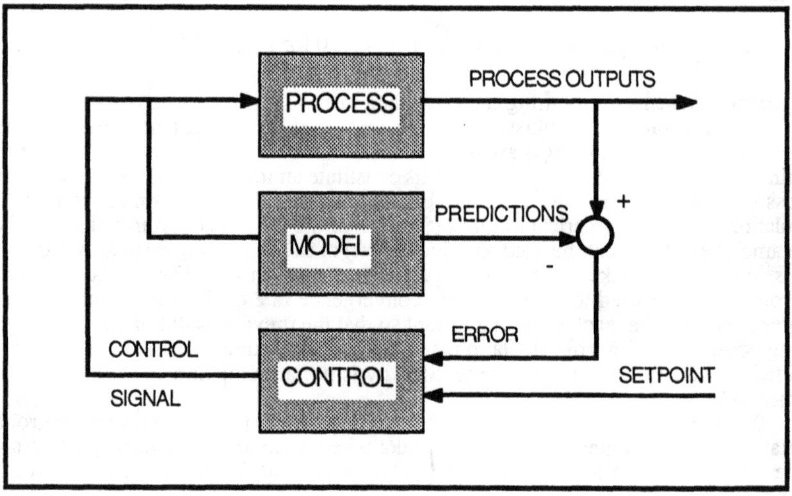

Figure 9.3: Block diagram for the parallel compensation scheme.

9.4 THE STABILTY ANALYSIS.

The stability analysis is first developed using the standard least squares algorithm with forgetting factors updated via Eq.9.10 and the controller is updated via Eq. 9.21. We use the following definition of stability.

DEFINITION: A controller for system 9.1 is said to be stable if it produces a sequence {u(t),t=1,2,3...} of controls so that

$$\limsup_{t\to\infty} \frac{1}{t}\sum_{i=1}^{t} \|y(t)-y(t)^*\|^2 \le k_1, \text{ and}$$

$$\lim_{t\to\infty} \sup \frac{1}{t} \sum_{i=1}^{t} ||u(t)||^2 \le k_2$$

where k_1 and k_2 are positive constants independent of $\{y(i),u(i),i=1,2,3,...\}$.

Instead of using the least squares estimates directly, we use a set of smoothed estimates

$$\theta_c(t) = \theta_c(t-1) + [\theta(t) - \theta_c(t)]/L, \quad L \in [1,\infty).$$

to compute the control transfer function. $\theta(t)$ is uniformly bounded and we may conclude

$$\lim_{L\to\infty} |\theta_c(t) - \theta_c(t-1)| = 0 \tag{9.24}$$

We have

RESULT 9.3: Suppose that det $L(z)\neq 0$ for $|z|\le 1$ and that Assumptionset 1 is satisfied. The controller based on equation 9.21 is then stable provided that L is chosen large. Moreover, for a constant output disturbance and a constant setpoint we have $\lim|y(t)-y^*| = 0$.

(This result holds true for the LP approach as well)

PROOF: The time varying closed loop system

$$y(t) = L(q^{-1})\, R(q^{-1},t)^{-1}[\{\, T(q) + \; S(q^{-1},t)\}y(t)^* - S(q^{-1},t)\upsilon(t)]$$

has bounded, slowly time-varying coefficients. Moreover it defines a "frozen" stable system for all t. The adaptive feedback system above can thus be made arbitrary close to a an exponentially stable, time invariant system provided L is large. Stability then follows since $\upsilon(t)$ is root mean square bounded according to Eq. 9.9 [25]. The controller includes an integrator and it is now straightforward to show that the second part of the result follows.

A similar result can be developed for the more general control law based on the optimization procedure.

9.5 APPLICATION STUDY USING EXTENDED HORIZON CONTROL.

The multi-step predictive control policy has been tested in a number of pilot plant and simulation studies of chemical process applications. We report here on the results from experiments performed on a MeOH distillation column situated in the Department of Chemical Engineering at the University of Massachusetts. The column has 12 bubble cap trays, is 12' tall and 8" in diameter. The adaptive controllers work in a cascade arrangement as shown in Fig 9.4. This arrangement was introduced to overcome a serious hysteris problem in the control valves. The control objective is simply to use the steam pressure and the reflux flow to adjust the top and bottom temperatures so that these follow predetermined set points. Disturbances were introduced by making large changes in the feed flow rate to the column. This is the standard dual temperature (composition) control problem which has attracted much academic research interest over the past several decades. An industrial application of this theory would most likely involve the use of a different set of control pairings.

The EHC was tested both in the SISO and the MIMO modes. In the MIMO mode two separate least squares estimators were used. One for the top and one for the bottom of

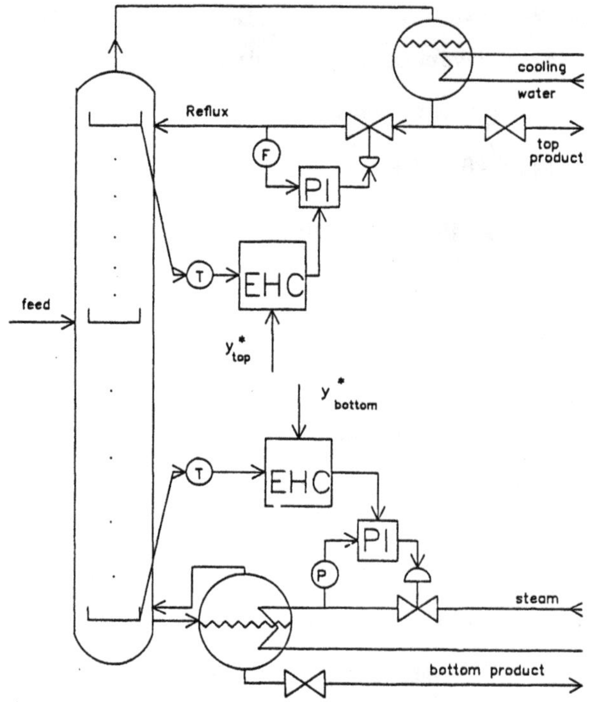

Figure 9.4: Schematic of distillation column.

the column. In all experiments we used a UDU^T factored version of the least squares estimator with forgetting factors updated as described earlier and the control law 9.21. The estimator deadzone was not necessary to implement in these particular experiments. We investigated the performance of the EHC controller when it was subjected to frequent set point and disturbance changes. This led to sufficient excitation and the problems associated with parameter drift were avoided. The indirect adaptive strategy was chosen. The control parameters are then based on estimating the parameters in the MISO model (9.3). The parameter projection procedure was not used in these experiments. A discussion on the use of the direct adaptive control law is given by Ydstie et al.[26]. It was genrally found that the indirect adaptive control approach gave better results. This conclusion was made after a number of simulation and pilot plant experiments were carried out. A satisfactory theory has not been developed yet that explain exactly why this is so. However, there are a couple of contributing factors that can be used to explain this observations. Firstly, the number of estimated parameters is less in the indirect control approach. Secondly, the regression vector in the indirect startegy contains more recent data. It is thought that both of these factors help producing more stable estimates and smaller prediction errors.

The MIMO extended horizon controller was found to be sensitive to start-up conditions in the column. It helped to bring the adaptive controller through a "learning phase" before start up. In our case this consisted of producing a few step changes in the manipulated variable with the estimator running but control set to "manual". We used a multirate sampling strategy. The top controller had a sampling rate of about 10 sec. and the bottom about 20 sec. Some leading coefficients of the elements of the $B(q^{-1})$ polynomial

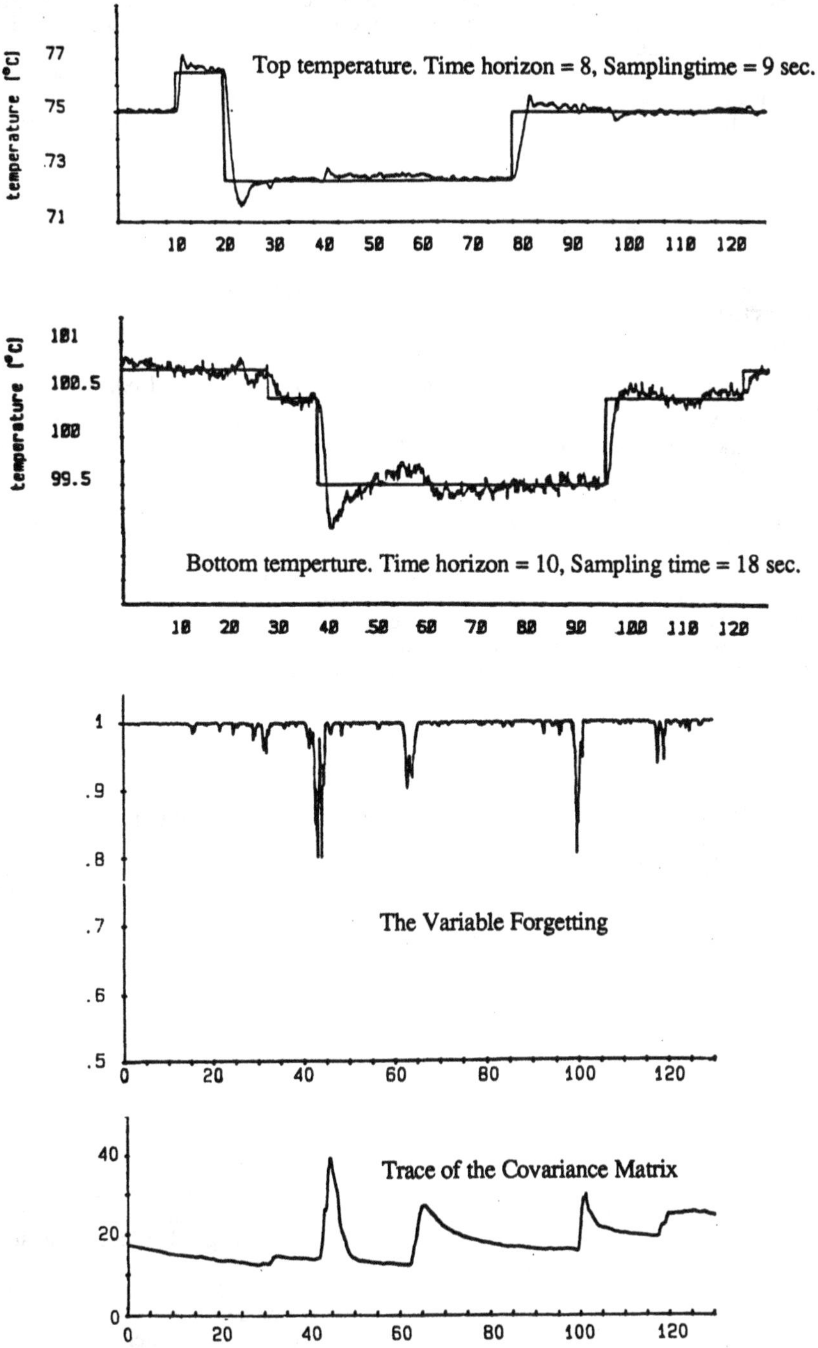

Figure9.5: Typical responses due to setpoint changes and disturbances.

matrix were set to zero to indicate the fact that there are large process delays involved in the distillation control problem. In particular, there was a large time delay associated with the interaction from the reflux to the temperature in the bottom of the column . Such process knowledge was used to reduce the size of the regressor. The number of parameters to be estimated and the size of the covariance matrix is then reduced accordingly and significant computational savings are realized. The input, output and set points were filtered. The expressions for the filters are

$$F(q^{-1}) = \begin{bmatrix} f_{11}(q^{-1}), f_{12}(q^{-1}) \\ f_{21}(q^{-1}), f_{22}(q^{-1}) \end{bmatrix}$$

where f_{ij} i, j = 1,2 are rational functions of the form $f(q^{-1}) = n(q^{-1})/d(q^{-1})$. In practice it suffices to use first order filters and set

$$d(q^{-1}) = 1 + (1-\alpha)\, q^{-1} \;,\; 0 < \alpha \leq 1 \;,\; n(q^{-1}) = \alpha$$

This gives unity steady state gains and a filter pole on $\mathbf{R}^+$ at $(1 - 1/\alpha)$. Decreasing α gives a slower response as the pole moves to the origin along the real axis. Setting $\alpha = 1$ removes the effect of the filter altogether since this places the pole at zero. This filter is the same as the model following filter used in the Model Predictive Controller [6]. Diagonal input and output filters were used. It was found advantageous to filter the off diagonal terms in the setpoint filter heavily. This reduces the effect of steady state interaction introduced by the EHC design procedure.

A typical result is shown in Fig. 9.5. It is seen that a 50% feed change disturbance goes by almost unnoticed. The parameters change noticeable, indicating that there are dynamic changes in the column. This is accompanied by a drop in the forgetting factor. Similarly, setpoint changes can be performed at either end of the column without significant interaction. The MIMO controller performed better than two independent SISO adaptive controllers for these types of disturbances.

9.6 SUMMARY AND DISCUSSION.

We have in this chapter discussed a new approach to multistep control based on adaptive parameter estimation and mathematical programming. The approach allows for the handling of process constraints. A new approach to the forgetting factor problem has also been proposed. Experiments indicate that the proposed adaptive control algorithm with the new forgetting factor update is not very sensitive with respect to the choice of the design parameters. The major design parameters are:

- The order of the process transfer functions.
- The control horizon.
- The prediction horizon.
- The sampling time.
- The constraints.
- The filter constants.

The adaptive controller has been tested in simulation and pilot plant experiments. The SISO version of the algorithm is particularly easy to implement since it requires few parameters to be specified by the user. Moreover, the algorithm appears to be relatively insensitive with respect to how these parameters are chosen. A good process simulator can be used as a help in the selection of these parameters. For example: In our study we found that simulation experiments performed on the linear transfer function model of a similar column taken from the literature gave relevant information about how to choose sampling time and prediction horizons for our experiments. Only minor adjustments were needed to fine tune the performance of the algorithm when it was applied to the real column. The most notable change needed was introduced by the necessity of filtering the input and output signals. Filtering of the input signal (u(t)) was especially important. The effect of small amplitude chaotic responses generated by nonlinearities can then be avoided or at least reduced without impairing performance noticably.

Acknowledgements: Support from the *National Science Foundation* Grant # CPE 84 10852 and the *US Department of Energy* Grant # DE-FG02-85ER1318 is gratefully acknowledged.

REFERENCES:

1. Åstrøm, K. J. and B. Wittenmark (1973) "On Self-tuning Regulators", Automatica **9**, p. 185.

2. Clarke, D. W. and P. J. Gawthrop (1975) " Self -tuning Controller" Proc. IEE Pt. D **122**, p. 929.

3. Seborg, D. E. S. L. Shah and T. F. Edgar (1986) "Adaptive Control Strategies for Process Control: A Survey" AIChE Journal **32** p. 881.

4. Kwon W. H., and A. E. Pearson (1977) "Qadratic Cost Feedback Stabilization", IEEE-T-AC-22, p.838.

5. Cutler, C. R. and B. L. Ramaker (1980) "Dynamic Matrix Control - A Computer Control Algorithm" JACC, San Francisco, CA.

6. Richalet, J., A. Rault, J. L. Testud and J. Papon (1978) "Model Predictive Heuristic Control: Applications to Industrial Processes", Automatica **14**, p. 413.

7. de Keyser, R. M. C. and A. R. Van Cauwenberghe (1982) "Simple Self-Tuning Multistep Predictors", 7th IFAC Symposium on Identification and System Parameter Estimation, Washington, D. C., p.1558.

8. de Keyser, R. M. C., Ph. G. A. Van de Velde and F. A. G. Dumortier (1985) "A Comparative Study of Self Adaptive Long Range Predictive Control Methods", 7th IFAC Symposium on Identification and System Parameter Estimation, York, U. K, p. 1317.

9. Lee, K. S. and W. K. Lee (1983) "Extended Discrete-Time Multivariable Adaptive Control Using Long Term Predictor", Int. J. Control **38,** p. 495.

10. Tung, L. S. (1983) "Sequential Predictive Control of Industrial Processes", American Control Conference, San Francisco, CA.

11. Casalino, G., F. Davoli, R. Minciardi and G. Zappa (1985) "Adaptive Finite Horizon Implicit LQ Controllers", 7th IFAC Symposium on Identification and System Parameter Estimation York, U.K., p. 1311.

12. Goodwin, G. C., L. Dugard and X. Xianya (1984) "The Role of the Interactor Matrix in Multivariable Stochastic Adaptive Control", Automatica, **20**, p. 701.

13. Ydstie, B. E. and Liu, L. K. (1984) "Single and Multivariable Control with Extended Prediction Horizon", the ACC, San Diego, CA. p.

14. Ydstie, B. E., L. S. Kershenbaum and R. W. H. Sargent (1985) "Theory and Application of an Extended Horizon Self-tuning Controller," AIChE Journal, **31**, p. 1771.

15. Clarke, D. W., P. S. Tuffs and C. Mohtadi, (1985) "Self-Tuning Control of a Difficult Process", 7th IFAC Symposium on Identification and System Parameter Estimation, p. 1009, York, U.K. p. 1009.

16.Tsakalis, K. S. and P. A. Ioannou (1987) "Adaptive Control of Linear Time-Varying Plants: A New Controller Structure". American Control Conference, Minneapolis, MI. p. 583.

17. Campo, P.J. and M. Morari (1986) " ∞-norm Formulation of Model Predictive Control Problems" American Control Conference, Seattle, WA. p. 339.

18. Golden, M.P., S. A. Chesna and B. E. Ydstie (1986) "Adaptive Nonlinear Model Control" Tech. Report Dept. Chem. E. Univ. of Mass. Amherst MA01003. Also to appear in Chem. E. Comm.

19. Fortescue, T. R., L. S. Kershenbaum and B. E. Ydstie (1981) "Implementation of Self Tuning Regulators with Variable Forgetting Factors", Automatica, **17**, p. 831.

20. Ydstie, B. E. and T. B. Co (1985) " Recursive Estimation with Adaptive Divergence Control" Proc. IEE pt. D, **132**, p 19.

21. Cordero, O. A. and D. Q. Mayne (1981) "Determininstic Convergence of a Self tuning Regulator with a Variable Forgetting Factor" Proc. IEE pt. D, **128**, p 19.

22. Ydstie, B. E. and R. W. H. Sargent (1986) "Convergence and Stability Properties of an Adaptive Regulator with a Variable Forgetiing Facrtor"Automatica **22**, p. 749.

23. Ydstie, B. E. and M. P. Golden (1987) "Chaos and Strange Attractors in Adaptive Control Systems", IFAC World Congress, Munich West Germany.

24. Garcia, C. and M. Morari (1982) "Internal Model Control: Part 1 A unifyieng Review and Some New Results" Ind. and Eng. Process De. and Dev. **21** p. 308.

25. Desoer, C. A., (1970) "Slowly Varing Discrete System $x_{k+1}=A_kx_k$", Electronic Letters, **6**, pp. 339-340.

26. Ydstie, B. E., A. H. Kemna and L. K. Liu (1987) " Robust Multivariable Predictive Control" Tech Report, Dept of Chem. Eng., Univ. of Mass., Amherst MA01003. Also to be published in Comp. in Chem Eng.

Chapter 10

Application of extended prediction self-adaptive control

R. M. C. DeKeyser

10.1. INTRODUCTION

During the last years a number of interesting control methods were published all having a common idea : the control action is based on a long-range prediction of the process output (Richalet, 1980; Cutler and Ramaker, 1980; Rouhani and Mehra, 1982; De Keyser and Van Cauwenberghe, 1981; Peterka, 1984; Ydstie, 1984; Greco et al., 1984; Clarke et al., 1986).

Self-adaptive long-range predictive control methods have some nice properties.

Compared to methods based on single-step cost indices (minimum-variance and dead-beat type control laws) they seem to be fairly robust w.r.t. wrong process structure and non-minimumphase process zeros. This is important in a practical application where a simple mathematical process model is only a crude approximation of the complex real world. Maybe of even greater importance to a practical situation is the fact that the closed-loop performance criteria realized by the algorithms can be easily understood by the user : they have a direct and simple physical interpretation. Compared to classical control laws (such as PID) long-range predictive control laws are extremely powerful if there is a process deadtime or if the setpoint is preprogrammed.

This paper deals with real-life applications of the EPSAC strategy, which is a specific long-range predictive control method (De Keyser and Van Cauwenberghe, 1985, 1986). The development of a real-life automatic control system is much more than just programming a control algorithm. Therefore it was decided not to focus solely on the adaptive methods, which can be found elsewhere in this book and in the references. Rather we pay some attention to the several aspects (hardware, software) which are important for the implementation and to the various phases that occur during the realization and testing of an adaptive automatic control system for industrial processes.

The first application describes the results of a demonstration project where it was the objective to illustrate by means of real-life experiments the feasibility of some concepts of modern control theory and digital instrumentation to low cost automation. A typical application field is in heating and ventilation control.

The second application describes the development of a self-learning computer-based automatic control system for a cutter suction dredger, a difficult process which never was controlled successfully by means of classical PID control algorithms.

10.2. EXTENDED PREDICTION SELF-ADAPTIVE CONTROL (EPSAC)

Long-range predictive control methods are based on the principle of Fig. 10.1 which illustrates the EPSAC method.

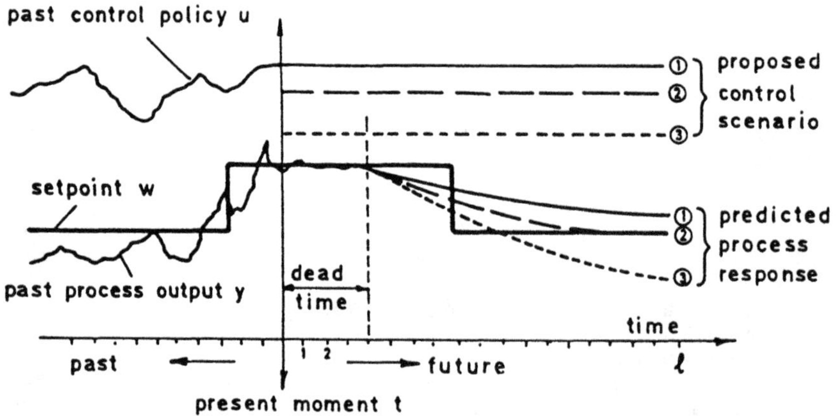

Fig. 10.1 Long-range predictive control

They are characterized by the following strategy :

- At each sampling period t, a forecast of the process output over a long-range time horizon (l sampling periods) is made in the control algorithm, based on an estimated mathematical model of the process dynamics. Moreover it is a function of the control scenario proposed to apply in the future.
- From several proposed control scenarios, the strategy is selected which brings the predicted process output back to the setpoint in the "best" way according to a suitable control objective. One possible form is (v = setpoint, k = deadtime) :

$$\min_{\Delta u(t)} \sum_{i=k}^{l} [v(t+i)-y^*(t+i/t)]^2 + \mu \Delta u^2(t) \qquad (10.1)$$

  where $\Delta u(t) = u(t)-u(t-1)$ is the current control input variation and where $y^*(t+i/t)$ denotes the predicted process output (the output $y(t+i)$ predicted at time t). Many other control objectives could be selected according to the specific application.
- The control action for the first sampling period of the best strategy is then applied as a control action to the real process input at the present moment. At the next sam-

pling instant the whole procedure is repeated, leading to an updated control action with corrections based on the latest measurements (receding-horizon strategy).

The algorithms to implement this strategy have been described in De Keyser & Van Cauwenberghe (1981, 1985, 1986). The Generalized Predictive Control (GPC) method of Clarke et al. (1986) is also closely related. The results of a comparative study of some long-range predictive control methods is described in De Keyser et al. (1987). Reference is also made to the contributions of Clarke, Mohtadi and Ydstie in this book.

10.3. LOW COST HEATING AND VENTILATION CONTROL

This demonstration project illustrates two important facts : first microprocessor-based digital hardware is today a valuable alternative besides classical electronics for low cost automation equipment; and second, as a result of this, the availability of advanced digital control algorithms is no longer a privilege of the larger process control projects.

10.3.1. System Hardware

The general structure of the Adaptive Microcomputer Controller (AMC) is shown in Fig. 10.2.

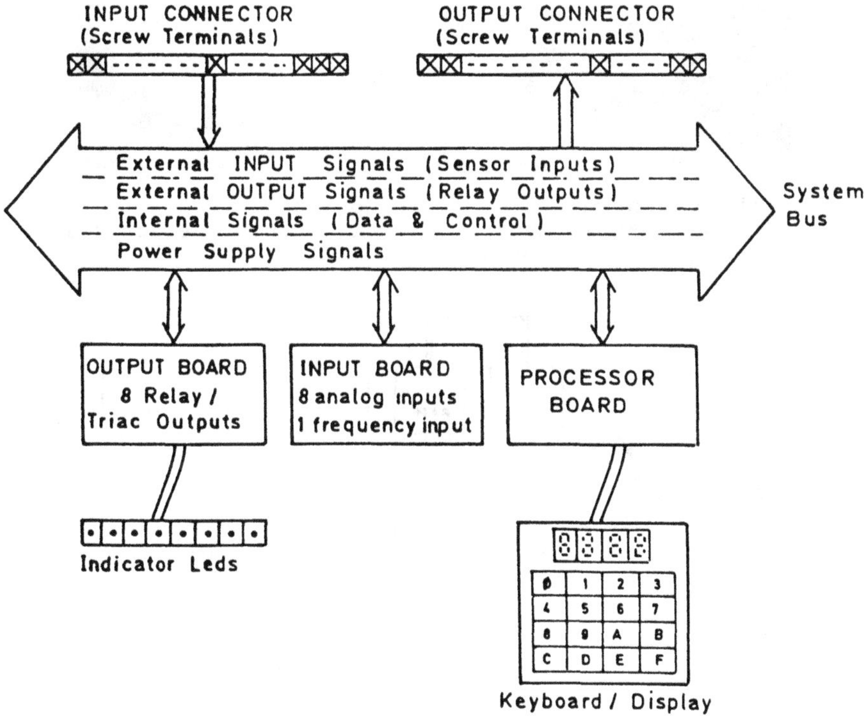

Fig. 10.2 Hardware structure of the AMC

Apart from the main board with system bus and power supply, there are 4 smaller boards which implement specific functions : 3 identically sized (12 cm x 7 cm) boards which can be plugged into any of the 4 bus slots (further called the Processor, Input and Output board) and a Keyboard/Display board which is mounted on the frontpanel for interaction with the operator. The whole system is very compact and is contained in a (11 cm x 13 cm x 26 cm) wall-mounted case. This is illustrated by the picture showing the AMC in a building heating application.

A block scheme of the processor board is given in Fig. 10.3. The well-known MC6809 8/16 bit microprocessor chip is surrounded by 4 other popular VLSI-chips.

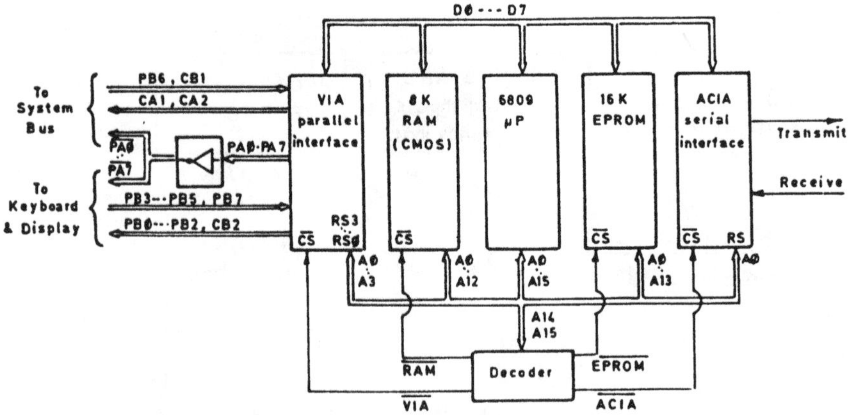

Fig. 10.3 Block scheme of processor board

The program code is stored in a 16 Kbyte eprom chip, the program data in a 8 Kbyte ram chip. The CMOS static RAM chip is plugged into a 28 pin DIP socket with a built-in CMOS watch function, a nonvolatile RAM controller circuit and an embedded lithium energy source. This provides a complete solution to problems associated with memory volatility and uses a common energy source to maintain time and date. There is no battery backup on the system level, so when there is a power shut-down the processor stops running. This is no severe drawback for the kind of applications that were envisaged (there is indeed a good chance that the power for the actuators will be off as well). However it is important that after power start-up the control system restarts autonomously, having the exact time-of-day available (because the operator setpoint, e.g. a heating schedule, may be a function of time) and with the same data as just before shut-down. This is especially true when using self-adaptive control algorithms, because the controller actions are based on the estimated numerical process model, which should not be lost.

The ACIA chip (Asynchronous Communications Interface Adapter, MC6850) allows for serial communication with any computer (usually a PC) which has a standard RS232 interface. In this way it is possible (albeit not necessary) to send all measured and internal data of the AMC system to a Personal Computer for supervision or further processing, or to receive data and instructions from the PC (e.g. for remote operation instead of local operation via the keyboard). This option has turned out to be a valuable feature, especially to coordinate from a PC at a hierarchically higher level the operation of several AMC systems in a larger application environment.

Perhaps the most important chip on the Processor Board is the VIA (Versatile Interface Adapter, SY6522) because it is the interface between the processor and all other system functions such as sensor inputs, relay outputs, keyboard input and display output. The control and data transfers for all these functions have been realized through a relatively low number of binary input-output lines. This was only possible when using the same signals for different functions and multiplexing them in time, resulting in a less straightforward but cheaper hardware design and an essentially more complex software. It was however decided from the beginning that the hardware cost should have a higher weight than the software cost in the total cost optimization exercise, to end up with a low cost product. The reason is mainly because low cost automation is only economically feasible when it leads to a mass product, in which case the software cost is spread over a large number of units. This is perhaps another difference with the design of industrial process control systems where the cost of software becomes more and more important.

The input board contains three main parts :

- the electronic circuits to translate the input signals coming from the external sensors into a suitable 0-5 Volt DC signal; as the system was initially built for climate control applications, the currently available input sen-

sors are temperature and relative humidity.
- an electronic multiplexer to select one channel out of the eight input channels;
- a voltage-to-frequency converter to convert the selected 0-5 VDC signal into a 0-5 kHz block wave; the frequency is directly measured by means of a counter in the VIA chip. In this way the analog sensor signals are converted into a digital value.

The output board contains eight output switches (relay or triac) that are controlled from the VIA binary output signals $\overline{\text{PA0}}$...$\overline{\text{PA7}}$ by means of the CA2 signal which acts as 'enable' input to an octal D-type latch. When the 'enable' is high, the latch outputs will follow the data inputs $\overline{\text{PA0}}$...$\overline{\text{PA7}}$. When the 'enable' is taken low, the outputs will be latched at the levels that were set up at the inputs. Each of the eight output switches has a corresponding LED indicating its state on the system front panel.

10.3.2. System Software

All software is developed on a 6809-microcomputer development system. The executable code is then stored in the 16 Kbyte EPROM chip of the target system. As the target system itself does not run under the supervision of an operating system, the software for the self-adaptive microprocessor controller has to include the low level procedures for the real-time control of all system hardware.

Normally this is done in a low level computer language such as assembler. In this project however all software, including the low level procedures such as timer control, interrupt processing and input/output port manipulation, is successfully programmed in Pascal. The available Pascal compiler has some extensions compared to the standard language which allowes this low level programming.

The heart of the AMC system is a timer in the VIA chip which is interrupting the processor every 20 milliseconds. The time critical tasks that are initiated by the timer tick are e.g. keyboard input, display refreshing, time-of-day updating, sensor measurement and control. The remaining part of the processor time - between the finishing of the interrupt service routine and the arrival of the next timer interrupt - is for eventual dialog with the user. We refer to this part as the 'background program', while the interrupt handling part of the software is called the 'foreground program'.

The background program consists of 2 parts : the initialization procedures and a loop executing the user interface (dialog) software which is running as a low priority task and which is interrupted every 20 ms by the foreground (high priority) tasks. At power on, the necessary RAM locations are initialized with default values, the VIA and ACIA chips are properly initialized and the time-of-day is read from the battery protected CMOS watch. From now on the VIA timer is allowed to interrupt the background program every 20 ms, while it is waiting for keyboard input from the user.

The main part of the foreground program contains the control algorithms, including the sensor measurements and the relay setting and also the communication procedure. This part is repeated every 5 seconds. It can take one or more seconds to execute depending on the number of measurements, the complexity of the control algorithms, whether or not PC communication is requested, etc... . Obviously the other parts of the interrupt service procedure (display refreshing, keyboard scanning, etc...) must be executed strictly every 20 ms (and take only some milliseconds to execute). Therefore the control part should start by allowing new interrupts thus resulting in a program structure with nested interrupts and several priority levels.

10.3.3. Applications

The system described until now is essentially a small general-purpose microcomputer controller. It offers some nice application possibilities for low cost automatic control and regulation in the field of climate control. This section contains a brief summary of some recent applications. The control of a building heating system will be described more in detail as a specific example.

10.3.3.1. Greenhouse climate control. Temperature and relative humidity are important parameters that should be kept close to a desired value in order to stimulate the growth and quality of the craps. The main disturbing effect is outside weather (solar radiation, temperature, rain, wind). The actuators that can be used for control are essentially heat supply and ventilation windows (eventually sprinkling). The popularity of mini- and microcomputers for the control of greenhouses is steadily growing. Typically in practice control procedures are improved by adding decisions. The combination of one or more AMC systems for basic temperature and humidity control with optionally a PC for monitoring and rule based expert control has obviously a promising future.

10.3.3.2. Incubator control. Temperature is here an extremely important parameter which should be kept within a range of +/- 0.1 ^{o}C around a reference value that slowly varies with time during the incubation period. Also humidity has to be controlled within some %RH. The air is ventilated around the eggs and is heated by means of electrical resistances. The temperature is kept within the specified range by switching the heating power on and off frequently by means of the AMC triac outputs (no mechanical wear). The air is moistened by means of blades turning in and out a water reservoir. The driving motor is controlled by the AMC system on the basis of the measured and desired humidity. The setpoint for temperature and relative humidity can be preprogrammed by the user for the whole incubation period and for several types of eggs (pheasant, duck, hen ...). The system also takes care of some secondary functions such as turning the eggs regularly.

10.3.3.3. Sty ventilation control. A stable sty temperature is important for low fodder consumption and fast fattening. The dense sty occupation which characterizes modern pig breeding leads to considerable internal heat production coming from the animals. Efficient ventilation is thus of major importance. Applying natural ventilation (which is usually the case) requires that window aperture is adapted in order to keep internal temperature at a specified level. It is well known that in natural ventilation systems the air freshening effect strongly depends on wind parameters such as speed and direction. They represent the main disturbing variables, while variations in outside air temperature (mainly day/night fluctuations) obviously introduce supplementary perturbations. The task of the AMC system is to control continuously the position of the ventilation windows in order to keep the sty temperature between the specified limits in spite of all disturbances.

10.3.3.4. Building heating control. Most central heating plants in small buildings and residences are either controlled by a central room thermostat as the most simple system or by a mixture valve controlled by a weather compensation circuit. Large public buildings are more and more controlled by computer based energy management systems (EMS) which are rather expensive and cannot be justified on smaller installations (Birtles et al., 1984). Thanks to its low cost aspects it is not unreasonable to apply the AMC system for more intelligent control in medium-sized installations. Compared to thermostatic or weather compensation control this leads to both a more comfortable and less energy consuming heating policy.

A typcial configuration that can be controlled by a single AMC system is given in Fig. 10.4. The residence is divided in 4 heating zones (each consisting of one or more rooms) which are controlled independently. Three of them (1-2-3) are floor-heated, the fourth being heated by a large-surface (low-temperature) radiator. The following instrumentation is available for the control of the heating system : temperature sensors for the outside air (T_o), the boiler water (T_b) and the supply water (T_s); a valve motor (VM) for the 4-way mixture valve; in each zone a temperature sensor for the room air (T_r) and an electrothermic valve (ETV) which can shut-off the circuit. The 7 temperature sensors are connected to the input board of the AMC system, while the 8 relays (triacs) of the output board are used to switch on/off the circulation pump, the boiler, the mixture valve (open/close) and the 4 shut-off valves.

The desired room temperature is preprogrammed as a function of time for each zone independently, which assures that a zone is only heated when occupied. This is the only information the user has to enter in order for the system to operate. Building specific characteristics (such as heating curves that have to be tuned by the user in current weather dependent control systems; or gain constants for a modulating mixture valve regulator) are totally absent. All parameters are identified and optimized by the control software itself from information in the measured data, resulting in a self-learning or self-adaptive control system.

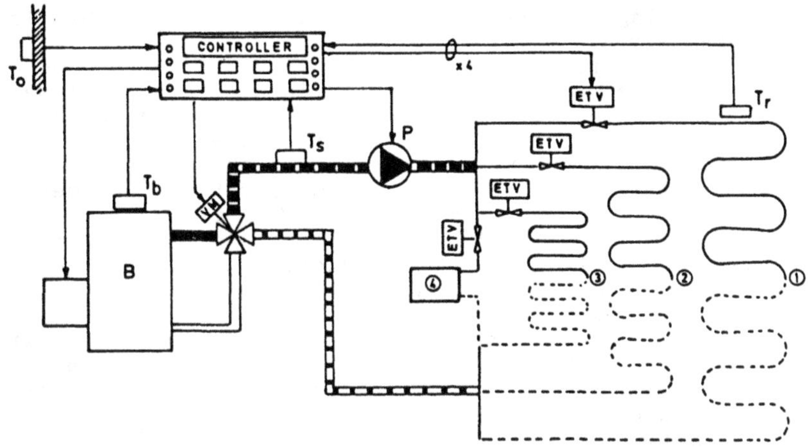

Fig. 10.4 Configuration of building heating control system

As an example let us focus on the calculation of the minimum supply water temperature T_s necessary to keep a zone temperature T_r at its desired value T_r^w. The heat balance equation when the zone temperature is in equilibrium at T_r with the supply water at T_s gives :

$$k_1(T_s - T_r) = k_2(T_r - T_o) \qquad (10.2)$$

where k_1 and k_2 are unknown heat conduction coefficients and T_o is the outside air temperature. The ratio $k = k_2/k_1$ can be estimated in real-time by means of a moving average filter :

$$\hat{k}(t) = \alpha\hat{k}(t-1) + (1-\alpha)\,\frac{T_s(t) - T_r(t)}{T_r(t) - T_o(t)} \qquad (10.3)$$

where α is the filter constant ($\alpha \lessapprox 1$) and t denotes discrete time. The minimum required supply water temperature is then computed from

$$T_s^W(t) = T_r^W(t) + \hat{k}(t).[T_r^W(t) - T_o(t)] \qquad (10.4)$$

This is the setpoint for the mixture valve regulator loop which controls the position of the mixture valve in such a way that the measured supply water temperature $T_s(t)$ follows the desired value $T_s^W(t)$. The boiler is then controlled such that its water temperature $T_b(t)$ is just a few degrees centrigrade higher. Although this strategy does not expect a single tuning parameter from the user, it assures that the whole system is always operating at its lowest possible heat output in order to realize the specified zone temperature, thus saving energy. Notice that there is only one mixture valve for the heat supply to four independently controlled zones. Possible conflicts can however be eliminated by shutting off one or more zones by means of the electrothermic valves.

The above strategy is able to keep the temperature in each zone close to its setpoint. It is overruled by a self-optimizing strategy for setpoint variation. It computes how many hours in advance the heating of a zone should start up in order to attain the new setpoint at the specified time. The underlying ideas were borrowed from the EPSAC method. The technique is based on : 1. on-line estimation of an input-output model of the heating plant dynamics for each zone; 2. use of the model for prediction of the future zone temperature as a function of the postulated heat supply to the zone; 3. use of the prediction facility for decision-making, i.e. to select the best heat supply to the specific zone at the current time instant. These steps are repeated the next sampling instant such that new measurement information can be used to improve previous decisions.

The zone dynamics are described by the incremental model

$$\Delta T_r(t)+a_1\Delta T_r(t-1)+a_2\Delta T_r(t-2) = b_1\Delta T_s(t-1)+b_2\Delta T_s(t-2)+e(t) \quad (10.5)$$

or in vector notation

$$\Delta T_r(t) = \underline{\phi}^T(t).\underline{\theta} + e(t) \quad (10.6)$$

with the data vector

$$\underline{\phi}^T(t) \equiv [\Delta T_r(t-1) \quad \Delta T_r(t-2) \mid \Delta T_s(t-1) \quad \Delta T_s(t-2)] \quad (10.7)$$

and the parameter vector

$$\underline{\theta}^T \equiv [-a_1 \quad -a_2 \mid b_1 \quad b_2] \quad (10.8)$$

The notation $\Delta T_r(t)$ denotes the increment $[T_r(t) - T_r(t-1)]$ and the signal e(.) represents zero-mean modelling errors.

At the current time instant t, given the estimated $\hat{\underline{\theta}}(t)$ and the postulated control policy $\{T_s(t+i/t);\ i=0 \ldots l-1\}$, it is possible to predict the behaviour of the future zone temperature $\{T_r(t+i/t);\ i=1 \ldots l\}$:

$$T_r(t+i/t) = T_r(t+i-1/t) + \underline{\phi}^T(t+i/t).\hat{\underline{\theta}}(t); \quad i = 1 \ldots l \quad (10.9)$$

where $\underline{\phi}^T(t+i/t) \equiv [\Delta T_r(t+i-1/t)] \quad \Delta T_r(t+i-2/t) \mid$

$\Delta T_s(t+i-1/t) \ \Delta T_s(t+i-2/t)]$

When using this strategy for deciding when to start preheating a zone, the postulated control policy $\{T_s(t+i/t);\ i = 0 \ldots l-1\}$ is normally the maximum possible supply water temperature. This corresponds to the strategy : start heating as late as possible and with maximum heat input into the zone. This rule is by no means limiting and could be replaced by others. The decision problem is twofold : the selection of the most appropriate time instant t_1 to switch-on heat supply to the zone; and at the end of the heating-up period the selection of the most appropriate time instant t_2 to switch from maximum supply water temperature to the temperature required to maintain the zone temperature in equilibrium at its desired value (Fig. 10.5).

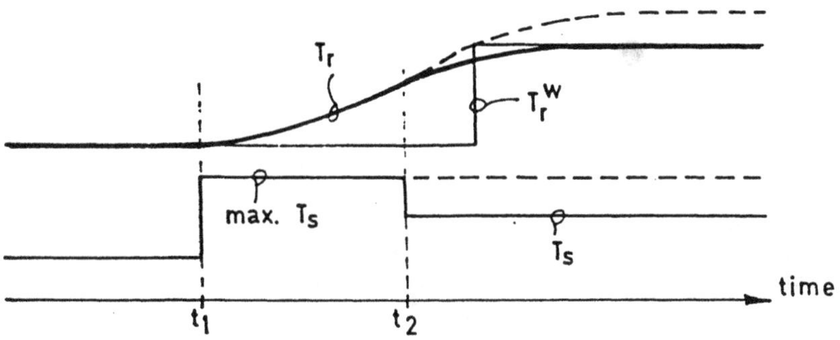

Fig. 10.5 Rule based control strategy

Summarizing this section, the control software for the building heating controller can be described as being a simple rule based expert control system (Sutton, 1984; Åström et al., 1986) in which an essential part of the knowledge is obtained by means of self-learning algorithms.

10.3.4. Real-life Experiments

The operation of the hardware and software of the AMC system has been thoroughly investigated by means of simulation studies and by real-life demonstration projects. Some results are described in De Keyser (1985). Some typical real-life results of the building heating control application are summarized in this section.

The upper part of Fig. 10.6 shows the variation of the outside air temperature T_o during a 1-day period. The day/night difference is over 10°C. The middle part shows the measured temperature in zone 2 during that day as well as the position (open/closed) of the zone shut-off valve. The desired temperature was 19°C all day long. Obviously the regulation performance can hardly be improved, the variations during the night (when no uncommon disturbances of sun or occupants occur) being +/- 0.1°C. The lower part of Fig. 10.6 illustrates how a setpoint increment (19°C ⟹ 20°C) is realized in zone 4.

10.4. ADAPTIVE CONTROL OF CUTTER SUCTION DREDGERS

The development of a self-learning computer-based automatic control system for a cutter suction dredger is an important step in the evolution of dredging. The system is based on the latest concepts of control theory, hardware and software. Commercial controllers of type PI, whether or not implemented in digital hardware, require careful tuning in order to operate well on this kind of dredger. Therefore in the new automatic controller the classical control algorithms were replaced by more powerful adaptive predictive control methods. The software has been implemented in an onboard multi-microcomputer system.

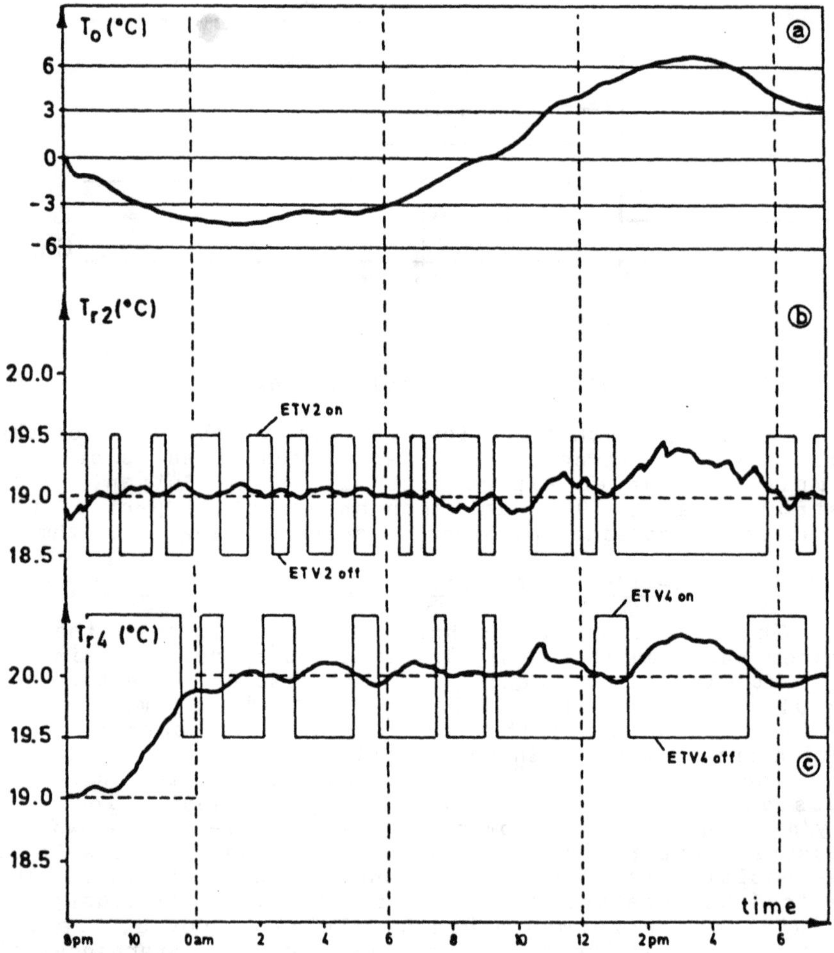

Fig. 10.6 Experimental results

10.4.1. Cutter Suction Dredgers

The cutter suction dredger is at the moment the most frequently used dredger : 41% of the world fleet of dredgers are cutter suction dredgers. They have the advantage that they can be used to work on the most varied soil types :

from silt to hard rock (with a tensile strength of concrete). Moreover it has the advantage of hydraulic transport.

Fig. 10.7 illustrates the main features of a cutter suction dredger (Claeys, 1978) :

- the cutter (1) to dislodge the soil
- the ladder (2) to fix the dredging depth. It holds the shaft for the cutter and also the suction pipe (and pump).
- the spud (3) to fix the length of the cut or step. The dredger swings around the spud and steps forward relatively to the spud at the end of each swing.
- the swing-winches (4), wires (5) and anchors (6) which, by moving the dredger, determine together with spud and ladder position the quantity of soil to be cut per unit of time.
- the pipeline (7) and the sandpumps (8) to transport the dislodged soil over the required distance and height.

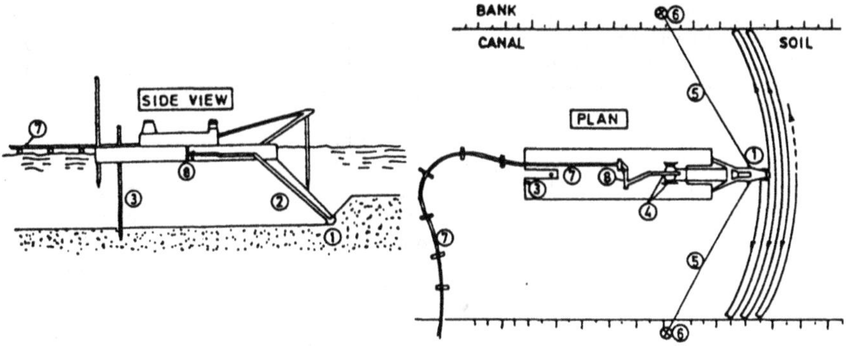

Fig. 10.7.a Cutter suction dredger, side view
b Cutter suction dredger, top view

A cutter suction dredger is a stationary dredger, i.e. it remains on the spot while operating. The dredging process can be divided into two parts :

- The mechanical part in which the soil is loosened with a rotating cutter. The movement of the pontoon is brought about by pulling one of the side winches. This causes the ship to rotate around a spud that is fixed into the ground.
- The hydraulic process in which the loosened material is diluted with water and suctioned at the cutter. Via a floating pipeline the species is pumped ashore and furtheron transported by a pipeline to the reclamation area.

Several process variables have to be controlled continuously either manually by the dredgemaster or by an automatic control system :

- the load of the cutter motor
- the concentration of the soil/water mixture in the pipeline
- the load of the swing-winch motors
- the velocity of the mixture in the pipeline
- the load of the pumps and their diesel engine drives
- the vacuum before the suction pump

- the tension in the side wires
- the total pressure after the pumps, etc.

10.4.2. Control problems

For the control of these parameters the dredgemaster (or the computer) can intervene in several ways : by changing the step length, the ladder depth, the number of revolutions of the cutter, the speed of the pumps. The most important factor for a continuous control is however the swing speed of the ship around the spud. This swing speed determines primarily the quantity of material dredged per time unit, i.e. the production. The swing speed is controlled by means of the current to the starboard and portside winch motors.

The aim of the control policy is to keep the swing speed and thus the production as high as possible. However, the electromotors of the side winches and the cutter may not be overloaded, the velocity of the mixture may not become too low in order to avoid deposition of the material resulting in the blocking of the pipeline and the pumps should not start cavitating.

From the above it is clear that the main task of an efficient control system is to adjust continuously the swing speed by controlling the side winch motors, so that at least one of the critical factors of the dredging process is at its nominal value. In this way the capacity of the ship is fully used; in other words, the production is at its maximum taking into account the available power of the different drives.

The realization of this objective is not obvious and asks for a continuous adaptation of the control mechanism. Commercially available cutter controllers (type PI) fail to realize this task efficiently due to the extremely random and time-varying nature of the process characteristics and loads. This type of vessel is used to dredge clay, sand as well as rock. The relationship between process input (swing velocity) and process outputs (cutter load, mixture concentration ...) is changing almost continuously. The soil characteristics do not remain constant along the swing and drastical variations in a stochastic sense are quite normal. This constitutes the whole problem in controlling this kind of process.

To give an idea of the irregularity of the process we refer to Fig. 10.8. The load of the cutter motor and the mixture concentration at the measuring-spot in the pipe (near the sandpumps 8 in Fig. 10.7) is illustrated under manual control by the dredgemaster.

The occurence of breaches is another illustration of the variation of process gains and time constants. At some spots along the cut it is possible that there is less material so that the production is very low even when hauling at the maximum swing speed. According to the control theory this means that the gain between the process input (= swing speed) and the output (e.g. concentration) is very low. At other places it is possible that due to breaches the production is very high, even when the ship is almost stopped.

This means an extremely high process gain.

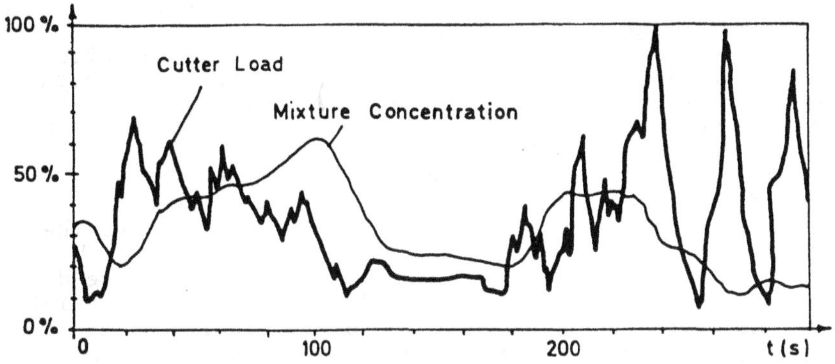

Fig. 10.8 Irregularity of the dredger process

These working conditions make it necessary for the skipper in manual control to be extremely attentive, if not the mean production will decrease. Likewise a good automatic control system has to adapt itself to the continuously changing situations. Because the control algorithms are based on the feedback principle, the control parameters, among which the gain, have to be adapted continuously to the stochastically fluctuating process parameters. This explains why commercially available control systems with pretuned parameters are not very successful : with increasing process gain the behaviour of the control loop will be oscillating (instability) and when the process gain is diminishing the control performance tends to become very slow (production loss).

10.4.3. Modelling, Identification and Simulation

In order to gain some experience with the control system before it was installed aboard of the ship, it was decided to build a computer simulation of the dredging process. A fairly comprehensive dynamic model was developed. The model comprises both the mechanical part (vessel, winches and electric drives, side wires, cutter) and the hydraulic part (pipeline, pumps and diesel drives) of the dredging process. The simulator was built from "white" models based on physical (mechanical, electrical, hydraulic) laws. However in order to obtain some lacking numerical parameters in these models the identification of simple "black-box" models was very helpful.

For modelling purposes the cutter suction dredger was structured into four submodels :

- the mechanical process : as input we have the currents of both electromotors of the sidewinches, resulting in a swing speed (angular speed around working spud). If we presume that the ladder remains at constant depth and the angular speed of the cutter does not vary, only the hauling speed will influence the dredging process (next to the soil characteristics).

- the cutting process : the swing speed determines the cutterload. When the swing speed increases, there is more soil to be cut so that it is clear that the torque on the cutter is higher. As a matter of fact the influence of the soil characteristics is very important : the behaviour in clay is very different from that in rock, even though the motor in both cases is running on full load.
- the suction process : the material that is cut is sucked. Depending on the soil characteristics, the shape of the cutter, the angular speed of the cutter and the swing speed, more or less material is sucked. This results in a specific concentration (density) of the mixture that is pumped.
- the hydraulic process : the density of the mixture determines all parameters of the hydraulic process, consisting of the pumps and their drives and the pipelines. An important parameter is the velocity of the mixture, which may not be too low in order to avoid deposits in the pipeline. If no action is taken, the pipeline will be blocked so that intervention from outside is necessary.

For all these subprocesses a numerical model was made. From the on-board measurements the unknown parameters were estimated with the maximum likelihood identification method. The goal of this analysis was not only to learn about the dynamic process but also to lay the foundation for the computer simulation of the dredging process.

The experiments for the identification were done aboard of the cutter suction dredger "Rubens" belonging to the fleet of the Belgian company "Dredging International N.V.". It is one of the biggest dredging companies operating worldwide and the tests were done while the ship was engaged in dredging operations in the Suez Canal (Ismailia, Egypt, 1980) and in the harbour of Lazaro Cardenas (Mexico, 1981).

Some major specifications of the vessel are :
- built in 1977
- overall length 101 m
- installed power 11436 KW
- maximum dredging depth 25 m

The following picture gives a view of the "Rubens", clearly showing the cutter which is normally under water.

We further concentrate on the mechanical subprocess to illustrate the identification phase. The identification of simple dynamical models describing the mechanical subprocess was done step by step :
- the model of the unloaded (autonomously running) winches
- the model of the vessel in an unloaded swing (i.e. without cutting)
- the model of the vessel in a loaded swing
- the model of the cutter load
- the model of the suction concentration

During the experiments with the unloaded winches and vessel, active testing with PRBS-type input sequences was allowed. The models under load conditions had to be identified from normal operating records.

The parameter estimation was done off-line with a computer package that contains several well-known identification methods. The maximum likelihood method estimates the parameters in the input-output model

$$A(q^{-1})y(t)=B(q^{-1})u(t-k)+C(q^{-1})e(t) \tag{10.10}$$

As an example let us describe briefly the results for the model of the vessel in an unloaded swing. The parameters were estimated with :

- input signal $u(t) = I_h(t) - I_v(t)$, i.e. the difference between the armature currents of hauling and veering swing winch motors (in A)
- output signal $y(t)$ = swing velocity $v_h(t)$ of the dredger (in m per minute)

Five time series (3 starboard swings and 2 portside swings) were processed and used for identification, giving the result :

$$y(t)-0.904y(t-1) = 0.017u(t-2)+0.007u(t-3)+ \\ +e(t)+0.36e(t-1) \tag{10.11}$$

This approximates the continuous-time model

$$\frac{v_h(s)}{I_h(s)-I_v(s)} = \frac{0.25\ e^{-4s}}{1+29.7\ s}\ (\frac{m/min}{A}) \tag{10.12}$$

The sampling period was here T=3s. (For a loaded swing the sampling period had to be T=0.5s. This is also the sampling period of the ultimate automatic control system aboard of the dredgers).

Fig. 10.9 gives the output of this model (simulation output, not prediction output!) for a portside swing showing excellent results taking into account the simplicity of the model.

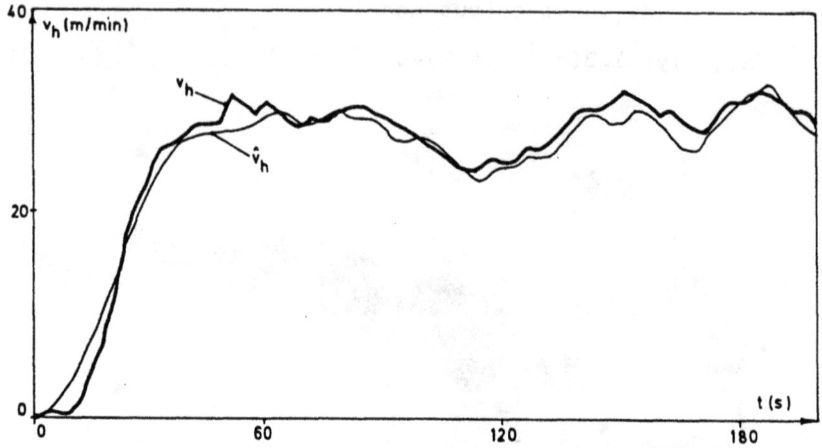

Fig. 10.9 Model output compared to real output (unloaded portside swing)

The identification of all parts of the dredger is described in detail in De Keyser et al. (1986). The numerical models were used to build a comprehensive simulation of the dredging process. As the dynamics of the total dredging process are quite complex, a fairly performing simulation system is needed. The simulator runs in a multi-microcomputer system which is linked to a dredging desk with instrumentation similar to that on board of a real cutter suction dredger. It allows for the study and visualization of dynamical effects that occur during the dredging operation.

10.4.4. Adaptive Predictive Control

Taking into account the complexity of the process dynamics the closed loop control of cutter suction dredgers is a real challenge for applying adaptive and self-learning strategies. Because of the multiple control loops and because of the numerical complexity of adaptive regulators, the realization was done by means of parallel operating microcomputers.

Although computer control of a dredging ship has several additional aspects and secondary control functions, the most important loops are depicted in Fig. 10.10. This is essentially a master/slave structure with a slave loop controller, that aims to haul the dredger with a certain swing speed and several master controllers, which compute the desired swing speed from the load of the different parts of the dredger.

All software runs on a parallel computer system. The principle is shown in Fig. 10.11.

It consists of a 68000 processor on a 16/32-bit bus (VMEbus) and a 6809 processor on an own local 8/16-bit bus (EURObus). The EURObus system is connected to the VMEbus through an interface board (VIOC = VMEbus/IOchannel convertor). This board maps the 64 Kbyte address space available

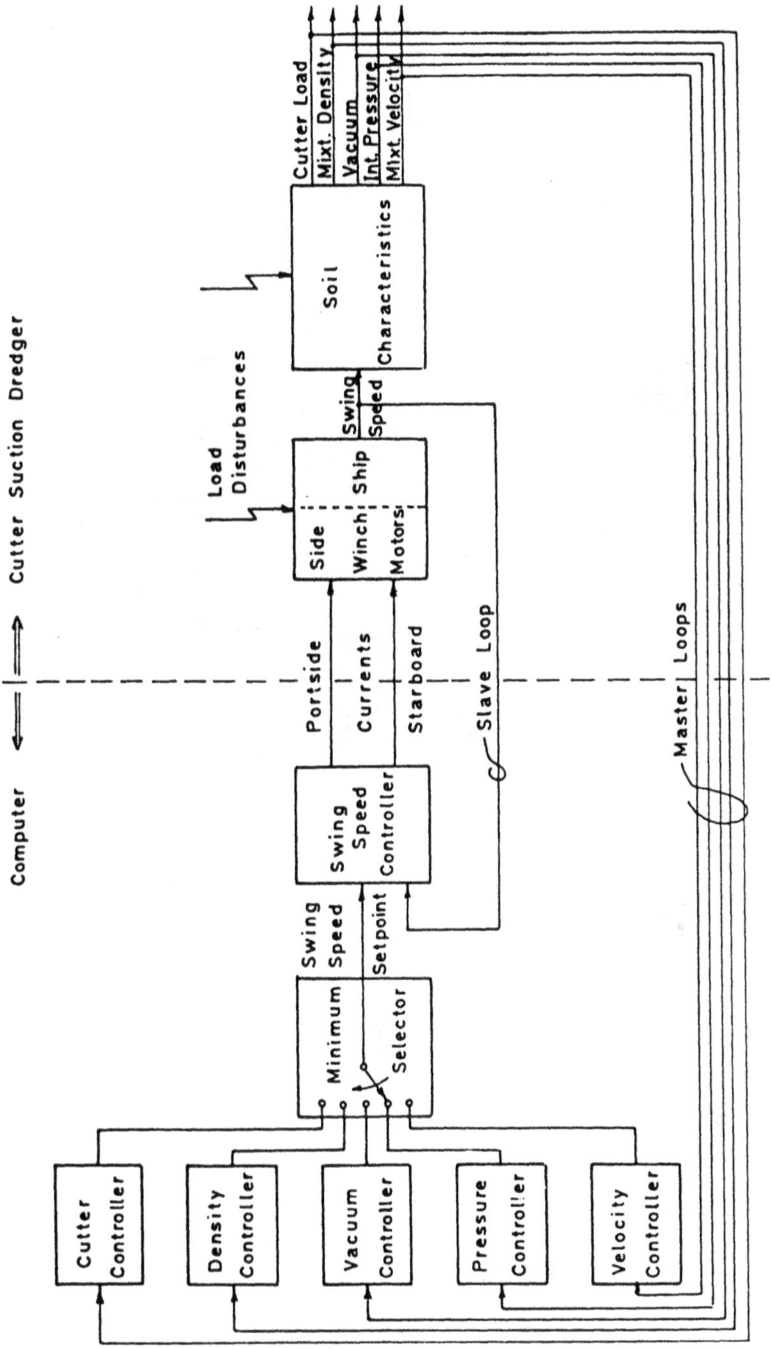

Fig. 10.10 Structure of the control loops

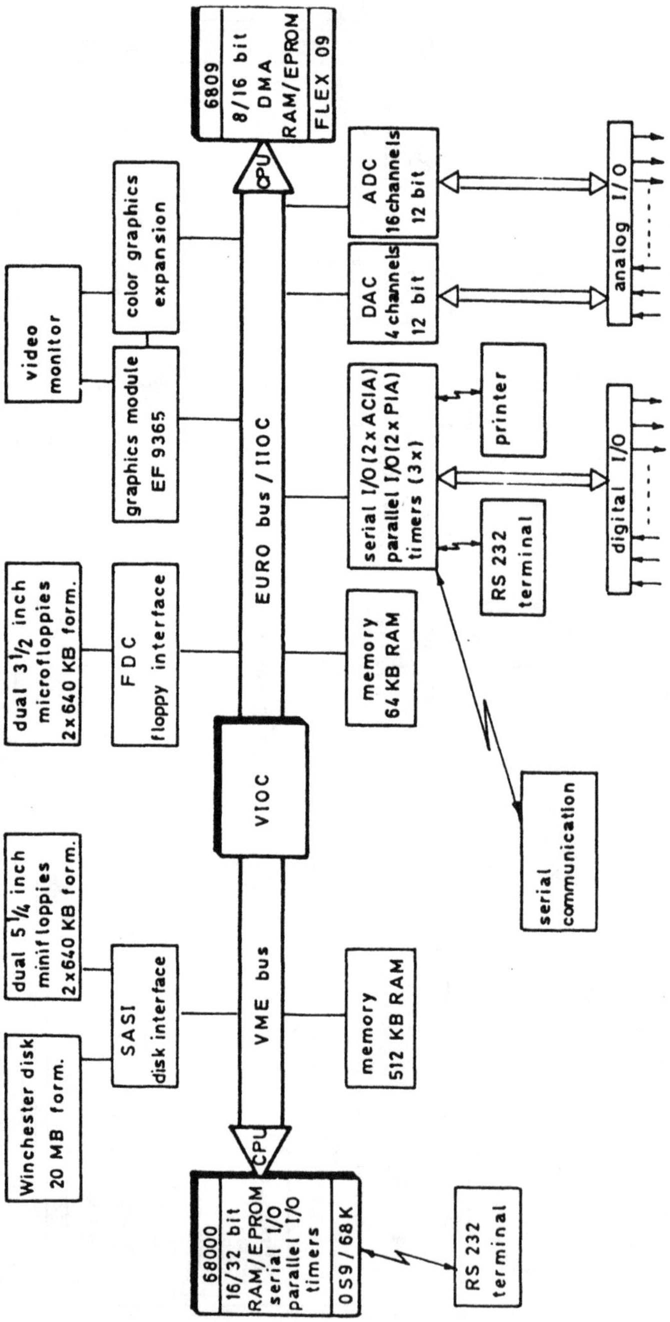

Fig. 10.11 VME-based multi-processor system for dredger controller

on the EURObus into the VMEbus memory. In this way the VME processor can access RAM and memory-mapped I/O peripherals on the EURObus as if they were part of its own memory map.

The configuration is very tempting for advanced real-time automatic control applications, where the fast "master" (68000) processor can perform the more complex tasks (e.g. optimization, complex calculations, identification and prediction algorithms, ...) whereas the "slave" (6809) processor does the more simple jobs (I/O, filtering and scaling, slave control loops). An additional advantage of splitting up the tasks in this way, is that all the hardware for real-world interfacing (analog and digital I/O) resides on the 8-bit side, resulting in less expensive equipment.

For the evaluation of the controller extensive measurements on board of the cutter suction dredger "Rubens" were performed. At first the slave loop controller was examined. Fig. 10.12 shows the behaviour of the dredger in steady state conditions but with strongly stochastic load.

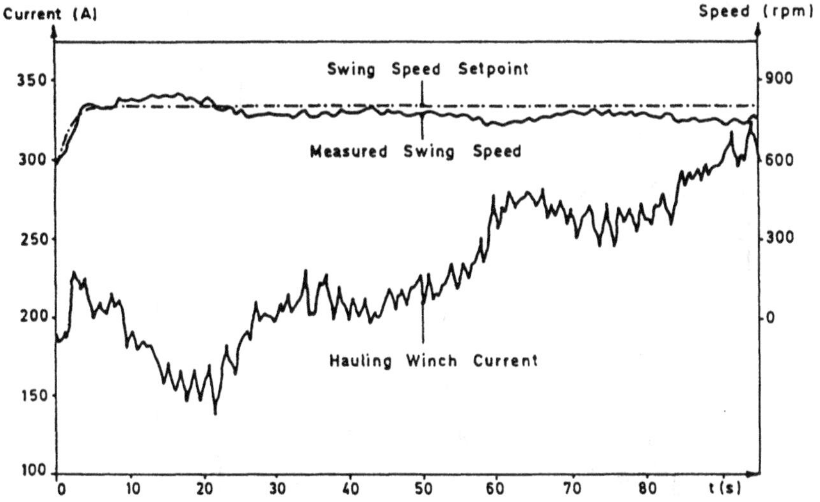

Fig. 10.12 Evaluation of slave loop controller

From this it is obvious that the adaptive automatic controller keeps the swing speed well around its setpoint. Furtheron the reaction speed to set point changes of the hauling speed is improved compared to manual control. Because of the fact that the current of the side winches is controlled directly, the hauling forces in the side wires are better kept below the limits so that breaking of a cable occurs less frequently.

As for the master controllers, Figs 10.13 and 10.14 show respectively the control of the cutter load and of the concentration. The outcome could be compared to the corresponding signals for manual control presented earlier in Fig. 10.8 (notice the scale difference). Especially the performance of the density regulation is unachievable by manual

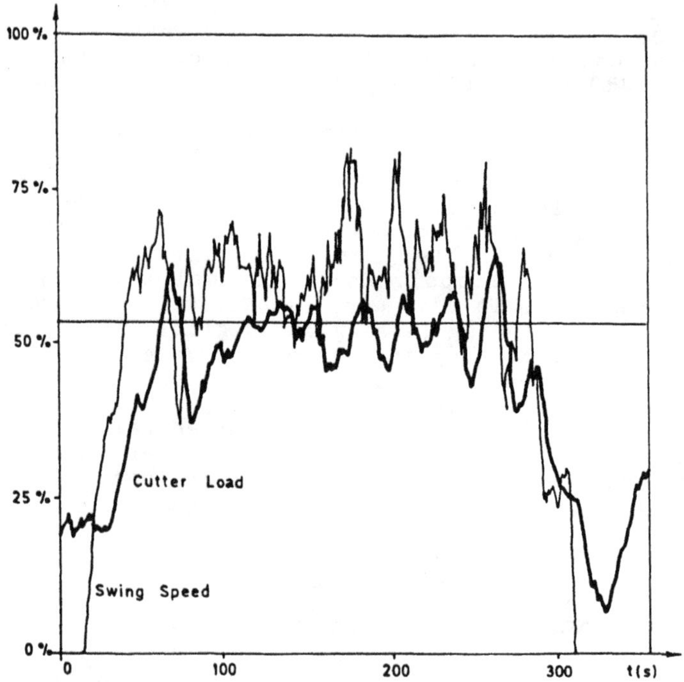

Fig. 10.13 Master controller of the cutter load

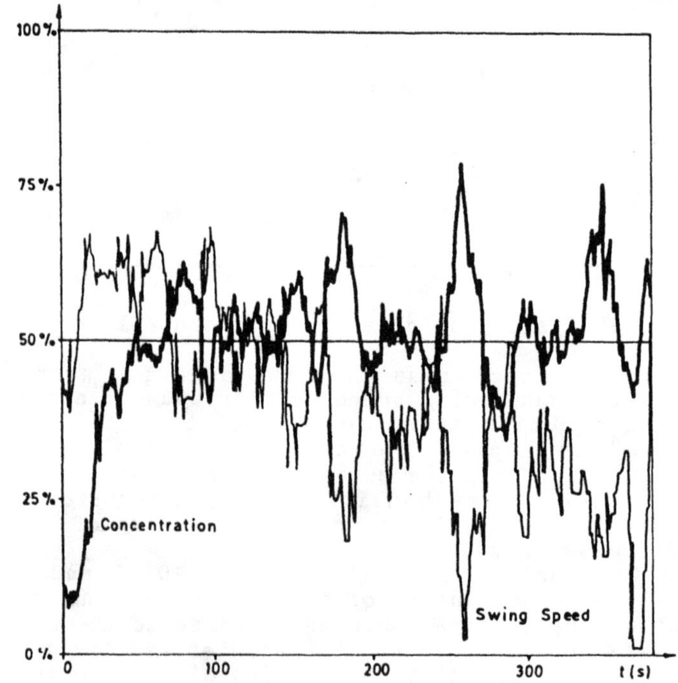

Fig. 10.14 Master controller of the concentration

control because of the important deadtime. Indeed this control loop is hindered by a severe transportation lag because the measuring instrumentation is located near the pumps (indicated by 8 in Fig. 10.7), which is about 50 m away from the suction inlet (indicated by 1 in Fig. 10.7). Moreover the deadtime changes due to the varying mixture velocity in the pipe (between 4m/s and 9m/s). The controller has to be robust w.r.t. the unknown, varying deadtime.

From the experiments it was also evident that the control system is reacting well to varying soil characteristics. It is here that classical PI controllers failed.

10.5. CONCLUSIONS

Two real-life applications of adaptive long-range predictive control have been described. Attention was paid mainly to the important implementation aspects which distinguish a real-life industrial application from a computer simulation application. For the underlying theory and algorithms reference was made to other contributions in this book.

In the first application, a self-adaptive microprocessor system for climate control, special attention was paid to the low cost aspect. The control algorithms are essentially a combination of simple rule based expert control and a self-learning feature to improve the controller's knowledge base.

The second application, a computer control system for large cutter suction dredgers, was an illustration of a difficult process where the use of commercial PID-type regulators was not successful. Because of the highly stochastic and time-varying nature of the dredging process and the presence of considerable deadtimes, it could be expected that adaptive predictive control might lead to a better overall performance. This was confirmed by several on-board experiments. The adaptive control system is now installed on the newly built dredgers of the company's fleet.

10.6. ACKNOWLEDGEMENTS

The development project of the adaptive automatic cutter controller is an initiative of the Belgian contractor "Dredging International n.v.". It was realized in collaboration between the consulting firm "IMDC n.v." and the Automatic Control Lab of the University of Gent.

10.7. REFERENCES

Åström, K.J., J.A. Anton and K.E. Arzén (1986). Expert control. Automatica 22 (3), 277-286.

Birtles, A.B., R.W. John and J.J. Smith (1984). Before and after study of the performance of an energy management system. CIB-W'79 Performance of HVAC Systems and Controls in Buildings, BRE Garston U.K.

Brouwer, K. (1985). The selection of dredging equipment. Marintec China '85 Conference, Vol. 5 : Ports, Dredging and Cargo Handling, Lloyd's of London Press, London.

Clarke, D.W., P.S. Tuffs and C. Mohtadi (1986). Generalized predictive control : A new robust self-tuning algorithm. In : I. Landau, L. Dugard (Eds), Commande Adaptative : Aspects Pratiques et Théoriques, Masson, Paris, 209-228.

Claeys C. (1978). Production limiting factors of cutter suction dredgers. 7th International Harbour Congress, Antwerp, paper 3.15, 10 pp.

Conte, G. and D. Del Corte (1985). Multi-processor systems for real-time applications. D. Reidel Publishing Company, Dordrecht.

Cutler, C.R. and B.C. Ramaker (1980). Dynamic matrix control - A computer control algorithm. JACC, San Francisco, paper WP5-B.

De Keyser, R.M.C., A.R. Van Cauwenberghe (1981). A self-tuning multistep predictor application. Automatica 17 (1), 167-174.

De Keyser, R.M.C. (1985). Adaptive microcomputer control of residence heating. CIB-W'79 Recent Advances in the Control and Operation of Building HVAC Systems. SINTEF Trondheim, 154-164.

De Keyser, R.M.C. and A.R. Van Cauwenberghe (1985). Extended prediction self-adaptive control. In : H.A. Barker, P.C. Young (Eds), Identification and System Parameter Estimation, Pergamon Press, Oxford, 1255-1260.

De Keyser, R.M.C. and A.R. Van Cauwenberghe (1986). Towards robust adaptive control with extended predictive control. In : Proceedings of the 25th IEEE Conference on Decision and Control, Athens, Greece.

De Keyser, R.M.C., J. Lefever and J. Van Ostaeyen (1986). Identification and modelling of a cutter suction dredging ship. In : Lopez P. (Ed.) Identification and Pattern Recognition, IASTED, Toulouse, 18 pp.

De Keyser, R.M.C., Ph.G.A. Van de Velde and F.A.G. Dumortier (1987). A comparative study of self-adaptive long-range predictive control methods. To appear in Automatica, November Issue.

Greco, C., G. Menga, E. Mosca and G. Zappa (1984). Performance improvements of self-tuning controllers by multistep horizons : the MUSMAR approach. Automatica 20 (5), 681-700.

Peterka, V. (1984). Predictor-based self-tuning control. Automatica 20 (1), 39-50.

Richalet, J. (1980). General principles of scenario predictive control techniques. JACC, San Francisco, paper FA9-A.

Rouhani, R. and R.K. Mehra (1982). Model algorithmic control - Basic properties. Automatica 19 (4), 401-414.

Sutton, R.W. (1984). Expert systems for process control. In : S. Bennett & A. Linkens (Eds). Real-time Computer control, Peter Peregrinus Ltd, London, 247-251.

Tokheim, R.L. (1983). Microprocessor fundamentals. McGraw-Hill, New York.

Van Ostaeyen, J.L.S. and R.M.C. De Keyser (1985). The design of an adaptive cutter controller. Marintec China '85 Conference, Vol. 5 : Ports, Dredging and Cargo Handling, Lloyd's of London Press, London, 25 pp.

Van Zutphen, A.C. (1983). The influence of comprehensive automation systems on cutter suction dredging operations. Europort Congress Dredging Days, Amsterdam, paper D6, 20 pp.

Ydstie B. (1984). Limited horizon adaptive control. 9th IFAC World Congress, Budapest, Hungary.

Chapter 11

Self-tuning and self-adaptive PIP control systems

P. C. Young, M. A. Behzadi and A. Chotai

11.1 INTRODUCTION

In this Chapter, we are concerned with the self tuning (STC) and self adaptive (SAC) control of dynamic systems using a particularly simple digital approach based on discrete-time, state variable feedback pole assignment. The design method can be considered as a direct development of a previous continuous-time servomechanism design method suggested by Young and Willems (1972). Its simplicity in the digital situation is achieved by the novel definition of a Non-Minimal State Space (NMSS) representation of the system, in which the state vector is composed only of the present and past sampled values of the input and output signals, all of which are directly measurable, together with an "integral-of-error" state, which introduces the required servomechanism (type 1) performance.

This particular state variable feedback design is notable because, in the deterministic situation, it does not require the introduction of a state reconstruction filter (or Observer), as normally proposed (see e.g. O'Reilly, 1983). Of course, it is well known that closed loop pole assignment for discrete-time systems is possible using simple linear feedback control involving only the sampled inputs and outputs of the controlled system; indeed this idea is the basis for most recent self-adaptive and self-tuning algorithms (e.g. see Astrom and Wittenmark, 1980; Wellstead et al., 1979). On the other hand, although the state variable feedback interpretation of such approaches is quite straightforward and provides valuable insight into the nature of the resulting control system designs, it seems to have received little, if any, exposure in the technical literature. Moreover, as we will see in this Chapter, the NMSS motivated designs have novel features which directly exploit the state-space formulation and help to provide a unified approach to fixed gain and self-adaptive/self-tuning control system design.

The new NMSS design procedure for SISO systems is attractive in practical terms not only because it is extremely simple to implement, but also because it provides a logical extension to direct digital control (DDC) algorithms of more conventional design. In particular, one of the simplest implementations of NMSS control system synthesis, for a first order system with unity time delay, has a conventional proportional-integral (PI) form; while higher order implementations retain the PI elements but introduce additional feedback terms. For this reason, we feel that the resulting control systems might be conveniently termed **Proportional-Integral-Plus (PIP)** controllers.

The self adaptive or self-tuning PIP controllers are implemented quite straightforwardly by a procedure of "identification and synthesis". Here, some form of recursive identification algorithm provides continuously updated estimates of the system model parameters which, together with the closed loop characteristic polynomial (CLCP) coefficients (defining the desired closed loop pole positions in the complex plane), then provide the inputs to a computational algorithm for the feedback gains. In the present PIP context, this algorithm simply involves the solution of a set of linear simultaneous equations of a specified structural form, as defined by the model parameter estimates and CLCP coefficients. Finally, the continuously updated control gains obtained in this manner are used in the synthesis of the self-tuned or adaptive feedback control system.

This concept of identification and synthesis goes back a long way (e.g. Kalman proposed an early system of this type in 1958), but the STC and SAC systems described in this Chapter have their origins in self adaptive systems designed and implemented in the middle to late nineteen sixties. The first of these (Young,1965), was based on a control law which can be considered either as a self adaptive PID system, or as a precursor of the minimum variance self-tuning systems that appeared in the nineteen seventies (Astrom and Wittenmark,1970), depending upon whether it is considered from a deterministic or a stochastic standpoint. The second (Young, 1980), however, is a continuous-time equivalent of the PIP system in which a pole assignment procedure is combined with continuous-time NMSS feedback to yield an adaptive type 1 servomechanism system.

11.2 THE SISO DYNAMIC MODEL and the NON-MINIMAL STATE-SPACE FORM

As we point out later in Section 11.8, the design procedures described in this Chapter can be applied to multivariable deterministic or stochastic systems. However, little is gained by introducing them in such a general and complex setting. Rather, for simplicity of exposition, we will restrict the main discussion to the control of single input, single output (SISO) systems. In particular, the analysis will be presented in the context of the following discrete-time model of a SISO system,

$$y(k) = \frac{B(q^{-1})}{A(q^{-1})} u(k) + \xi(k) \qquad \text{(i)}$$

or,

$$y(k) = x(k) + \xi(k) \qquad \text{(ii)} \qquad (11.1)$$

where,

$$x(k) = \frac{B(q^{-1})}{A(q^{-1})} u(k) \qquad \text{(iii)}$$

Here, $u(k)$ and $y(k)$ are, respectively, the input and output of the system measured at the kth sampling instant; $x(k)$ is that part of $y(k)$ causally related to $u(k)$ (i.e. the "noise free" output); and $\xi(k)$ represents additional disturbances affecting the system. $A(q^{-1})$ and $B(q^{-1})$ are polynomials in the backward shift operator q^{-1} (i.e., $q^{-1}y(k) = y(k-1)$) of the following general form,

$$\begin{aligned} A(q^{-1}) &= 1 + a_1 q^{-1} + \ldots\ldots + a_n q^{-n} \\ B(q^{-1}) &= b_1 q^{-1} + \ldots\ldots + b_m q^{-m} \end{aligned} \qquad (11.2)$$

No prior assumptions are made about the nature of the Transfer Function $B(q^{-1})/A(q^{-1})$, which may be marginally stable, unstable or possess non-minimum phase characteristics. However, if the input-output behaviour of the system is characterised by any pure time (transport) delay effects of δ samples, then these are accommodated by assuming that the first δ-1 coefficients of the $B(q^{-1})$ polynomial, i.e. b_1, ... $b_{\delta-1}$, are all zero. In this situation, the "noise-free" part of the system model, as given by equation (11.1)(iii), can be rewritten in the following alternative form, which explicitly shows the presence of the delay,

$$x(k) = \frac{B^{+}(q^{-1})}{A(q^{-1})} u(k-\delta) \qquad \text{(iii)} \qquad (11.1A)$$

where,

$$B^{+}(q^{-1}) = b_{\delta} + b_{\delta+1}\, q^{-1} + \ldots\ldots + b_m\, q^{-(m-\delta)}$$

The disturbances $\xi(k)$ can be either deterministic or stochastic but, if they are the latter, then it is assumed they can be modelled as a general AutoRegressive-Moving Average (ARMA) process of the following form,

$$\xi(k) = \frac{D(q^{-1})}{C(q^{-1})}\, e(k) \qquad (11.3)$$

where e(k) is a zero mean, serially uncorrelated sequence of random variables (discrete white noise) with variance σ_e^2 ; while $C(q^{-1})$ and $D(q^{-1})$ are the following polynomials in q^{-1},

$$C(q^{-1}) = 1 + c_1\, q^{-1} + \ldots\ldots + c_p\, q^{-p}$$
$$D(q^{-1}) = 1 + d_1\, q^{-1} + \ldots\ldots + d_q\, q^{-q} \qquad (11.4)$$

In contrast to the usual assumptions about ARMA processes (e.g. Box and Jenkins,1970), it is assumed that the process may be unstable, i.e. the roots of the C polynomial may lie on or inside the unit circle in the complex plane. This can create theoretical difficulties but it is a useful assumption in certain practical situations.

For the moment, let us consider the deterministic case of the SISO model in equation (11.1), where $\xi(k)$ is either a deterministic, unmodelled disturbance or identically zero. It is well known that, if such a dynamic system is controllable , then a linear, state variable feedback (SVF) control law will, in theory, allow for arbitrary assignment of the poles of the closed loop system (e.g. Popov,1964; Wonham,1967; Young,1972). This powerful result applies to both continuous and discrete-time systems and is the basis for both pole assignment and optimal control system designs.

One limitation of such SVF control, which has tended to discourage its use in practice, is the apparent need either to measure all of the state variables or to generate surrogate state variables by means of a state reconstruction filter (or Kalman filter/state estimator in the stochastic case). In the discrete-time situation considered here, however, it is possible to avoid these difficulties by selecting a "Non-Minimal State Space" (NMSS) model form, in which the state variables consist of the present and past values of the system input and output signals, all of which are clearly available for direct measurement and utilisation in DDC terms.

In the present context, the NMSS model is an unusual state space representation of the discrete-time TF model (11.1) of the following general form,

$$\begin{aligned} \underline{x}(k) &= F\,\underline{x}(k-1) + \underline{g}\,u(k-1) + \underline{d}\,y_d(k) \\ y(k) &= \underline{h}\,\underline{x}(k) \end{aligned} \qquad (11.5)$$

where the state transition matrix F, input vector $\underline{g}$, and output vector $\underline{h}$ are defined below,

$$F = \begin{bmatrix} -a_1 & -a_2 & \cdots & a_{n-1} & a_n & b_2 & b_3 & \cdots & b_{m-1} & b_m & 0 \\ 1 & 0 & \cdots & 0 & 0 & 0 & 0 & \cdots & 0 & 0 & 0 \\ 0 & 1 & \cdots & 0 & 0 & 0 & 0 & \cdots & 0 & 0 & 0 \\ \cdot & \cdot & & \cdot & \cdot & \cdot & \cdot & & \cdot & \cdot & \cdot \\ 0 & 0 & \cdots & 1 & 0 & 0 & 0 & \cdots & 0 & 0 & 0 \\ 0 & 0 & \cdots & 0 & 0 & 0 & 0 & \cdots & 0 & 0 & 0 \\ 0 & 0 & \cdots & 0 & 0 & 1 & 0 & \cdots & 0 & 0 & 0 \\ 0 & 0 & \cdots & 0 & 0 & 0 & 1 & \cdots & 0 & 0 & 0 \\ \cdot & \cdot & & \cdot & \cdot & \cdot & \cdot & & \cdot & \cdot & \cdot \\ 0 & 0 & & 0 & 0 & 0 & 0 & & 1 & 0 & 0 \\ a_1 & a_2 & \cdots & a_{n-1} & a_n & -b_2 & -b_3 & \cdots & -b_{m-1} & -b_m & 1 \end{bmatrix} \qquad (11.6)$$

$$\underline{g}^T = [\; b_1 \quad 0 \quad \cdots \quad 0 \quad 1 \quad 0 \quad 0 \quad \cdots\cdots \quad 0 \quad -b_1 \;]$$

$$\underline{h} = [\; 1 \quad 0 \quad \cdots \quad 0 \quad 0 \quad 0 \quad 0 \quad \cdots\cdots \quad 0 \quad 0 \;]$$

$$\underline{d}^T = [\; 0 \quad 0 \quad \cdots \quad 0 \quad 0 \quad 0 \quad 0 \quad \cdots\cdots \quad 0 \quad -b_1 \;]$$

with the state vector $\underline{x}$ defined as,

$$\underline{x}^T = [x(k)\; x(k-1) \ldots x(k-n+1)\; u(k-1) \ldots u(k-m+1)\; z(k)] \qquad (11.7)$$

where z_k is the "integral of error" state defined by the following equation,

$$z(k) = z(k-1) + \{\, y_d(k) - x(k) \,\} \qquad (11.8)$$

and $y_d(k)$ is the "desired output", i.e. command input to the servomechanism.

This integral-of-error state is not an essential component of the NMSS description and it is possible to proceed with the subsequent analysis without including it in the NMSS equations (see Wang and Young, 1987[1]). However, its presence has the advantage of ensuring inherent "type 1" servomechanism behaviour for the resulting SVF control system. This is clear from equation (11.8), where we see that, if $y_d(k)$ is constant and the closed loop system is stable, then $x(k) = y(k) = y_d(k)$ in the steady state, as desired for type 1 servomechanism performance. In practical terms, of course, the feedback of this state variable simply introduces the required and familiar integral action into the SVF control law.

The non-minimality of the state-space model (11.5) is obvious: the state has been extended from its minimum dimension n so that it is able to include not only the n sampled output signal values $x(k)$, $x(k-1)$,.., $x(k-n+1)$, but also the m-1 past sampled values of the input $u(k-1)$, $u(k-2)$,.., $u(k-m+1)$, and the integral-of-error state $z(k)$. The importance of this particular non-minimal definition of the state lies in its implications as regards state variable feedback (SVF) control: if such an NMSS representation is used as the basis for SVF control, then it is clear that the resulting control law involves only the directly measurable input and output signals over a given time interval, and does not need to resort to the complication of a state reconstructor (i.e. observer). It remains to demonstrate, however, that such a NMSS control law retains the most notable property of the minimal state SVF law; namely its ability to arbitrarily assign the poles of the closed loop system.

11.3 THE PROPORTIONAL-INTEGRAL-PLUS CONTROL SYSTEM

From the definition of the observation vector $\underline{h}$ in equations (11.6), we see that $x(k)=y(k)$. As a result, the non-minimal SVF control law can be written as,

$$u(k) = -\underline{v}^T \underline{x}(k) \tag{11.9}$$

$$= -f_0\, y(k) - f_1\, y(k-1) - \ldots - f_{n-1}\, y(k-n+1) - g_1\, u(k-1) - \ldots$$
$$\ldots - g_{m-1}\, u(k-m+1) - k_I\, z(k)$$

where,

$$\underline{v}^T = [\, f_0 \;\; f_1 \;\ldots\; f_{n-1} \;\; g_1 \;\ldots\; g_{m-1} \;\; k_I \,] \tag{11.10}$$

is the SVF control gain vector for the NMSS model form.

The recursive nature of the above control law is clear if we note from equation (11.8) that,

$$z(k) = \frac{1}{1 - q^{-1}} \;[\, y_d(k) - x(k) \,]$$

and substitute this in the SVF control law (11.9) which,

following simple manipulation, can be written as,

$$u(k) = u(k-1) - k_I \{y_d(k) - x(k)\} - f_0 \nabla x(k) - \ldots$$
$$\ldots - g_{m-1} \nabla u(k-m+1) \qquad (11.11)$$

where ∇ is the differencing operator, i.e., $\nabla x(k) = x(k) - x(k-1)$.

It is easy to consider the control system defined by the control law (11.9), or equivalently (11.11), in block diagram terms, as shown in Fig. 11.1. This reveals that, in addition to the proportional action and the standard, delay-free integrator arising from the feedback of z_k, the system is characterised, in these more conventional block diagram terms, by feedback and forward path, discrete time filters which are the consequence of SVF terms arising from the remaining state variables (i.e. the past values of the input and output).

Fig. 11.1 also demonstrates that this particular NMSS control system can be considered as a logical extension of the well known proportional-integral (PI) controller. For this reason, we will refer to it as a "Proportional Integral Plus" or PIP control system. Of course, having introduced the general concept of the NMSS, it is possible to think of other NMSS forms and their corresponding transfer function interpretations; for example, as we pointed out above, an NMSS model without the integral-of-error state is discussed by Wang and Young (1987[1]) and may be more appropriate in certain circumstances.

The general PIP control algorithm, and the proof of its efficacy in pole assignment terms, can be developed in two ways (see Young et al,1987[1]): by state-space analysis using an approach similar to that proposed previously for continuous-time systems (Young and Willems,1972; Young,1972); or, alternatively, by straightforward polynomial algebra, based on the block diagram model form (Fig. 11.1) suggested by the NMSS analysis. The former provides greater insight into the nature of the solution but the latter is simpler, and it may well appeal more to the practicing control systems designer.

Having justified the form of the basic PIP block diagram in Fig. 11.1, it is straightforward to reduce the diagram to a single closed loop block and equate the coefficients of the closed loop characteristic polynomial (the denominator of the closed loop TF) with those of the desired closed loop polynomial. This results in the following set of linear equations in the m+n control gains.

$$\Sigma \cdot \underline{v} = \underline{\beta}$$

P - **Proportional Control** (gain f_0)
I - **Integral Control** (gain k_I)
F_0 - **Feedback Filter**
$$f_1 q^{-1} + f_2 q^{-2} + \cdots\cdots + f_{n-1} q^{-n+1}$$
$$G = 1 + g_1 q^{-1} + \cdots\cdots + g_{m-1} q^{-m+1}$$

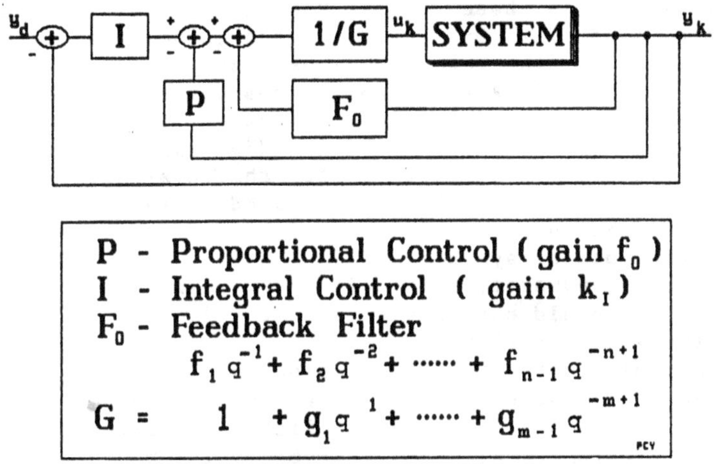

Fig. 11.1   The Proportional-Integral-Plus (PIP)
Servomechanism Control System

Here Σ is a matrix of dimension $(m+n) \times (m+n)$ of a specified form (see Young et al,1987[1]) while $\underline{v}$ is the SVF control gain vector defined in equation (11.10), and $\underline{\beta}$ is the following vector,

$$\underline{\beta}^T = [\ \beta_1 \quad \beta_2 \quad \cdots\cdots\cdots \quad \beta_{m+n}\ ]$$

with,

$$\beta_i = d_i - (a_i - a_{i-1})\ ; \quad a_0 = 1.0\ ; \quad a_i = 0 \text{ for } i > n$$

where d_i, $i=1,2,\ldots m+n$, are the coefficients of the desired characteristic polynomial $d(q^{-1})$, i.e.,

$$d(q^{-1}) = 1 + d_1 q^{-1} + d_2 q^{-2} + \cdots\cdots + d_{m+n} q^{-(m+n)}$$

Clearly, the conditions for the non-singularity of the matrix Σ in the above equation are linked to the controllability of the NMSS model (11.5). The necessary (but not sufficient) conditions for the controllability of this model are provided by the following Theorem:

<u>Theorem (Wang and Young,1987[1])</u>

Given a single input, single output system described by equation (11.1), a necessary condition for the controllability of the NMSS representation (11.5), as defined by the pair [F, $\underline{g}$], is that the polynomials $A(q^{-1})$ and $B(q^{-1})$ are coprime.

The proof of this theorem is straightforward but lengthy: it is based on the PBH eigenvector test and is omitted here for brevity. The coprimeness condition is equivalent to the rather obvious requirement that there should be no pole-zero cancellations between the $A(q^{-1})$ and $B(q^{-1})$ polynomials. It is interesting to note that the characteristic polynomial of the NMSS representation (11.5) is the product of the basic system characteristic polynomial $A(q^{-1})$ and another part, contributed by the delayed input variables and integral of error state, which does not depend upon the basic system. The above condition is not sufficient, however, since it is clear that the $B(q^{-1})$ polynomial should not have a zero at q=1, which would cancel with the integrator pole.

An alternative SVF solution to the pole assignment problem for the system (11.1) could be obtained with the help of a state reconstruction filter (Observer). In this alternative approach, the observer design would follow from the definition of an appropriate minimal state-space description, such as one of the well known canonical forms (see O'Reilly, 1983), and the SVF law would be defined in terms of the observer outputs. However, this solution is relatively complex and the alternative NMSS concept proposed here not only simplifies the structure of the control system and the computational algorithms for the feedback gains, but also nicely avoids the need to define the gains and associated dynamic behaviour of the observer system itself (see Wang and Young, 1987[2]).

Finally, it will be noted that the NMSS approach allows automatically for any pure time delay effects. If a time delay of δ sampling intervals is present, then the specification that $b_1 = \ldots = b_{\delta-1} = 0$ in the $B(q^{-1})$ polynomial will ensure that the PIP system allows for the presence of this time delay. It is interesting to note that, when applied in this manner, the PIP system can be considered as a generalised Smith Predictor for sampled data systems, as discussed by Chotai and Young (1987).

11.4 THE DESIGN PROCEDURE

The previous Sections of this Chapter have outlined the theoretical background to the PIP control system. It is now necessary to show how this theory can be utilised in the design and implementation of PIP-based self-tuning and self-adaptive controllers.

Although it is inherently more complex when applied to higher order and time delay systems, we have pointed out that the PIP controller is structurally related to the ubiquitous PI and PID controllers which have dominated control engineering design for so many years. But this superficial resemblance tends to obscure an underlying and quite fundamental difference in design philosophy. In particular, it can be argued (Young,1987[1]) that the PIP control system design heralds a new era of "**True Digital Control**" (TDC) systems; systems which overtly acknowledge the advantages of discrete-time signal processing and control system synthesis and are not merely based on the mechanical digitisation of their continuous-time forbears.

The TDC system utilises a control sampling interval (or intervals in a multi-rate design) which is chosen on the basis of both the frequency response characteristics of the system, as reflected in the discrete-time model; and the desired performance of the closed loop system, as defined in the present PIP context by the desired closed loop characteristic polynomial (CLCP). This usually results in a sampling interval which is relatively coarse when compared with digitised continuous-time control systems. Certainly there is no attempt in TDC to sample as rapidly as possible, since it is well known that this will result in poor estimation of the discrete-time model parameters and consequent deficiencies in any STC or SAC system design that relies on such parameter estimation.

The approach to TDC system design used in developing self-adaptive and self-tuning versions of the PIP controller is inherently digital in nature and depends strongly on one of the most powerful of digital concepts, the recursive algorithm (Young;1987[1],1984). In digital computer terms, the recursive algorithm is a FOR...NEXT or DO loop, in which some variable - in the present context a control signal, a parameter estimate or a state estimate - is updated at each recursion through the loop. By exploiting such a recursive formulation, we are able to consider all aspects of the design procedure, from initial data analysis to control system implementation, in a completely digital manner, with only minimal reference to continuous-time concepts, except where these may prove advantageous because of physical considerations.

The proposed TDC design procedure is composed of five major steps: experiment design; data analysis; model identification and parameter estimation; PIP control system design and evaluation; and, finally, STC/SAC system design and implementation. The details of the latter three stages in this design process are outlined in the next three Sections of this Chapter. By decomposing the design in this manner, we are emphasising our belief that self-tuning and self-adaptive system synthesis requires extensive **off-line** investigation prior to the final, on-line implementation of the design. This is rarely stressed in the literature on self-adaptive control, but it is felt that such a rigorous procedure can lead to much more reliable self-adaptive systems.

Of course, it could be argued that, by emphasising the need for prior off-line design studies, we are being overly conservative. Certainly, such an approach is not always necessary and the appealing idea of a "universal" SAC system, which will adapt itself to any dynamic process, will remain a long term aim. But, at this point in time, we believe that it is unwise to proceed with STC or SAC implementation without first ensuring that the **structural** nature of the control system is well suited to the process being controlled. For example, such incorrect structural specification can easily lead to poor on-line parameter (or control gain) estimation, and consequent deficiencies in the controlled system performance. There is clearly no need to risk such deficiencies, however, given the accessibility of numerous, off-line CAD packages, such as the **microCAPTAIN** program discussed in subsequent Sections of this Chapter, which allow for prior identification, estimation and control studies. Such studies may sometimes prove redundant, of course, particularly if the controlled process is dynamically very simple, but they will always ensure greater discipline in the design process and greater confidence in the resulting STC/SAC design.

11.5 PRIOR MODEL IDENTIFICATION AND PARAMETER ESTIMATION

There are two basic methods for deriving mathematical models of dynamic systems: the **physically-based** or mechanistic approach, in which the mathematical equations are obtained by the application of physical principles and laws; and the **data-based** approach, where the model equations are obtained directly from experimental data using methods of time-series analysis and data reduction. These approaches are not mutually exclusive and should normally complement each other.

The physically-based model is usually defined in terms of continuous-time differential or partial differential equations, since physical laws are usually postulated using differential or integral calculus. These continuous-time model formulations have two disadvantages in TDC system design terms. First, they can sometimes rely too much on the perception of the model builder and not sufficiently on the reality of the practical control situation. This can give rise to excessive complexity in model formulation and the derivation of models which are of too high a dynamic order for use in control system design , although they may well be useful later as simulation models for evaluating control system performance. Secondly, they clearly need to be converted into discrete-time form if they are to be a useful vehicle for TDC system design. One way of obviating these disadvantages is to carry out exercises in "model reduction" where, in the DDC situation, one obtains a lower dimensional, discrete-time model by some form of mathematical analysis applied either to the original high order continuous-time model equations, or the response of these equations to simulated input disturbances (see e.g. Young,1987[1]).

The alternative data-based approach assumes the existence of experimental sampled data from the dynamic process and relies on the application of computer-based methods of discrete-time model identification and parameter estimation. This has the obvious disadvantage of requiring the acquisition of sampled data, either during the normal operation of the process or from specially designed dynamic experiments. Also, there is a risk of considering the resulting model in purely "black-box" terms, so losing a physical appreciation for the nature of the process and risking deficiencies in the control system design; deficiencies which can result from problems such as over or under-parameterisation of the model.

Although the practical examples considered later in Section 11.7 will demonstrate how easy it can be to succumb to the dangers of such a black-box mentality, we believe that the data-based approach provides the most satisfactory general framework for discrete-time modelling. For this reason, we will restrict further discussion on dynamic modelling to the data-based methods, assuming that the user of these methods will take care to fully investigate the physical implications of his model studies before employing them in TDC system design.

11.5.1 Data Based Modelling : Recursive Estimation of the SISO Model

Even if data-based modelling is to be used only in an off-line situation, there are still advantages in exploiting recursive procedures. For example, the recursive estimates provide an extra dimension to the modelling studies, allowing for improved model structure identification and the investigation of parametric non-stationarity arising either from changing system dynamics or nonlinearity (see Young,1984). In the present STC/SAC context, however, the use of recursive methods is almost mandatory: if recursive estimation is to be employed subsequently for self tuning or self adaptive purposes, then there are obvious advantages to using similar procedures in the prior modelling phase of the design study.

There are many sophisticated methods of recursive parameter estimation for input-output systems, but discussion will be limited here to the recursive instrumental variable (IV) approach [e.g. Young (1965,1970,1984); Soderstrom and Stoica (1983)]. This method has the unique advantage that, in its basic form, it can yield consistent estimates of the parameters in transfer function models such as (11.1) without simultaneous estimation of a model for the disturbance $\xi(k)$; while in its optimum or "refined" form [Young (1976,1984); Young and Jakeman (1979,1980,1981,1983); Soderstrom and Stoica (1983)] it is able to generate consistent and asymptotically efficient (minimum variance) estimates if this disturbance has rational spectral density and so can be modelled as an ARMA process such as equation (11.3).

The basic and refined IV methods for the discrete-time model (11.1) have been described elsewhere in great detail [Young and Jakeman (1979,1980)] and we will not repeat this description here. In both cases, the recursive algorithms can be implemented in fully recursive or recursive-iterative form. The latter approach is the preferred one for off-line modelling (in contrast to on-line adaptive control applications), since iteration improves the statistical efficiency of the estimates for relatively small data sets. In either case, the refined IV algorithm generates a recursive estimate $\hat{\underline{a}}(k)$ of the model parameter vector $\underline{a}$, where,

$$\underline{a}^T = [\ a_1\ a_2 \ \ldots\ldots\ a_n\ b_1\ \ldots\ldots\ b_m\ ] \tag{11.12}$$

The algorithm takes the following form,

$$\hat{\underline{a}}(k) = \hat{\underline{a}}(k-1) + P(k-1)\hat{\underline{x}}^*(k)[1+\underline{z}^{*T}(k)P(k-1)\hat{\underline{x}}^*(k)]^{-1} \{y^*(k)-\underline{z}^{*T}(k)\hat{\underline{a}}(k-1)\} \tag{I}$$

$$P(k) = P(k-1) - P(k-1)\hat{\underline{x}}^*(k)[1 + \underline{z}^{*T}P(k-1)\hat{\underline{x}}^*(k)]^{-1}\underline{z}^{*T}P(k-1)$$

where,

$$\underline{z}^*(k) = [-y^*(k-1)\ -y^*(k-2)\ldots\ldots-y^*(k-n)\ u^*(k-1)\ldots\ldots u^*(k-m)]^T$$

and,

$$\hat{\underline{x}}^*(k) = [-\hat{x}^*(k-1)\ -\hat{x}^*(k-2)\ldots\ldots-\hat{x}^*(k-n)\ u^*(k-1)\ldots\ldots u^*(k-m)]^T$$

Here, $x(k)$ is the instrumental variable, which can be considered as an estimate of the "noise-free" output of the system $x(k)$ and is generated by an adaptive auxiliary model of the form [cf equation (11.1)(iii)],

$$\hat{x}(k) = \frac{B(q^{-1})}{A(q^{-1})} u(k) \qquad (11.13)$$

in which the hat indicates that the polynomial parameters are based on estimates obtained, in the recursive-iterative case, from the previous iteration. Note again that a pure time delay of δ sampling intervals can be accomodated by assuming that the first δ coefficients of the $B(q^{-1})$ are zero, in which case the auxiliary model (11.13) could be written in the alternative form of equation (11.1A)(iii), with the polynomial coefficients replaced by their estimates.

The star superscript is used in algorithm I to denote that the associated variable has been pre-filtered by an adaptive filter $F(q^{-1})$ of the form,

$$F(q^{-1}) = \frac{\hat{C}(q^{-1})}{\hat{D}(q^{-1})\ \hat{A}(q^{-1})} \qquad (11.14)$$

As in the case of the adaptive auxiliary model (11.13), in the recursive-iterative algorithm, the hat indicates that the polynomials are based on the parameter estimates obtained from the previous iteration. The details of this adaptive updating and the associated recursive algorithm for the noise model parameters (as required for the above adaptive pre-filtering operations) are given in Young (1976,1984). It will suffice here to say that the recursive estimation of the noise model parameters can be achieved by either: assuming an AR model and using the RLS algorithm; employing the Approximate Maximum Likelihood (AML) or Prediction Error Recursion (PER) methods, if ARMA models are preferred (see e.g. Young,1984); or using the AR model estimation as a first step in ARMA model estimation (see e.g. Young,1985). Also, a simplified refined IV (SRIV) algorithm (Young,1984), which has adaptive pre-filtering but no simultaneous noise model estimation, can be advantageous in certain applications; for example, in the modelling of systems from impulse or step response data or in "model reduction" (Young,1987[1],1987[2]).

An important aspect of modelling from time-series data is the "identification" (the term used by statisticians) of an appropriate model structure; i.e. the order of the transfer function polynomials $A(q^{-1})$, $B(q^{-1})$, $C(q^{-1})$ and $D(q^{-1})$ and the presence of any pure time delay δ in the system transfer function equation (11.1). In the present context, the various methods of stochastic model identification , such as the Akaike AIC and Parzen CAT(see e.g. Priestley,1981), can be used to identify the polynomial orders p and q in the noise model equation (11.3). Such criteria can also be applied to identification of the most appropriate structure of the system transfer function equation (11.1), but our experience suggests that they do not work so well in this input-output situation, since they do not overtly utilise the additional information provided by the presence of the input signal.

An alternative approach to the identification of the structure of the system transfer function from input-output data is described in Young et al (1980). This approach, which exploits the IV nature of the recursive estimation algorithms, involves the use of two statistical measures : a "coefficient of determination", R_T^2; and an "error variance norm", EVN, which together help to define the model which combines good explanation of the data (a relatively high value of R_T^2 close to its maximum of 1.0) and well defined parameter estimates (a relatively low value of the natural logarithm of EVN). A single statistic which combines aspects of these two previous criteria is currently being evaluated and seems to provide a rather simple yet effective approach to the problem. This statistic is defined as follows,

$$YIC = \log_e[\sigma_\xi^2 / \sigma_y^2] + \log_e[NEVN] \qquad (11.15)$$

where,

σ_ξ^2 is the sample variance of the model residuals

σ_y^2 is the sample variance of the system output y(k) about its mean value

while NEVN is the normalised EVN defined as,

$$NEVN = \frac{1}{m+n} \sum_{i=1}^{m+n} \sigma_\xi^2 \, [\, p_{ii}(N) \,]/[\hat{a}_i(N)]^2 \qquad (11.16)$$

Here p_{ii}, i= 1,2,..., m+n, are the diagonal elements of the P(N) matrix obtained from the estimation algorithm at the completion of estimation. In the refined IV case, it can be shown that, when multiplied by σ_e^2, these elements provide a measure of the model parameter estimation error variance.

It can be seen that the first term in (11.15) provides a normalised measure of how well the model explains the data; while the second term is a normalised measure of how well the parameter estimates are defined for the (m+n)th order model. Thus the model which minimises the YIC provides a good compromise between model fit and parametric efficiency: as the model order is increased, so the first term tends always to decrease; while the second term tends to decrease at first and then to increase quite markedly as the model becomes over-parameterised. In this IV context, the YIC criterion can also be justified in purely numerical analysis terms: the second term is a sensitive measure of the conditioning of the "instrumental product matrix", which increases sharply when serious ill-conditioning is encountered because the model is of too high an order (Young et al, 1980).

It must be stressed that this new criterion for model identification needs to be evaluated further, both in theory and practice, before its efficacy is firmly established. However, its use will be demonstrated in the case studies presented in subsequent Sections of the Chapter.

11.6 THE SELF-TUNING PIP CONTROLLER

The PIP controller can be implemented in fixed gain or self tuning/adaptive form. In the latter cases, the simplest mechanisation and the one used in the examples discussed here and elsewhere [Young et al (1987[1]); Behzadi et al(1987)], is to utilise the conventional recursive least squares (RLS) or instrumental variable (IV) parameter estimation algorithms to generate continuously updated estimates of the unknown system model parameters required by the control law.

The recursive version of the basic IV algorithm is identical in form to the refined IV algorithm I, but without the adaptive pre-filtering on the variables y(k), $\hat{x}(k)$ and u(k). In other words, the star superscript is removed from the variables in algorithm I and they are re-defined as,

$$\underline{z}(k) = [-y(k-1) \; -y(k-2) \; \ldots \; -y(k-n) \; u(k-1) \; \ldots \; u(k-m)]^T$$

$$\hat{\underline{x}}(k) = [-\hat{x}(k-1) \; -\hat{x}(k-2) \; \ldots \; -\hat{x}(k-n) \; u(k-1) \; \ldots \; u(k-m)]^T$$

The RLS algorithm is simply this basic IV algorithm with the instrumental variable $\hat{x}(k)$ replaced everywhere by the measured output y(k), so that $\hat{\underline{x}}(k)=\underline{z}(k)$.

When employed for self-tuning, under the implied assumption that the model parameters are sensibly constant, these IV and RLS algorithms are used either in the form described above or with simple modifications to encourage initial convergence. For example, a common device is to introduce an exponential-weighted-past (EWP) or "variable forgetting" modification, in which the exponential weighting is made short at the initiation of the algorithm, but is automatically extended to become infinitely long as time progresses [see e.g. Ljung and Soderstrom (1983), page 279 et seq.; Young (1984), page 60 et seq.]

In the Self Adaptive Control (SAC) situation, EWP algorithms with fixed exponential time constant can be used, but we prefer an alternative and more flexible approach [Young and Yancey (1971); Young (1980,1984)]. For slow changes in the system dynamics this reduces to the well known technique, in which the parameter variation is modelled as a random walk. When the model parameters may change more rapidly, however, the system employs a more sophisticated stochastic model for parameter variations which exploits a priori information on the changing dynamics to improve the parameter tracking ability [see Young (1984), page 84 et seq.]. A practical example of this approach is the heated bar SAC system outlined in Section 11.7.3.

11.6.1 Full Adaption vs Adaptive Gain Control (AGC)

Often, when self-tuning or adaptive implementation follows prior model identification and estimation, it is only the system gain that is subject to uncertainty. For example, in some nonlinear systems, the basic dynamic behaviour is relatively invariant, but the gain is a function of the output level. In such situations, the designer can choose between full adaption or Adaptive Gain Control (AGC). Here, the model parameters are fixed at their a priori estimated values and only the scalar gain is estimated recursively.

There are various ways in which this can be accomplished, but the one preferred here is to assume that the model can be written in the form (cf equation (11.1),

$$y(k) = g \frac{B(q^{-1})}{A(q^{-1})} u(k) + \xi(k) \qquad (11.17)$$

where $A(q^{-1})$ and $B(q^{-1})$ are known and g is the unknown gain. It is then clear that, in the case where $\xi(k)$ is white noise, this equation represents a regression relationship in the gain g, so that an estimate of g can be obtained from a scalar recursive least squares algorithm, with the regressor defined by $[B(q^{-1})/A(q^{-1})]u(k)$. If $\xi(k)$ is not white, then a generalised least squares solution would be more appropriate.

A related but not so satisfactory solution to gain estimation is to re-arrange equation (11.17) and consider that $[A(q^{-1})\ y(k)]/[B(q^{-1})\ u(k)]$ is a direct observation of g and then obtain an estimate from the recursive algorithm for estimating the mean of a random variable (see Young,1984). This is a somewhat simpler solution, but the division inherent in constructing the observation and the nature of the additive noise clearly cause problems (see Behzadi, 1987).

11.6.2 Low Frequency Noise Considerations

One advantage of the off-line IV and refined IV algorithms is that they are inherently designed to allow for noise on the system signals. The concept of instrumental variables, for instance, was introduced originally simply as a convenient modification to the least squares equations in order to reduce noise-induced estimation bias. And the refined IV method is a straightforward development of the IV algorithm which reduces the error variance associated with the parameter estimates and so induces greater statistical efficiency. We might conclude, therefore, that if STC and SAC systems can be designed bearing in mind some of the lessons learnt from the use of these off-line IV algorithms, they should be more resilient to noise effects than those which are based simply on the deterministic concepts of linear least squares.

In this Chapter, where we are concentrating primarily on the deterministic or low noise situations, we will restrict attention to the resolution of the most common "noise" problem that affects STC and SAC system design, even under these desirable low noise conditions; namely the problem of low frequency noise in the form of "bias" or "drift" on the input and output signals of the system. The ever present possibility of such low-level, but troublesome, noise clearly make it desirable to pre-filter the signals in some manner, prior to their use in recursive estimation. Normally, however, only the simplest solutions, such as differencing the data, are discussed and a unified approach to the problem is rarely presented.

Fortunately, the refined IV algorithm actually prescribes the theoretically optimal nature of the data filter required for model parameter estimation, in the form of the pre-filter $F(q^{-1})$ defined by equation (11.14). The potential practical utility of this pre-filter becomes clear if we suppose that the system model is of the form given in equation (11.1), with $\xi(k)$ defined as a random walk to allow the possibility of low frequency "drift" or bias on the signals, i.e.,

$$\xi(k) = \frac{1}{1 - q^{-1}} \quad e(k)$$

Since $D(q^{-1})=1.0$ and $C(q^{-1})=1-q^{-1}$ in this model, the optimal pre-filter $F(q^{-1})$ takes the form,

$$F(q^{-1}) = \frac{1 - q^{-1}}{\hat{A}(q^{-1})} \qquad (11.18)$$

where, in theory, the estimated denominator polynomial of the system transfer function, $\hat{A}(q^{-1})$, should be chosen either on the basis of prior modelling studies, or updated adaptively, as in the refined IV algorithm.

Normally, of course, complicated pre-filtering such as (11.18) is difficult to justify and simpler alternatives are recommended. One alternative to the optimal pre-filtering becomes apparent when we note that $F(q^{-1})$ defined in this manner is simply a dc-blocking (or low-pass filtered differencing) operation, which makes obvious sense in practical terms. This suggests that simpler implementations can be obtained by approximating $\hat{A}(q^{-1})$ by a first or second order polynomial with an appropriately defined parameters chosen to provide satisfactory high frequency noise filtration, while passing those frequencies within the bandpass of the process.

The practical design of such DC blocking filters requires some care. For example, in the noisy SAC situation, it is important to take ensure that low pass or "smoothing" element in the filter (i.e. the denominator) does not deleteriously affect the initial convergence of the algorithm. One approach is to introduce a device similar to the variable forgetting factor mentioned above in the context of EWP estimation. Suppose, for instance, that a second order polynomial $\hat{A}(q^{-1})=[1 - \hat{a}_1 q^{-1}]^2$ is chosen in (11.18), then the parameter $\hat{a}_1$ can be made time variable by defining it as follows,

$$\hat{a}_1(k) = \alpha\, \hat{a}_1(k-1) + (1-\alpha)\, a_1 \qquad (11.19)$$

Here, $\hat{a}_1(0)$ and α can be considered as design variables, with typical values of 0 and 0.95, respectively. In this manner, the second order pre-filter starts as a simple differencing operation ($\hat{a}_1(0)=0$) and approaches a DC blocking filter with second order denominator, after an exponential transient having a time constant of $-1/\log_e(\alpha)$ samples (e.g. 20 samples for $\alpha = 0.95$). The second order low-pass characteristics of this DC blocking filter are defined by the double pole at a_1 [the steady state solution of (11.19)] in the complex q plane; this ensures the removal of the offending low frequency signal components together with attenuation of the higher frequency noise effects which would otherwise be amplified by the differencing operation.

Of course, complications such as those discussed above are not required in the situation where any higher frequency noise on the data is at a low level. Then the $\hat{A}(q^{-1})$

polynomial in the pre-filter (11.18) can be set permanently to unity, so that the pre-filter reduces to the common device of simply differencing the input and output signals prior to their use in the recursive algorithm. Furthermore, in such a low noise environment, the RLS algorithm is clearly appropriate for parameter estimation and is somewhat simpler than either the IV or any other alternative recursive estimators, such as the Approximate Maximum Likelihood [or Extended Least Squares (ELS)] and Prediction Error Recursion (PER) Methods [see e.g. Ljung and Soderstrom(1983); Young(1984) and Norton(1986)]

11.7 PRACTICAL EXAMPLES

The PIP approach to digital control system design has a number of attractions : it is simple yet widely applicable; it exploits the power of SVF pole assignment, without resort to complications such as state reconstruction; and it provides a logical successor to the conventional PI and PID controllers. The examples presented here are all associated with control problems in glasshouse systems and they are intended to demonstrate some of the attractions of the PIP concept within a practical self-adaptive (SAC) or self-tuning (STC) context. The first example is discussed in some detail and considers all steps of the design procedure outlined in Section 11.4, including practical implementation.

11.7.1 PIP Control of an NFT System

The Nutrient Film Technique (NFT) is a method of soil-less ("hydroponic") cultivation used in horticulture for growing plants, such as tomatoes and cucumbers, in glass-houses. The system is quite complex in dynamic terms: the nutrient mixture is delivered to the plants, which are suspended in a series of long, parallel channels, by a pump-driven circulatory flow system, with the mixture returned to the pump via a catchment tank or "trench". This process is characterised by long and variable transport delays, changing dispersive characteristics, and the over-riding influence of the positive feedback introduced by the circulatory flow system (Young et al.,1987[2]). The process is dynamically similar in flow terms to the system used by Astrom (1984) for certain adaptive control studies.

<u>11.7.1.1 Experiment design and data analysis.</u> The first two of the five stages in the design procedure outlined in Section 11.4 are concerned with experimental design and data analysis. In this NFT example, most of the experimental studies on the NFT system have been carried out on a pilot

plant constructed at Lancaster. This is "dynamically similar" in flow terms to the full scale process, but it has much shorter dominant time constants and so is simpler to use for control experiments. Naturally, the pilot plant can be described by a similar discrete-time model to the full scale plant but the sampling interval of 25 seconds corresponds to an "equivalent" sampling interval of 15 mins on the full scale process. The pilot plant was constructed following difficulties experienced during initial experiments on the full scale NFT system: in particular, the extremely slow dynamics of the full scale system necessitated unacceptably long experimental times, which interfered with the normal glass-house operation and seriously limited the nature of the control studies.

A diagram of the pilot plant is shown in Fig. 11.2, where it will be seen that the long and variable pure time delays are introduced by considerable (70 metre and 40 metre) lengths of plastic pipe, which simulate the advective transport delays associated with the growth channels and the trench in the full scale NFT system. The pump rate can be varied to simulate major variation in these pure time delays; and a sinusoidal leak of water from the system is incorporated to simulate the diurnal losses of nutrient that would occur on the full size NFT plant, due to uptake by the growing plants. The nutrient itself is simulated by a black dye: the dye delivery system, which constitutes the main control signal to the system, is in the form of an electrically driven peristaltic pump. The concentration of dye, which represents the system output, is measured by a photoelectric sensor positioned as shown in Fig. 11.2.

Both open and closed loop experiments can be conducted on the pilot plant, with data aqcuisition and subsequent digital control based on a BBC microcomputer, using a multi-tasking operating system. The system input (the microcomputer generated signal to the peristaltic pump), and output (the dye concentration measured by the photoelectric system) provide the main data used in subsequent analysis and modelling; but various other data, such as the sinusoidal leak control signal and the flow rate of water through the system, are normally monitored throughout the experiments.

The data from the experiments are processed in two ways. First, they are stored on disk in the BBC micro and presented graphically on the monitor, both for on-line monitoring of the experiment and, following a "screen dump" to the printer, for later reference. Second, they are transferred to an IBM-PC AT, by Kermit, for later analysis and modelling studies using the microCAPTAIN program. Since

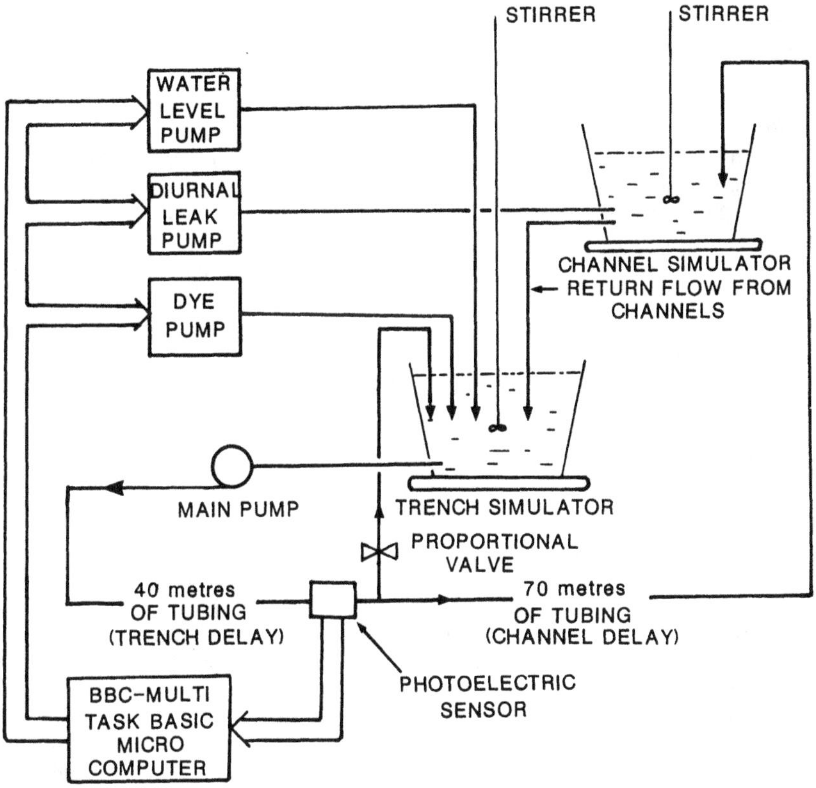

Fig. 11.2 The NFT flow pilot plant

on-line recursive estimation algorithms are programmed in the BBC for self-tuning/adaptive control studies, it is also possible to monitor the parameter estimation results during open or closed loop experiments.

<u>11.7.1.2 Model identification and parameter estimation.</u> The third stage of the design involves off-line exercises in both physical and data-based modelling. The physical-based simulation modelling of the full scale NFT system is described in Behzadi (1987) and Young et al (1987[2]), and was carried out prior to the pilot plant construction and experimentation. In this simulation model, which was initially implemented on an APPLE microcomputer, each of the

16 growing channels is represented by four first order "Aggregated Dead Zone" (ADZ) elements (Beer and Young,1983) connected in series; and the settling trench by eight first order ADZ, also in series. In systems terms, each ADZ element is a first order low pass filter similar to the well known chemical engineering model of a continuous stirred tank (CSTR) with associated pure time delay ("plug flow"). At the 15 min sampling interval (25 secs for the pilot plant), this high order model can be reduced straightforwardly (see Behzadi,1987) to a 12th order discrete-time model. A further exercise in model reduction using the SRIV algorithm is described in Young (1987[1,2]), where the model is reduced still further to a 4th order form at a sampling interval of 45 mins (1.25 mins for the pilot plant); and, finally, to the simplest first order form at the longest sampling interval of 3 hours (5 mins for the pilot plant).

Although the physical-based simulation modelling studies are important, it is the data-based modelling which is of major importance in control system synthesis. Consequently we will consider in detail a typical example of the data-based modelling studies carried out on the NFT pilot plant. These confirm the results of the physical-based simulation studies, but suggest certain simplified model structures that are useful in the subsequent fixed gain and adaptive control studies. All the modelling analysis was carried out using the microCAPTAIN computer program (Young and Benner,1986) and Fig. 11.3 is the medium resolution microCAPTAIN printout of the results obtained in the example.

The input-output data, shown in the top left hand block of Fig. 11.3(b), consist of 160 input (filename : PIL2160.MCI) and output (filename : PIL2160.MCO) pairs [i.e. $u(k);y(k)$] obtained at a sampling interval of 1.25 mins from the pilot plant during a closed loop experiment. Here, following start-up, the command input was peturbed up and down three times in a repeated step fashions during the 200 minute run. The water pump flow rate was maintained constant throughout the experiment, but the sinusoidal leak introduces a disturbance to the system which, as we will see, produces a noticable effect on the system behaviour.

Fig. 11.3(a) shows the results for a discrete-time TF model with a 5th order denominator, scalar numerator and time delay δ of 2 sampling intervals [i.e. (5,1,2)]; while the other three blocks in Fig. 11.3(b) show the model output $\hat{x}(k)$ compared with the measured output $y(k)$; the modelling error [i.e. the estimated noise $\hat{\xi}(k) = y(k)-\hat{x}(k)$] between $y(k)$ and $\hat{x}(k)$; and the model impulse response plotted over 40 samples (i.e. 50 mins in the pilot plant and 30 hours in the full scale system). We can see that the data are represented well by this (5,1,2) model, which has R_T^2 = 0.979; i.e. 97.9% of the output is explained by the model.

It is interesting to observe that the unexplained noise signal in the lower right hand block of Fig. 11.3(b) has a distinct temporal pattern. After removing the very long period components by recursive smoothing (see Young,1987[1]), univariate analysis of the remaining "detrended" signal (based on AIC identification) suggests a 17th order Autoregressive [AR(17)] model. The spectrum associated with this AR(17) model has a dominant peak at the 40 minute period (24 hours on the full scale NFT system), so revealing that it is associated with the sinusoidal leak introduced to model the diurnal uptake effects.

As might be expected, the impulse response of the model in the lower left hand block is similar to that of the physical-based simulation model and exhibits the major dynamic characteristics of the NFT flow system. The long term response is dominated by a first order mode with a very large time constant. This arises from the low losses in the system: except for the sinusoidal leak, all dye entering the system is retained, so that the system is always quite close to being an integrator. The shorter term behaviour is more interesting and is dominated by the circulatory flow system, with the three decaying peaks on the impulse response indicating that the initial impulsive input is being transported and progressively dispersed around the system.

Although it explains the data well, the (5,1,2) model is not the best identified model when viewed from a statistical standpoint. This is shown in Table 11.1 which provides the structure identification results, in the form

TABLE 11.1 Structure identification NFT pilot scale plant

Model No	MODEL	R_T^2	YIC
1	(1,2,2)	0.965	-8.00
2	(2,5,2)	0.954	-1.19
3	(3,4,2)	0.984	-3.89
4	(4,3,2)	0.985	+0.10
5	(5,1,2)	0.979	-3.03
6	(6,2,2)	0.975	-1.26
7	(5,1,2) SUBSET	0.979	-8.69

```
             IMMMMMMMMMMMMMMMMMMMMMMMMMMMMMMMMMMMMMMMMMMMMMMMMMMMMMMMMMMMMMM;
 DATAFILE:  :  microCAPTAIN        microCAPTAIN        microCAPTAIN    :   MODE:
O¦C:PIL2160 :  microCAPTAIN        microCAPTAIN        microCAPTAIN    :    I-O
I¦C:PIL2160 HMMMMMMMMMMMMMMMMMMMMMMMMMMMMMMMMMMMMMMMMMMMMMMMMMMMMMMMMMMMMMM<
                              Final iteration results

Coeff. of Detn:  0.97937       ln EVN: -3.70507       H-R Criterion: 4.17125
   Output mean:  196.956     Variance:  2595.14     Noise variance: 53.5158

               Parameter Estimates                  Time Delay:   2

           A 1: -0.79316 +/- 0.13376                 B 2:  0.07683 +/- 0.00797
           A 2:  0.17364 +/- 0.21054
           A 3: -0.13170 +/- 0.21350
           A 4:  0.05720 +/- 0.17873
           A 5: -0.28319 +/- 0.08818
```

(a)

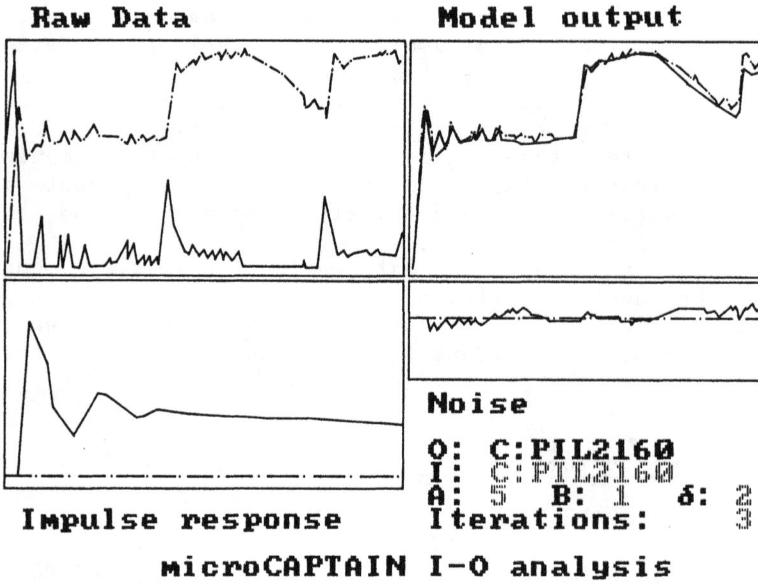

(b)

Fig. 11.3 MicroCAPTAIN results for the NFT Pilot Plant: (a) print-out of final estimation results; (b) data and model plots.

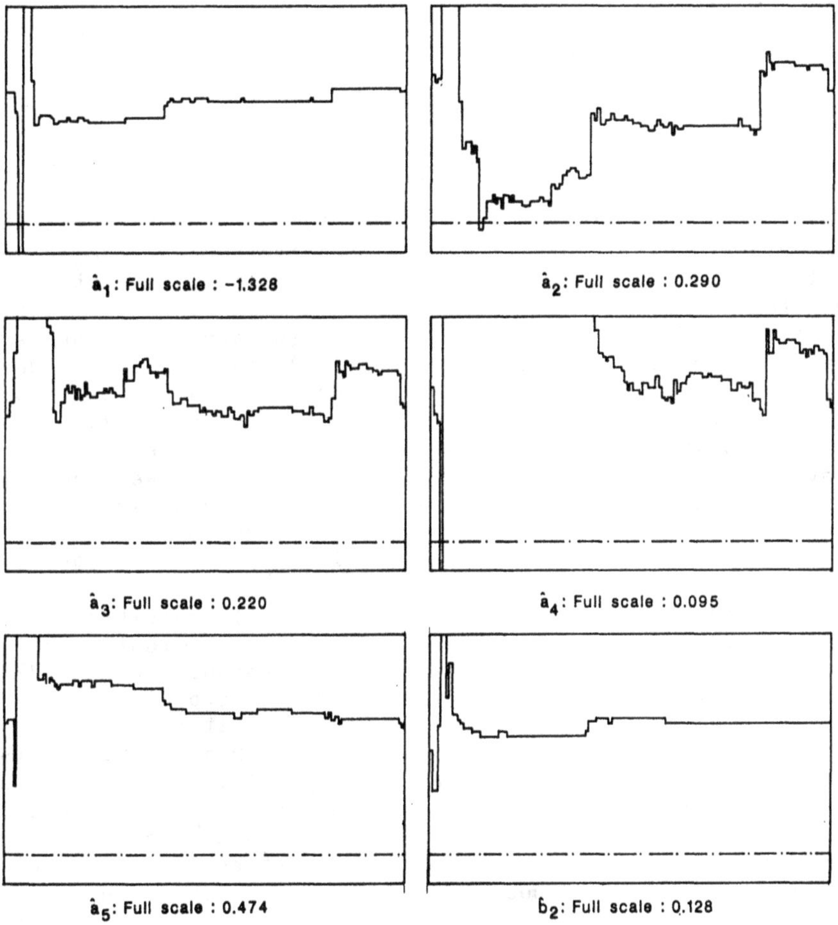

(c)

Fig. 11.3 continued. MicroCAPTAIN results for the NFT Pilot Plant: (c) plots of recursive parameter estimates.

of the best identified models from 1st to 6th order (n=1 to 6). The model shown in Fig. 11.3 is given as No.5 on this Table. Although it is clearly not the best identified on a YIC basis, we see that it is, in fact, closely related to the best identified model 7, which has the following "subset" (5,1,2) form,

$$y(k) = \frac{0.0805}{1 - 0.6904q^{-1} - 0.2857q^{-5}} u(k-2) \qquad (11.20)$$

In other words, the model has the (5,1,2) structure but the a_i parameters, i=2,3,4, are all constrained at zero value. This structure is suggested by the fact that the estimates of these constrained parameters are all insignificantly different from zero when their estimated standard errors are taken into account (see Fig. 11.3(a) where the standard errors are shown as +/- values adjacent to the estimates). This is confirmed by the fact that the constrained model has virtually the same R_T^2 as the unconstrained model, despite having three fewer parameters. Also, as we see in Fig.11.3(c), the recursive estimates of the parameters a_2, a_3 and a_4 show much more instability than those of a_1 and a_5, a clear indication of their poor definition in statistical terms.

Model No.1 in Table 11.1 is also well identified: its YIC is only slightly less than model 7. More significantly, however, it has a reduced R_T^2 at 0.965 compared with 0.979. This reduced descriptive ability is easily explained: since the (1,2,2) has only first order dynamics, it can only model the long term mode of behaviour and is unable to charaterise the shorter term, circulatory effects. This is illustrated in Fig. 11.4, which compares the impulse responses of models 1,3,4 and 7: the model 4 response is shown because, although it has a poor YIC, it has the highest R_T^2 of all the models. However, this good descriptive ability is obtained at the cost of considerable imprecision in the parameter estimates.

Some additional control studies (Young et al,1986;1987[1]; Young, 1987[1])) have been carried out at the very coarse sampling interval of 5 mins (3 hrs. on the full scale system). The most recent of these have used a simplification of the above subset (5,1,2) model, as obtained by identifying the model which best explains the step response of the model (11.20) at this coarse sampling interval. This yields the model,

$$y(k) = \frac{0.147}{1 - 0.9563q^{-1}} u(k) \qquad (11.21)$$

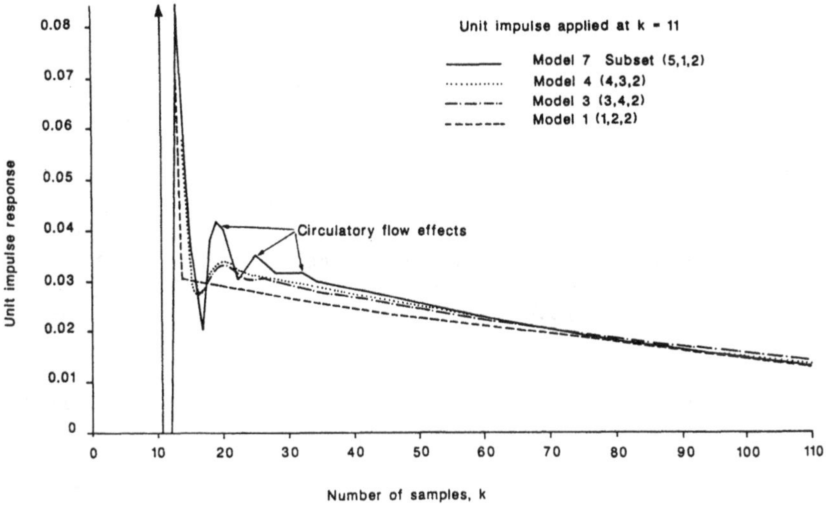

Fig. 11.4 Impulse responses of selected models from Table 11.1

This has the advantage of yielding a simple fixed gain PI controller which, because of its long sampling interval, is insensitive to changes in the system time delay. However, in the present Chapter we will limit attention to the more complex self adaptive controllers associated with model (11.20), and direct the reader to the other references for further discussion on the coarse sample case.

<u>11.7.1.3 PIP Control System Design.</u> The fourth stage in the design procedure is concerned with the off-line design and evaluation of the PIP controller in fixed gain and STC/SAC form. This is carried out using the physically-based simulation model and the best identified models from the previous stage 3. To conserve space, we will not discuss this stage in detail, since it merely confirms the efficacy of the various STC and SAC versions of the PIP controllers, the practical implementation of which is described below.

<u>11.7.1.4 STC/SAC System Design and Implementation.</u> The noise level on both the full scale NFT system and the pilot plant is relatively low, as we see in Fig. 11.3(b). For this reason, it is only necessary to use a RLS algorithm for parameter estimation in STC and SAC system design. The PIP controller defined above is implemented in the on-line BBC micro and the control gains are updated continually on the

basis of the RLS estimates of the associated model parameters. In the two examples below, the input/output signals were pre-filtered, prior to their use in the RLS algorithm, by a high pass filter $F(q^{-1})$ [see equation (11.18)], with $\hat{A}(q^{-1}) = (1-0.5\ ^{-1})^2$. In all cases, the RLS algorithm is initiated in the normal manner, with the $\underline{\hat{a}}(0)$ parameter vector set to zero and the $P(0)$ matrix set $\bar{I} \times 10$. For the first n+m samples, the self-tuning control gains are set to the values obtained in 11.7.1.3 and this maintains reasonable control over this initiation period. Although not absolutely necessary, such an approach is clearly advisable in practical terms.

Figs. 11.5 and 11.6 are typical of the STC results for the pilot plant. Fig. 11.5 shows an STC system in which all parameters in a (5,3,1) model are estimated in an unconstrained manner : this is based on the identified (5,1,2) model, but compensates for any possible changes in the pure time delay δ by allowing the additional b_i parameters $(i=1,3)$ to provide flexibility in the definition of the time delay characteristics. We see from the final estimated values, however, that only b_2 is significant, indicating that the delay was, indeed, two sampling intervals. It is also clear that a_2 and a_3 are insignificant, although a_4 is estimated as -0.109 and appears significant during this run.

Fig. 11.6 is an example of the simple gain tuning method discussed in Section 11.6.1. Here, the subset (5,1,2) model is used and the parameters are fixed at their values from 11.7.1.3. Clearly the convergence of the self tuning gain is rapid and effective and yields similar performance to that shown in Fig. 11.5. It is worth noting, however, that the control signal in Fig. 11.5 is smoother than in Fig. 11.6: in both cases, the control signal can be seen to compensate well for the sinusoidal leak, but the the former control would obviously be more acceptable in practical terms.

11.7.2 Temperature Control in a Glasshouse

Here we consider a simple implementation of the PIP system for a glasshouse temperature control system. The discrete-time model in this case is first order for a sampling interval of 10 minutes, with a one period time-delay. As in the simplest NFT case, therefore, the PIP design reduces to a conventional PI controller but, in contrast to the PI controller used normally in the glasshouse, the PIP implementation is a true digital design. The sampling interval of 10 minutes was selected following model identification studies carried out both at Lancaster and at the Agriculture and Fisheries Research Council,

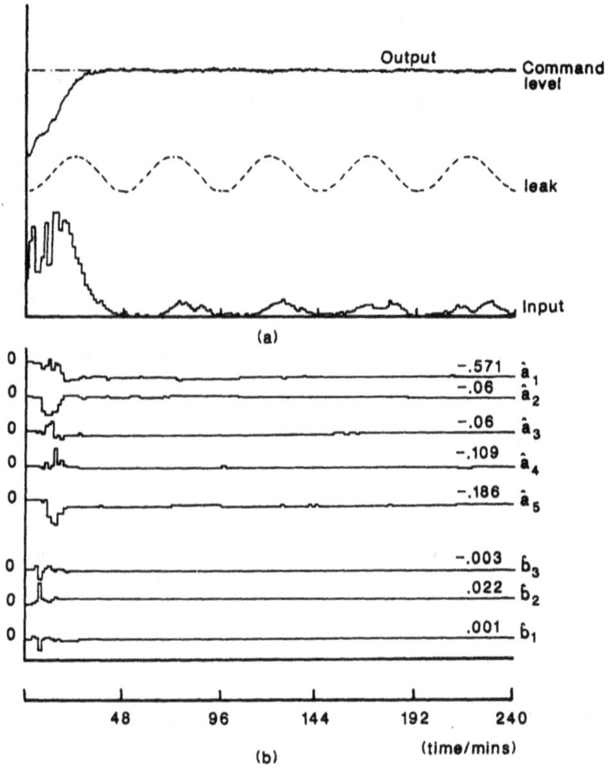

Fig. 11.5 Self tuning control of NFT pilot plant: full self tuning with (5,3,2) model

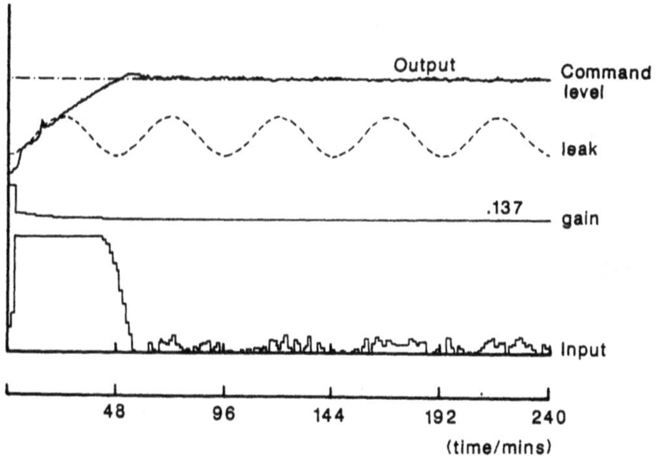

Fig. 11.6 Self tuning control of NFT pilot plant: simple adaptive gain control (AGC)

Institute of Engineering Research, Bedfordshire, England (Davis,1986) using instrumental variable estimation procedures. Such a sampling interval provides the model with the best defined parameter estimates and also allows, most conveniently, for the dead-time characteristics of the process.

Unlike the PIP system, the conventional controller used in the glasshouse is a digitised continuous-time PI system operating on the much smaller sampling interval of one minute. It is typical of some commercially available controllers and, although it was probably not tuned optimally, the control parameters were set at reasonable values. The performance of the two controllers is compared in Fig. 11.7, which shows them responding to similar, but not identical, set point changes over a period of some 55 hours. The failure of both systems to maintain the set points during parts of day time is due to the nature of the system: only heating action is available so, during the day when natural temperatures rise because of solar irradiance, the controllers provide their minimum zero input most of the time. However, at other times, and particularly during the cold night hours, the controllers are fully operative. It is here that we see the superiority of the PIP system, which quickly compensates for the disturbances and maintains the desired temperature level. Fig. 11.7 is provided by Drs. P.Davis and A.Hooper of the AFRC Institute of Engineering Research (see Davis, 1986).

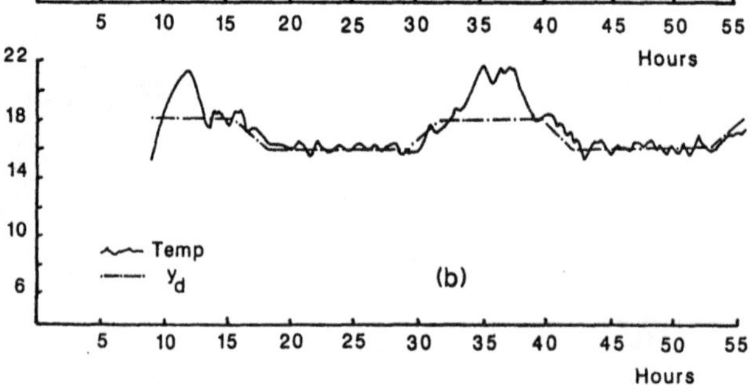

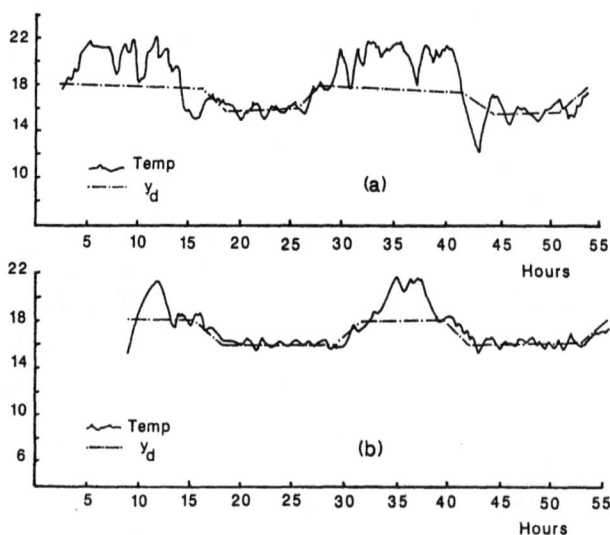

Fig. 11.7 Comparison of (a) conventional PI; and (b) PIP glasshouse temperature control systems

The conventional PI system, despite its more rapid sampling frequency, is quite sluggish and oscillatory in its response to both set point changes and disturbances. This apparently paradoxical behaviour is quite easy to understand. First, the rapid sampling rate makes the system more sensitive to the time-delays in the system and it is well known that conventional, continuous-time integral action leads to oscillatory behaviour when significant time delays are present. Second, the pole assignment design in this case is intended to provide a closed loop characteristic polynomial of $1-0.5q^{-1}$, which ensures that the response is rapid and critically damped.

11.7.3 Self-Adaptive Control of a Nonlinear Heated Bar System

The PIP approach can also form the basis for the design of controllers for certain nonlinear systems characterised by piecewise linear or bilinear behaviour. This is demonstrated by recent research on the adaptive temperature control of a nonlinear heated bar system. Here, a novel PIP controller is utilised to implement pulse width modulation of the "bang-bang" control signals to either the electric heater or the cooling fan (which provide the mechanism for controlling the bar temperature); and adaption is based on a novel recursive algorithm capable of tracking the extremely rapid variation of model parameters that occur when the control signal changes sign.

This new SAC system is based on an earlier adaptive system for an airborne vehicle (Young,1980), It is a computationally efficient solution, in the sense that a single recursive estimator is able to track the changes in the parameters of what are, in effect, the two, quite different dynamic systems: namely, the bar temperature when controlled rather sluggishly by the heater, and the same temperature controlled much more rapidly by the fan. Clearly an SAC system based on two separate recursive estimation algorithms would provide alternative but computationally somewhat slower and more complex solution.

Fig. 11.8 shows a typical run using the SAC system based on a single recursive estimator: here the system is responding to a "staircase" type input, chosen to move the bar temperature over a fairly wide range. In this manner, the adaptive action is required not only to allow for the changing dynamics associated with the control action but also to compensate for the longer term changes in the system dynamics which arise because of the increased heat losses at higher bar temperatures and the lower cooling effect of the fan at lower temperatures.

In Fig. 11.8, the model parameter estimates used to

update the PIP controller are shown in the bottom three traces. Based on its prior knowledge that the dynamic characteristics change radically when the input signal changes sign, the single recursive estimator automatically switches between parameter values appropriate to the state of the system (i.e. whether the switch is from heater to cooler or vice versa) and then continues to update these estimates until switching again takes place with the next change in the sign of the input signal. The switching

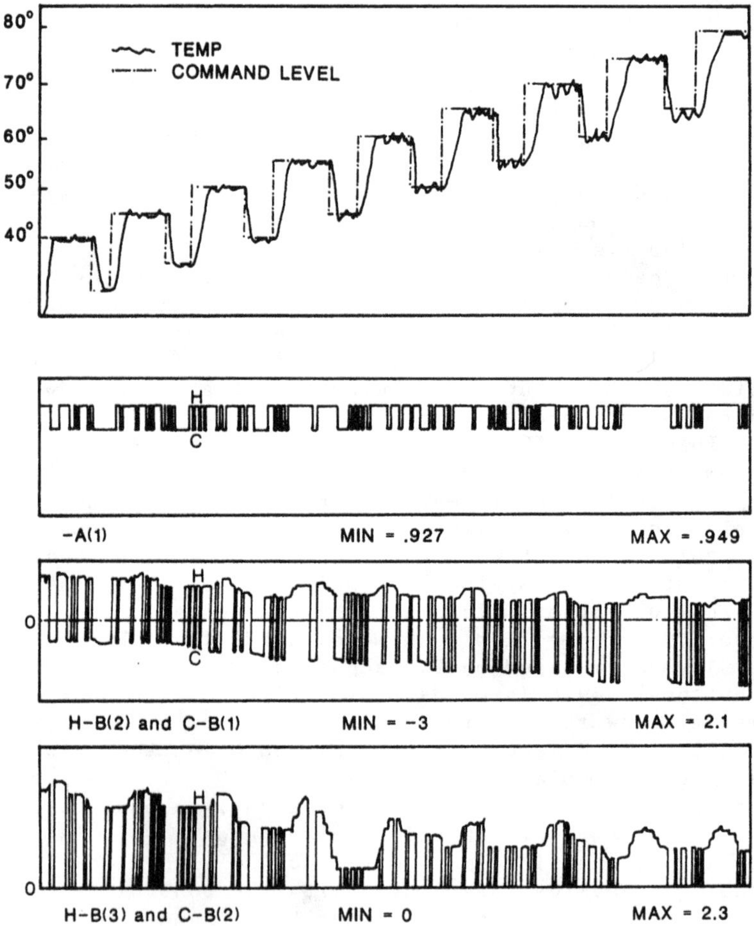

Fig. 11.8 Self adaptive control of the temperature of a metal bar

between levels is clearly apparent on the traces and the upper and lower "envelopes" illustrate the changes occurring in the parameter estimates over both the short and long term (when control switching is frequent, the estimates are frozen to avoid problems).

It is interesting to note that the sampling intervals for both modelling and control are different for the heater and fan "sub-systems" (namely 7.5 secs for the heater and 2.5 secs for the more rapidly responding fan); thus the plots of the parameter estimates should be interpreted with this in mind.

The top trace shows that the estimates of the negative of the a_1 parameter for both heater (upper envelope) and fan (lower envelope) sub-systems have been frozen at their a priori values, as estimated by the SRIV algorithm in microCAPTAIN. This proved possible because these parameters did not appear to change significantly over the operating regime of the system. The bottom two traces show the adaptive estimates of the parameters associated with the models of the two subsystems: the heater sub-system model has two numerator coefficients (b_2 and b_3, indicating the presence of a two period time delay); while the fan sub-system has only one significant numerator coefficient (b_1, indicating a single period delay). Here b_2 is set to zero for the whole of the run, as shown by the lower envelope being always coincident with the horizontal axis. This constraint was again suggested by prior estimation and modelling experiments.

Although this example is not strictly associated with glasshouses, it has features in common with glasshouse environmental control and is being used as a convenient test example to evaluate advanced adaptive concepts. It is hoped that these will later be applied, in modified form, directly to the control of multivariable glasshouse systems. Further details of the heated bar control system are given in Behzadi (1987).

11.8 EXTENSIONS

Perhaps the most intriguing aspect of the NMSS concept is that it opens the way to a whole new class of control systems, of which the PIP controller for SISO systems is only the first example. Because of its inherent state-space formulation, for example, the NMSS design method discussed here can be applied to multivariable and stochastic systems. Multivariable exploitation of the NMSS idea is quite obvious: it is possible to develop related NMSS design procedures for multiple-input, single output (MISO) pole assignment design (Wang and Young,1987[1]) ; full MIMO pole assignment designs with "static" decoupling derived from the inherent integral action (Wang and Young,1987[2]); or complete MIMO designs incorporating pole assignment and full dynamic decoupling (Wang and Young,1987[3]). These multivariable extensions follow straightforwardly from the SISO approach described in this Chapter, although they are naturally more complicated and have not yet been evaluated in practical terms.

Theoretically, in the linear, stochastic situation, a Kalman filter or its equivalent is required so that the "separation theorem" can be invoked. But since the whole of the state is directly observed in the NMSS case (i.e. the dimension of the observation vector is, effectively, equal to that of the state vector), state variable estimation reduces to the simple filtration of these directly measured states and does not require the complication of simultaneous state variable reconstruction. Within this stochastic setting, it is clearly possible to formulate "optimum" NMSS design procedures: for example, the well known "Linear-Quadratic-Gaussian" (LQG) approach or the less well known "Model-in-the-Performance-Index" discussed by Young and Willems (1972).

In STC or SAC application of these stochastic extensions, it is necessary to incorporate a recursive parameter estimation algorithm that is capable of handling the inherent closed loop noise effects. General discussion on identification in the closed loop is given in a number of publications [e.g. Norton (1986);page 212 et seq]. In the present context, however, we will restrict attention to an instrumental variable-type of solution. It is well known that, provided a reasonably active command input signal is present, then the system will normally be identifiable and its parameters can be estimated by any of the well known recursive estimators that allow for noise. A rather obvious IV-type solution is that suggested by Young (1970), in which the auxiliary model is expanded to represent the whole closed loop system, including the known feedback control elements, so that instrumental variables can be generated for the system input and output variables, both of which are contaminated by the noise circulating within the closed loop. For complete stochastic self adaptive control, however, such a solution could become quite complex, since it would require simultaneous noise model estimation using a simplified form of the refined IV algorithm.

11.9 CONCLUSIONS

The design procedure presented in this Chapter provides a simple yet flexible method of synthesising self-tuning and self adaptive control systems which are inherently digital in form and can be considered as logical successors to more conventional digital systems, such as the ubiquitous PI and PID controllers. Unlike these more traditional systems, which are often derived by the mechanical digitisation of continuous-time designs, this new class of controller offers "True Digital Control"(TDC). In the TDC design, the sampling rate is chosen to enhance the performance of both the closed loop dynamic system and the estimation algorithms employed in the self-tuning/self-adaptive mechanisation. As a result, it can be quite slow in relation to the alternative digitised continuous-time alternatives and gives rise to the concept of "coarse sampling" control systems (Young et al,1986).

The practical examples presented in the Chapter emphasise the advantages of the true digital PIP designs in relation to their more conventional forbears. For instance, all of the examples involve processes that are difficult to control using continuous-time or digitised continuous-time designs, since they are characterised by relatively long pure time delays. Also, the PIP system appears quite robust to the application of constraints on the control input, since the method of implementing the TDC law ensures that the integral action is suspended when the constraints are encountered, so counteracting problems such as integral "wind-up". The results from other experiments confirm these advantages: fixed gain and self-tuning/self-adaptive PIP systems have been developed by colleagues at Sunderland Polytechnic for the control of both a "ball-and-groove" Laboratory experiment and a fluid-flow valve system. In both cases, the PIP design is at least as good as other, more conventional designs, and normally out-performs them in difficult situations (Boucher, Cox and Young,1987).

The PIP concept can be extended in various directions. A forthcoming paper (Wang and Young,1987[3]) will report initial research on multivariable systems, which shows how the additional degrees of freedom provided by the multiple control inputs can be exploited to allow for simultaneous pole assignment and dynamic decoupling. Finally, because of the state-space setting, it is clearly straightforward to consider the approach to control system design presented here in optimal and stochastic contexts, although such systems will naturally be more complex.

11.10 ACKNOWLEDGEMENTS

Some of the research reported in this paper was carried out with the support of a SERC CASE studentship in association with the AFRC Institute of Engineering Research, Silsoe, Bedfordshire. The authors are grateful for this support and would like to express their special appreciation to Drs. Paul Davis and Bernard Bailey of the AFRC for all their help and encouragement.

11.11 REFERENCES

Astrom,K.J., and Wittenmark,B. (1973) On self-tuning regulators, Automatica, 9, 185-199.

Astrom,K.J., and Wittenmark,B. (1980) Self Tuning controllers based on pole-zero placement, IEE Proc.,127,120-130

Astrom,K.J. (1984) in I.D.Landau, M.Tomizuka and D.M.Auslander (eds), Adaptive Systems in Control and Signal Processing, Pergamon Press : Oxford, 137-146.

Beer,T., and Young,P.C. (1983) Longitudinal dispersion in natural streams, American Soc. of Civil Eng., Jnl. of Env. Eng., 109, 1049-1067.

Behzadi,M.A., Young,P.C., and Chotai,A. (1986) Modelling and coarse sampling control of a nutrient film system, Int. Conf. on Systems Science, Wroclaw, Poland.

Behzadi,M.A., Young,P.C., and Chotai,A. (1987) Self adaptive control in glasshouses, I.E.E Workshop on Self-Tuning and Adaptive Control, Sommerville College, Oxford.

Behzadi,M.A. (1987) Adaptive Control in Glasshouses, Ph.D. Thesis, Dept. Environ. Science, University of Lancaster, England, 1987.

Boucher,R., Cox,C., and Young,P.C. (1987) Implementation and evaluation of the PIP controller, in preparation.

Box,G.E.P., and Jenkins,G.M. (1970) Time Series Analysis Forecasting and Control, Holden-Day, San Francisco.

Chotai,A., and Young,P.C. (1987) A generalised Smith Predictor for discrete-time systems, in preparation.

Davis,P. (1986) Pers. Comm. relating to confid. paper by Davis.P, and Hooper,A.W., NIAE, Bedfordshire, England.

Jakeman,A.J., and Young,P.C. (1984) Recursive filtering and the inversion of ill-posed causal problems, Utilitas Math., **25**, 351-376.

Kalman,R.E. (1958) Design of a self-optimizing control system, ASME Trans.,Jnl. Basic Eng., **80**, 468-478.

Ljung,L., and Soderstrom,T. (1983) Theory and Practice of Recursive Estimation, MIT Press, Cambridge, Mass.

Norton,J.P.(1986) An Introduction to Identification, Academic Press, London.

O'Reilly (1983) Observers for Linear Systems, Academic Press, London.

Popov,V.M. (1964) Hyperstability and optimality of automatic systems with several control functions, Rev. Roum. Sci. Tech., Ser. Electro. Tech. Energ., 9, 629-690.

Priestley,M.B. (1981) Spectral Analysis and Time Series, Academic Press, London.

Soderstrom,T., and Stoica,P. (1983) The Instrumental Variable Approach to System Identification, Springer-Verlag Lecture Notes Series, Berlin.

Wang,C.L., and Young,P.C. (1987[1]) Direct digital control by input-output, state variable feedback : theoretical background, accepted for publication in the Int. Jnl. of Control.

Wang, C.L., and Young, P.C. (1987[2]) on the pole assignment of linear discrete-time servomechanism systems via state variable, input-output feedback. submitted for publication.

Wang,C.L., and Young,P.C. (1987[3]) Dynamic decoupling and pole assignment control of multivariable systems by input-output feedback, Dept. of Environ. Science, Univ. of Lancaster, Rep. No. TR 50, 1987.

Wellstead, P.E., Prager, D. and Zanker, P. (1979) Pole assignment self-tuning regulator, Proc. I.E.E., **126**, 781-787.

Wonham,W.M. (1967) On pole assignment in multivariable, controllable linear systems, I.E.E.E. Trans. Auto. Control, AC-12, 660-665.

Young,P.C. (1965) Process parameter estimation and self-adaptive control. Proc. IFAC Symp. on Theory of Self Adaptive Systems, (appears in P.H.Hammond (ed.) Theory of Self Adaptive Systems, Plenum Press, New York, 1966).

Young,P.C. (1970) An instrumental variable method for real-time identification of a noisy process, Automatica, **6**, 271-287.

Young,P.C. (1972) Dept. of Eng. Univ. of Cambridge, Rep. No. CUED/B- Control/TR 24.

Young,P.C.(1980) A second generation adaptive autostabilization system for airborne vehicles, Automatica, 17, 459-470.

Young,P.C. (1984) Recursive Estimation and Time-Series Analysis, Springer-Verlag, Berlin.

Young,P.C. (1985) Recursive identification, estimation and control, in E.J.Hannan, P.R.Krishnaiah and M.M.Rao (eds), Handbook of Statistics 5 : Time Series in the Time Domain, North Holland, Amsterdam.

Young,P.C. (1985) The instrumental variable method: a practical approach to identification and system parameter estimation, in H.A.Barker and P.C.Young (eds) Identification and System Parameter Estimation, Pergamon Press, Oxford.

Young,P.C. (1987[1]) Recursive estimation forecasting and adaptive control, to appear in C.T.Leondes (ed.) Control and Dynamic Systems : Vol XXVIV, Academic Press, Florida.

Young,P.C. (1987[2]) A refined instrumental variable approach to model reduction, in preparation.

Young,P.C., and Benner,S. (1986) microCAPTAIN Handbook : Version 1.0, Dept.of Environ. Science, Univ. of Lancaster, Rep. No. TR 37

Young, P.C. and Jakeman, A.J. (1979-80) Refined instrumental variable methods of recursive time-series analysis, Parts I, II and III, Int. Jnl. Control, 29, 1-30 and 621-644; 31, 741-764.

Young,P.C., and Wang,C.L. (1987) Identification and estimation of multivariable dynamic systems, in J.O'Reilly (ed) Multivariable Control for Industrial Applications, Peter Perigrinus, London.

Young,P.C., and Willems,J.(1972) An approach to the multivariable servomechanism problem, Int. Jnl. Control, 15, 961-979.

Young,P.C., and Yancey,C. (1971) A second generation adaptive pitch autostabilization system for a missile or aircraft, Tech. Note 404-109, Naval Weapons Center, China Lake, California.

Young,P.C., Jakeman,A.J., and McMurtrie,R. (1980) An instrumental variable method for model order identification, Automatica, 16, 281-294.

Young,P.C., Behzadi,M.A., Chotai,A., and Davis,P. (1987), Chapter in J.A.Clark et al (eds), Computer Applications in Agricultural Environments, Butterworths : London.

Young,P.C., Behzadi,M.A., Wang,C.L., and Chotai,A (1987) Direct digital and adaptive control by input-output, state variable feedback pole assignment, accepted for publication in Int. Jnl of Control.

Chapter 12

LQG adaptive autopilots

J. Byrne and M. R. Katebi

12.1 INTRODUCTION

The autopilot on a ship is required to fulfil two very different modes of operation; course-changing, where the ship must follow a desired path, and course-keeping, where the vessel is required to stay on a preset course.

The control objective in course-changing is to perform the manoeuvre quickly and accurately. The manoeuvre should have a clear start, to indicate to other ships the intention of the turn and, to avoid collision it should be completed without overshoot. In this control situation fuel economy plays a minor role.

In course-keeping the control objective is to maintain a fixed heading whilst minimising the rudder activity which hence minimises the fuel consumption with respect to steering. In this situation the controller must offset low frequency disturbances on the ship such as wind loading and sea current effects. At the same time high frequency components acting on the ship must not be compensated for but must be attentuated to minimise rudder activity and wear and tear of the steering machine.

With two entirely different control objectives ideally there should be two different controller structures. However, the conventional autopilot is usually based on a simple proportional-integral-derivative (PID) controller the parameters of which are usually tuned to give good manoeuvring performance at a given nominal ship speed (Kallstrom et al,1979). A fixed setting is therefore often used which takes no account of the course-keeping requirements and is also inappropriate for energy minimisation. Furthermore, the PID controller is too simple to compensate satisfactorily for the effects of bad weather conditions and the variations in the ship dynamics and seaway parameters.

These difficulties in providing adequate course-keeping performance can be removed by the inclusion of adaptive features in the controller design. Adaption has been incorporated in PID autopilot by Oldenburg (1975) and Sugimoto and Kojima (1978). Stochastic adaptive systems have been developed by Merlo and Tiano (1975). Model reference techniques have been used by van Amerongen (1982) and the Clarke and Gawthorp (1975) type of self-tuner based on a conditional cost function has been employed by Lim and Forsythe (1983). The Astrom and Wittenmark (1973) minimum variance self-tuning algorithm has also been used by Kallstrom (1979) to design an adaptive autopilot and LQ optimal state-space based designs have been proposed by Saelid and Jenssen (1982).

The new generation of adaptive autopilots are based on the optimisation of a cost function which represents the energy used in maintaining a set heading. This cost function should ideally represent "the added resistance due to steering" and the elongation of distance sailed caused by the sway and yaw motions. However, the cost functions used in many of the existing adaptive autopilots contain very poor models of the added resistance and their validity has been questioned by Clarke (1982) and Reid (1982), who concludes that designs which minimise these cost functions may actually increase fuel consumption.

Another major problem with existing non-adaptive autopilots is their poor performance in rough seas (Blanke, 1981). One cause is the improper modelling of the sea wave environment and as a consequence high frequency measurement components are fed back to the ship's rudder.

In this chapter an autopilot is proposed which will minimise the added resistance due to steering and in addition will significantly improve the course-keeping performance in all weather conditions. To achieve these control objectives a new dynamic cost function is formulated which will be optimised in a stochastic linear quadratic gaussian (LQG) framework using a polynomial systems approach. An explicit self-tuning scheme is included to provide parameter adaption to varying weather conditions.

The modelling of the ship and the seaway disturbance is discussed in Section 12.2 and the proposed cost criterion is defined in Section 12.3. The LQG control design stategy is developed in Section 12.4. The adaptation for course-keeping is analysed in Section 12.5 and the identification scheme in described in Section 12.6. The simulation results are presented in Section 12.7 and the conclusions in Section 12.8.

12.2 MATHEMATICAL MODELLING

12.2.1 Ship Model

The dynamics of a ship can be analysed by considering the ship as a rigid body with six degrees of freedom corresponding to translation along and rotation about three axes. In course keeping control problems only motions in the horizontal plane, i.e. surge,sway and yaw are considered. These motions can be represented by the following normalised non-linear equations:

$$\dot{u} = X_1u^2 + X_2vr + X_3\delta^2 + X_4r^2 + X_t + X_{ext} \qquad (12.1)$$

$$\dot{v} = Y_1uv + Y_2ur + Y_3v|v| + Y_4r|r| + Y_5\delta + Y_{ext} \qquad (12.2)$$

$$\dot{r} = N_1uv + N_2ur + N_3v|v| + N_4r|r| + N_5\delta + N_{ext} \qquad (12.3)$$

where u denotes the surge velocity, v the sway velocity, r the yaw rate and δ the rudder angle. The coefficients $X_1 \ldots X_5$, etc. are the ship's hydrodynamic derivatives which are normally computed from full scale trails and tank tests. X_t denotes the propeller thrust and X_{ext}, Y_{ext} and N_{ext} represent the external disturbances on the ship such as wind and wave forces and moment.

In addition, the heading angle , ψ can be written as:

$$\dot{\psi} = r \tag{12.4}$$

For course-keeping control design, a linearised ship model is required to predict the ship motion. The non-linear equations (12.1),(12.2) and (12.3) can be linearised about the steady-state operating point v=r=δ=0 and u=U. The linearised sway and yaw model becomes:

$$\begin{bmatrix} \dot{v} \\ \dot{r} \\ \dot{\psi} \end{bmatrix} = \begin{bmatrix} a_{11} & a_{12} & 0 \\ a_{21} & a_{22} & 0 \\ 0 & 1 & 0 \end{bmatrix} \begin{bmatrix} v \\ r \\ \psi \end{bmatrix} + \begin{bmatrix} b_{11} \\ b_{21} \\ 0 \end{bmatrix} \begin{bmatrix} \delta \end{bmatrix} + \begin{bmatrix} Y_{ext} \\ N_{ext} \\ 0 \end{bmatrix} \tag{12.5}$$

The autopilot translates the heading angle measurements from the gyrocompass to the rudder command angle. Hence the system output can be defined as:

$$y = \psi \tag{12.6}$$

and the system input as:

$$u = \delta \tag{12.7}$$

The polynomial systems approach to linear control design requires a discrete-time transfer function relating the system output to the system input. The first stage in this transformation is to express the state-space model of equation (12.5) as a transfer function in the s-domain:

$$\frac{y(s)}{u(s)} = \frac{k_r(1 + s\tau_r)}{s(1 + sT_1)(1 + sT_2)} \tag{12.8}$$

The second stage transforms this model to the discrete-time domain to obtain:

$$\frac{y(q^{-1})}{u(q^{-1})} = \frac{B(q^{-1})}{A(q^{-1})} = \frac{b_1q^{-1} + b_2q^{-2} + b_3q^{-3}}{1 + a_1q^{-1} + a_2q^{-2} + a_3q^{-3}} \tag{12.9}$$

The relationship between the sway velocity and the yaw rate will be used in the development of the dynamic cost function. This relationship can be expressed in the Laplace domain as:

$$\frac{v(s)}{r(s)} = \frac{k_v(1 + s\tau_v)}{(1 + s\tau_r)} \tag{12.10}$$

and in discrete-time as:

$$\frac{v(q^{-1})}{r(q^{-1})} = \frac{k(1 + \theta_v q^{-1})}{(1 + \theta_r q^{-1})} \qquad (12.11)$$

12.2.2 Wave Model

The response of a ship in a seaway is a complex process. The seaway is assumed to be unidirectional, resulting from the linear superposition of elementary sinusoidal waves with random phase angles, the amplitudes of which are determined from the International Ship Structure Congress (ISSC) energy spectrum. The spectrum is given by:

$$\Phi_{\epsilon\epsilon}(w) = A \exp(-B/w^4)\,/w^5 \qquad (12.12)$$

where $A = 172h_{1/3}^2T^4$ and $B = 691/T^4$

and $h_{1/3}$ and T denote the significant wave height and average wave period respectively.

The wave forces on a moving ship are however, dependent upon the wave encounter spectrum and the ship's receptance function $H(w,\beta)$. The receptance function is itself dependent upon the wave encounter angle, β and is defined as the transfer function between the input wave signal and the output yawing moment on the ship. Thus, the spectrum representing the wave induced yawing motion can be written as:

$$\Phi_{nn} = |H(w,\beta)|^2\Phi_{\epsilon\epsilon}\,(dw/dw_e) \qquad (12.13)$$

where w_e is the wave encounter frequency and is given by:

$$w_e = w - w^2\bar{u}g\cos\bar{\mu} \qquad (12.14)$$

and $\bar{u}$ and $\bar{\mu}$ are the mean values of the ship speed and wave direction relative to the ship respectively. The acceleration due to gravity is denoted by g.

Studies carried out by Reid et al (1982) show that Φ_{nn} can be represented for most sea states and encounter angles by a second order transfer function driven by white noise. Hence,

$$\Phi_{nn} = Y_n(q^{-1}).Y_n(q) \qquad (12.15)$$

where $Y_n(q^{-1}) = W_n(q^{-1})\epsilon_n \qquad (12.16)$

and $W_n(q^{-1})$ is a rational transfer function of the form:

$$W_n(q^{-1}) = \frac{C_2(q^{-1})}{A_n(q^{-1})} = \frac{c_{21}q^{-1}}{1 + a_{n1}q^{-1} + a_{n2}q^{-2}} \qquad (12.17)$$

and ϵ_n is a unity variance white noise source.

There is some debate in the literature regarding the point at which the wave model should be represented in the system. An output wave model is employed here (Fig. 6.1) since the control objective during course-keeping will be to minimise variations in the output signal y rather than the observation signal, z. That is, low frequency variations in heading are to be limited and this requires the use of the disturbance model structure used here.

12.2.3 The wind model

The wind forces on a vessel are usually separated into two components: mean wind and turbulence. The mean wind excitement, although random and non-stationary, gives steady forces over periods much longer than the ship's time constants. The turbulence can be modelled as white noise with intensity proportional to the wind speed (Kallstrom,1979). Hence the wind-induced yawing motion can be simulated as a slowly varying force of the form:

$$Y_d = N_m \sin 2\sigma (1 + \epsilon_d) \tag{12.18}$$

where N_m is a random variable representing the maximum wind amplitude, σ is the relative wind angle and ϵ_d is white noise of zero mean and unity variance. However, for control design purposes the following transfer function will be suffice to approximate the wind disturbance:

$$Y_d = W_d(q^{-1})\epsilon_d \tag{12.19}$$

where

$$W_d(q^{-1}) = \frac{C_1(q^{-1})}{A_d(q^{-1})} = \frac{c_1}{1 - q^{-1}} \tag{12.20}$$

12.3 THE COST CRITERION

The cost criterion proposed for autopilot designs are normally based on the concept of "added resistance due to steering". This is derived form the equation of motion in the surge direction (equation 1) by noting that for a steady speed the drag force on the vessel due to the ship's forwrd speed, $X_1 u^2$ must be cancelled by the thrust produced by the propeller, X_t ,i.e. $X_1 u^2 = - X_t$. Approximating the remaining drag forces to be minimised by those due to the low frequency rudder induced motions, the average added resistance force can be written as (Grimble et al, 1984):

$$J = \overline{\Delta X} = \lim_{T \to \infty} \frac{1}{2T} E\{ \int_{-T}^{T} (X_2 v_1 r_1 + X_3 \delta^2) dt \} \tag{12.21}$$

where the r^2 term in (12.1) is neglected. The low frequency sway and yaw velocities are denoted by v_1 and r_1 respectively.

This approximation can be justified by noting that the high frequency (HF) ship motions are filtered by the ship's dynamics and hence the contributions of the external high frequency forces to the sway and yaw motions are often small in comparison with the effects of the rudder variations. Further justification comes from the observation that any attempt to minimise HF drag forces will require HF rudder variations which will cause wear and tear in the steering gear.

Unfortunately the criterion (12.21) is not non-negative definite and cannot be used in a standard LQG control design. Note that X_2 is negative and X_3 is positive. This cost function has been used in an approximate form by many authors (Grimble et al, 1984). A common approach was to obtain a positive semi-definite cost index by using the approximation:

$$v = -kr \ , \quad k \leq 0 \tag{12.22}$$

in the integrand of (12.15). However, it is very difficult to find an adequate representative value of k. In general k is frequency dependent and cannot be approximated satisfactorily by a constant. The alternative approach taken here is to derive a dynamic criterion using the relation (12.10). Transforming (12.15) into the frequency domain and using (12.10), the dynamic cost function can be written as:

$$J = \frac{1}{2\pi j} \oint_{|z|=1} (F(q)F(q^{-1})\Phi_{\psi_1\psi_1} + X_3\Phi_{\delta\delta}) \frac{dz}{z} \tag{12.23}$$

where $F(q^{-1})$ is a Hurwitz spectral factor defined using:

$$F(q)F(q^{-1}) = \frac{kX_2[(\theta_r+\theta_v)q^{-1}+(1+\theta_r\theta_v)+(\theta_r+\theta_v)q](1-q^{-1})(1-q)}{(1+\theta_r q^{-1})(1+\theta_r q)} \tag{12.24}$$

The low frequency heading angle deviations ψ_1 are measured relative to an origin in earth co-ordinates. Hence, additional terms may be introduced into the cost function to directly cost the elongation of distance travelled. An extended version of the cost function can be written as (Katebi et al,1985):

$$J = \frac{1}{2\pi j} \oint_{|z|=1} ([P + F(q)F(q^{-1}) + \frac{I}{(1-q)(1-q^{-1})}]\Phi_{\psi_1\psi_1} + X_3\Phi_{\delta\delta}) \frac{dz}{z} \tag{12.25}$$

The scalar term, P provides a mechanism to penalise the added resistance due to steering against tight heading control while the second term $I/(1-q)(1-q^{-1})$ introduces

integral action into the controller.
The scalar quantity X_3 is represented in polynomial form by:

$$X_3 = \frac{B_r(q^{-1})B_r(q)}{A_r(q^{-1})A_r(q)} \tag{12.26}$$

where $B_r(0) = \sqrt{X_3}$ and $A_r(0) = 1$.

12.4 CONTROL PHILOSOPHY

Using equations (12.6),(12.7),(12.25) and (12.26) the optimal control problem now becomes the minimisation of the following cost function:

$$J = \frac{1}{2\pi j}\oint_{|z|=1} \left(\frac{B_q(q^{-1})B_q(q)}{A_q(q^{-1})A_q(q)}\Phi_{yy} + \frac{B_r(q^{-1})B_r(q)}{A_r(q^{-1})A_r(q)}\Phi_{uu}\right)\frac{dz}{z} \tag{12.27}$$

subject to the following dynamics (Fig. 6.1)

Ship:	$y_1(t) = A^{-1}(q^{-1})\, B(q^{-1})\ u(t)$	(12.28)
Wind:	$y_d(t) = A_d^{-1}(q^{-1})C_1(q^{-1})\ \epsilon_d(t)$	(12.29)
Waves:	$y_n(t) = A_n^{-1}(q^{-1})C_2(q^{-1})\ \epsilon_n(t)$	(12.30)
Reference:	$r(t) = A_d^{-1}(q^{-1})\ E(q^{-1})\ \epsilon_r(t)$	(12.31)
Observations:	$z(t) = y_1(t) + y_d(t) + y_n(t)$	(12.32)
Output:	$y(t) = y_1(t) + y_d(t)$	(12.33)

Define the strictly Hurwitz control spectral factor, D_c and filter spectral factor, D_f using:

$$D_c^*D_c = B_2^*B_2 + A_2^*A_2\ ;\quad B_2 = B_qA_rB\ ;\ A_2 = B_rA_qA \tag{12.34}$$

$$D_f^*D_f = [(E^*E + C_1^*C_1)A_n^*A_n + C_2^*C_2A_d^*A_d]A^*A \tag{12.35}$$

The following diophantine equations must be solved for the minimal degree solution (H_o,G_o,F_o) with respect to F:

$$\bar{D}_cG + FA_3 = \bar{B}_2B_qD_f \tag{12.36}$$

$$\bar{D}_cH - FB_3 = \bar{A}_2B_rD_f \tag{12.37}$$

where $g = \max(n_{d_c}, n_{b_2}, n_{a_2})$

$$\bar{D}_c = D_c^*q^{-g}\ ;\ \bar{B}_2 = B_2^*q^{-g}\ ;\ \bar{A}_2 = A_2^*q^{-g}$$

and $A_3 = A_dA_nA_qA$, $B_3 = A_dA_nA_rB$. A solution must also be obtained for the optimal minimal degrees solution (L_o, P_o) with respect to P of the equation:

$$\bar{D}_{fc}L + PA_4 = \bar{B}_4 B_q C_2 \qquad (12.38)$$

where $h = \max(n_{d_{fc}}, n_{b_4})$, $\bar{D}_{fc} = D_f^* D_c^* q^{-h}$, $\bar{B}_4 = B_4^* q^{-h}$

and $A_4 = A_n A_q$, $B_4 = B_2 C_2 AA_d$.

The optimal LQG controller follows as:

$$C_o = \frac{G_o A_r - (L_o A_d A_r)A}{H_o A_q + (L_o A_d A_r)B} \qquad (12.39)$$

Stability of the closed-loop control system is determined by the implied diophantine equation which follows from (12.36) and (12.37):

$$AA_q H_o + BA_r G_o = D_c D_f \qquad (12.40)$$

and represents the closed-loop characteristic equation.

A detailed analysis of the solution of the above optimal control problem is given by Grimble (1986).

12.5 ADAPTIVE CONTROL PHILOSOPHY

Adaptation can provide several benefits in autopilot design. It can account for changes in the ship's speed and loading conditions to ensure that the appropriate drag forces are minimised and it can also prevent the rudder responding to changes in the seaway environment. Two different explicit adaptive schemes are proposed here to identify variations in the ship model parameters and wave model respectively.

The changes in the ship model parameters when sailing in open seas are mostly due to changes in the ship's speed. Velocity scheduling of the low frequency ship model can therefore be employed. This dependence on ship speed, U can be described by the following (Kallstrom, 1979):

$$\begin{bmatrix} \dot{v} \\ \dot{r} \end{bmatrix} = \begin{bmatrix} a'_{11}\cdot\frac{U}{L} & a'_{12}\cdot U \\ a'_{21}\cdot\frac{U}{L^2} & a'_{22}\cdot\frac{U}{L} \end{bmatrix} \begin{bmatrix} v \\ r \end{bmatrix} + \begin{bmatrix} b'_{11}\cdot\frac{U^2}{L} \\ b'_{21}\cdot\frac{U^2}{L^2} \end{bmatrix} [\,\delta\,] \qquad (12.41)$$

where L is the length between the ship's perpendiculars.

The parameters of the cost function (12.27) and the ship model (12.28) can then be computed off-line using the above relationship. This gain scheduling scheme has two advantages. Firstly, it reduces the computational load by performing a number of calculations offline and secondly reduces estimation errors. Alternatively to account for variations in ship loading conditions the low frequency model parameters could be identified using the identification scheme proposed in the next section. This identification would only need to be carried out at the start of a voyage and the parameters found could be used as the base values for the velocity scheduling.

The wave model $W_n(q^{-1})$ is not so predictable and must be estimated continuously to account for changes in the sea spectrum and wave encounter angle. Hence, the wave model is identified on line to take into account all the variations in the sea wave environment.

The solution to the optimal control problem requires the ship model and cost function from the velocity scheduling scheme together with the disturbance models from the estimation procedure. To compute the optimal LQG controller one spectral factorisation must be performed (the other spactral factor is identified in the parameter estimation), and three diophantine equations must be solved. Due to the complexity of the algorithm, the computational requirements to generate the controller every sampling interval are very large and might in a few cases be prohibitive. In this case, the dynamics of the ship and the related seaway environment vary relatively slowly. Thus it is unnecessary and a waste of computing power to update the controller every sample. This has economic benefits as savings in computing power can be translated to possible savings in computer hardware. The controller is thus calculated and updated periodically.

The following discusses the implementation steps taken for the identification/control update.

1) The initial tuning-in phase involves applying the optimal control, without the identified filter spectral factor, and identifying the wave model parameters in the background. Once one update period has elapsed the parameters should have converged and the estimated polynomials can be included in the next controller calculation. This simple procedure ensures that the possibly detrimental effect of the initial parameter tuning-in phase on the ship's heading is minimised.

2) When the initial set of converged model parameters have been obtained the optimal controller is calculated and fixed. When the quality of the control begins to decrease, the control and identification update procedure can then be continued. In the example shown here after the initial controller was identified and fixed the identification of the disturbance models was continued in the background. Every 200 samples the controller was updated using these parameters.

3) Calculating the controller in this periodical manner enables one of the implementational problems of parameter estimation schemes to be avoided. Since the identification scheme does not need to produce a converged set of parameters until 200 samples the parameter covariance matrix can be re-initialised every controller update. This ensures that the estimation algorithm will always be able to track changes in the disturbance spectrum and hence prevents the

algorithm from going to 'sleep'.

The proposed adaptive scheme is shown in Fig. 6.2.

12.6 IDENTIFICATION SCHEME

With an explicit self-tuning scheme the plant and disturbance models are identified separately and the estimates are then used to generate the controller polynomials. Hence, the first problem considered here is to construct an identification scheme from which the unknown subsystems can be computed.

The estimation equation, which will be used as the basis of this scheme, is expressed in terms of an innovations signal with a unit spectral density white noise source e(t):

$$e_o(t) = \frac{C}{D} e(t) - \frac{B}{A} u(t) \qquad (12.42)$$

where $C^*C = [(E^*E + C_1^*C_1)A_n^*A_n + C_2^*C_2A_d^*A_d]$ and $D = A_nA_d$

and $D_f^*D_f = C^*C.A^*A$

The identification scheme used here is based on a version of the "Recursive Prediction Error Method" (RPEM) described by Ljung and Soderstrom (1983). The method can be summarised as follows.

Let,

$$\underline{\theta} = \begin{bmatrix} \underline{\theta}_1 \\ \underline{\theta}_2 \end{bmatrix} \; ; \; \underline{x} = \begin{bmatrix} \underline{x}_1 \\ \underline{x}_2 \end{bmatrix} \; ; \; \underline{y} = \begin{bmatrix} \underline{y}_1 \\ \underline{y}_2 \end{bmatrix}$$

For $A(q^{-1})$, $B(q^{-1})$ estimation:

$$w(t) = \frac{B}{A}(-u(t)) \qquad (12.43)$$

$$\underline{x}_1(t) = [-w(t-1),\ldots,-w(t-n_a);u(t-1),\ldots,u(t-n_b)]^T \qquad (12.44)$$

$$\underline{y}_1(t) = (D/AC)\underline{x}_1(t) \qquad (12.45)$$

$$\underline{\theta}_1(t) = [a_1,\ldots,a_{n_a} ; b_1,\ldots,b_{n_b}]^T \qquad (12.46)$$

For $C(q^{-1})$, $D(q^{-1})$ estimation:

$$v(t) = \frac{C}{D} e(t) = e_o(t) - w(t) \qquad (12.47)$$

$$\underline{x}_2(t) = [-v(t-1),\ldots,-v(t-n_d);s(t-1),\ldots,s(t-n_c)]^T \qquad (12.48)$$

$$s(t) = v(t) - \underline{x}_2^T(t)\underline{\theta}_2(t) \qquad (12.49)$$

$$\underline{y}_2(t) = (1/C)\underline{x}_2(t) \qquad (12.50)$$

$$\underline{\theta}_2(t) = [d_1,\ldots,d_{n_d} ; c_1,\ldots,c_{n_c}]^T \qquad (12.51)$$

$$\epsilon(t) = e_o(t) - \underline{x}^T(t)\underline{\theta}(t-1) \qquad (12.52)$$

The estimator and the gain algorithm can now be written as:

$$\underline{\theta}_1(t) = \underline{\theta}_1(t-1) + R_1^{-1}(t-1)\underline{y}_1(t)\epsilon(t) \quad (12.53)$$

$$R_1(t) = R_1(t-1) + [\underline{y}_1(t)\underline{y}_1^T(t) - R_1(t-1)] \quad (12.54)$$

$$\underline{\theta}_2(t) = \underline{\theta}_2(t-1) + R_2^{-1}(t-1)\underline{y}_2(t)\epsilon(t) \quad (12.55)$$

$$R_2(t) = R_2(t-1) + [\underline{y}_2(t)\underline{y}_2^T(t) - R_2(t-1)] \quad (12.56)$$

It is interesting to note that the only interconnection between the identification scheme for the ship and the disturbance models is through the prediction error, $\epsilon(t)$. This method is the most suitable for the adaptive scheme proposed, since either $\underline{\theta}_1$ or $\underline{\theta}_2$ or both $\underline{\theta}_1$ and $\underline{\theta}_2$ can be updated using (12.53) and (12.55) respectively.

The identification scheme can thus be very easily changed from identification of the ship parameters to identification of the disturbances and vice versa. The algorithm is in fact a type of instrumental varaiable method (Young,1976).

Since both the wind and wave disturbances are present, some knowledge about one of these must be available, because the identification algorithm cannot separate the disturbance effects. Therefore, to identify the wave model, the wind model is assumed to represent a very low frequency model and the reference polynomial $E(q^{-1})$ is set to zero. Identification of the $D(q^{-1})$ polynomial gives the denominator of the wave model $A_n(q^{-1})$ since $A_d(q^{-1})$ is assumed known. The wave model numerator polynomial $C_2(q^{-1})$ can be calculated from:

$$C^*C = C_1C_1^*A_nA_n^* + C_2C_2^*A_dA_d^*$$

Further, if the $C_2(q^{-1})$ polynomial is a scalar quantity its value can be found from the following:

$$c_2c_2^* = \left.\frac{C^*C}{A_dA_d^*}\right| \text{ at zeros of } A_nA_n^*$$

12.7 SIMULATION RESULTS

A computer simulation package (Byrne and Katebi,1986) was used to examine the course-keeping performance of the proposed adaptive autopilot. The ship and the associated disturbances were modelled as described in Section 12.2 with numerical values appropriate to a roll-on roll-off passenger ferry. The values used in the simulations are shown in Table 6.1.

The simulation results presented will consider three situations. The first will demonstrate the initial tuning-in phase of the controller when there is no previous knowledge of the disturbances available to the controller. The second test will show how the controller performs when the parameters identified from the tuning-in period are included in the controller. In the third simulation the effects of continuously updating the controller every sample time will be investigated.

In each of the simulations the cost function of equation (12.25) will be used as the basis for controller design. The values used in the tests are shown in Table 1. The discrete-time dynamic cost function polynomials are as follows:

$$B_q(q^{-1}) = 4.98658 - 11.8623q^{-1} + 9.63392q^{-2} - 2.62223q^{-3}$$

$$A_q(q^{-1}) = 1 - 1.66010q^{-1} + 0.66010q^{-2}$$

$$B_r(q^{-1}) = 3.0$$

$$A_r(q^{-1}) = 1.0$$

The first simulation run demonstrates the importance of correctly modelling the disturbancs acting on the ship. In this experiment the controller has no previous knowledge of the disturbances so the controller calculated assumes that the wave model can be modelled by white noise,

$$\text{i.e.} \quad \hat{C}_2(q^{-1}) = \hat{A}_n(q^{-1}) = 1.$$

The optimal controller computed is as follows:

$$C_o(q^{-1}) = \frac{-0.49622 + 1.42061q^{-1} - 1.48821q^{-2} + 0.67279q^{-3} - 0.11004q^{-4}}{1 - 2.78561q^{-1} + 2.81014q^{-2} - 1.21040q^{-3} + 0.18597q^{-4}}$$

The plot of rudder angle with this controller (Fig. 6.3) shows that the high frequency wave components are being fed back through the controller to the rudder angle. Such large oscillations on the rudder command signal would cause excessive wear and tear of the rudder's hydraulic system.

During this simulation the parameter identification scheme identified the following polynomials (Fig. 6.6):

$$\hat{C}\ (q^{-1}) = (2.5870 - 2.8551q^{-1} + 1.0341q^{-2})\text{x}10^{-2}$$

$$\hat{A}_n(q^{-1}) = 1 - 0.5998q^{-1} + 0.7855q^{-2}$$

The estimate for the wave model numerator polynomial was calculated at the end of the simulation run as:

$$\hat{C}_2(q^{-1}) = 0.0190q^{-1}$$

These polynomials were then included in the computation of the optimal controller to give:

$$C_o(q^{-1}) = \frac{-0.25717 + 1.04667q^{-1} - 2.07160q^{-2} + 2.507188q^{-3} - 1.85856q^{-4} + 0.78501q^{-5} - 0.16459q^{-6} + 0.01223q^{-7}}{1 - 3.86287q^{-1} + 6.27971q^{-2} - 5.50817q^{-3} + 2.72532q^{-4} - 0.70376q^{-5} + 0.06842q^{-6} + 0.00134q^{-7}}$$

This controller was then used in a simulation trial which replicated the first. The plots of this second experiment (Fig. 6.4) show a considerable reduction in the rudder activity. The graph of rudder angle shows that the high frequency components have been successfully removed by the controller's filtering action leaving the necessary low frequency control action to compensate for the effects of the wind disturbance. The plot of yaw angle shows that there is no degradation in the output response.

The improvement in the noise rejection properties of the autopilot can be explained graphically by comparing the autopilot's frequency response before and after tuning-in. The solid line in (Fig. 6.7) shows the amplitude versus frequency plot for the controller after the disturbance models have been identified while the broken line shows the response of the controller prior to tuning-in. The graph shows that the identified models allow the controller response to be shaped in the frequency range of interest. The post tuning-in controller has an attenuation of -16dB at 0.3rad/sec compared to approximately 0dB for the original controller. Note the response of each controller falls initially by -20dB/decade indicating the presence of integral action.

The third simulation demonstrates the one of the disadvantages of continuously updating the controller every sampling period. The plot of heading angle (Fig. 6.5) shows that the system has a very large overshoot. This is due principally to the transient response of the parameter estimates. The estimated C polynomial is used to calculate the filter spectral factor, D_f = CA, and hence the parameters identified directly influence the poles of the closed loop system. This transient behaviour may be acceptable in industrial processes such as paper-making where the maximum damage likely to be incurred is paper jammed between rollers. However, the effects of such a large overshoot in a ship's heading ,especially when in the vicinity of other ships, could be potentially disastrous.

12.8 CONCLUSIONS

Some of the difficulties in providing adequate course-keeping performance by conventional control have been overcome by the development of an explicit adaptive LQG design. The new scheme has several interesting features. An output wave model has been adopted so that only low frequency heading deviations were minimised. A dynamic cost function weighting has been introduced to enable the added resistance due to steering to be minimised. The polynomial systems approach to linear quadratic control provides a neat solution to the optimal control problem.

The proposed autopilot has been used in a simulation trial to demonstrate the improvements that can be attained in the disturbance rejection properties.

REFERENCES

Amerongen, J. van (1982). Adaptive steering of ships. PhD Thesis, Husidrukkerij, Delft University of Technology, The Netherlands.

Astrom, K.J. and B. Wittenmark (1973).On self-tuning regulators. Automatica,9,185-199.

Blanke, M. (1981). Ship propulsion losses related to automatic steering and prime mover control. PhD Thesis, Technical University of Denmark.

Byrne J. and M.R. Katebi (1986). Application of LQG polynomial approach to the design of a fin roll stabilisation control system. SERC Vacation School on Control System Design- Robustness and Polynomial Methods, University of Strathclyde, Glasgow, U.K.

Clarke, D. (1982). Do autopilots save fuel. Trans IMarE (c),Vol. 94.

Clarke, D.W. and P.J. Gawthorp (1975). Self-tuning controller. Proc. IEE, Vol. 122,Pt-D.

Grimble, M.J. (1986). Controllers for LQG self-tuning applications with coloured measurement noise and dynamic costing. Proc. IEE, Vol.133, Pt-D.

Grimble, M.J., M.R. Katebi and J. Wilkie (1984). Ship steering modelling and control design. Proc. 7th Ship Control Systems Symposium, Bath, U.K.

Kallstrom, C.G., K.J. Astrom, N.E. Thorell, J. Eriksson and L. Sten (1979). Adaptive autopilots for tankers. Automatica,15,241-254.

Kallstrom, C.G. (1979). Identification and adaptive control applied to ship steering. PhD Thesis, Lund Institute of Technology, Sweden.

Katebi, M.R., M.J. Grimble and J. Byrne (1985). LQG adaptive autopilot design. Proc. 7th IFAC Symposium on Identification and System Parameter Estimation, York, U.K.

Ljung, L. and T. Soderstrom (1983). Theory and practice of recursive identification. MIT Press.

Lim, C.C. and W. Forsythe (1983). Autopilot for ship control. Proc. IEE, Vol.130, Pt.D,No.6.

Merlo, P. and A. Tiano (1975). Experiments about computer controller ship steering. Semana International Sobre la automatica en la Marina, Barcelona, Spain.

Oldenburg, J. (1975). Experiments with a new adaptive autopilot intended for controlled turns as well as for straight course keeping. Proc. 4th Ship Control Systems Symposium, The Hague, The Netherlands.

Reid, R.E. (1982). Identification and Minimisation of Propulsion Losses Related to Ship Steering. Report No. MA-RD-940-82038, University of Illinois at Urbana-Champaign, U.S.A.

Reid, R.E., A.K. Tugcu and B.C. Mears (1983). The use of

wave filter design in Kalman Filter state estimation of the automatic steering problem of a tanker in a seaway. IEEE Trans. on Automatic Control, Vol Ac-29, No. 7.
Saelid, S. and N.A. Jenssen (1982). Adaptive autopilot with wave filter. Proc. 6th IFAC Symposium on Identification and System Parameter Estimation, Washighton D.C., U.S.A.
Sugimoto, A. and T. Kojima (1978). A new autopilot system with condition adaptivity. Proc. 5th Ship Control Systems Symposium, Annapolis, Maryland, U.S.A.
Young, P.C. (1976). Some observations on instrumental variable methods of time series analysis. Int. Jour. of Control,23.

TABLE 6.1 Simulation Data

Continuous-Time Ship Model Parameters		
State-space	Laplace Transfer Function	Cost Function
$a_{11} = -0.0344$	$k_r = -0.131412$	$U = 10.8$ m/sec
$a_{12} = 0.5230$	$\tau_r = 9.63$	$L = 150.0$ m
$a_{21} = -0.0022$	$T_1 = 24.47$	$X_2 = -0.5$
$a_{22} = -0.2193$	$T_2 = 4.698$	$X_3 = 9.0$
$b_{11} = 0.3470$	$k_v = -61.61$	$P = 0.10$
$b_{21} = -0.0111$	$\tau_v = 4.933$	$I = 0.01$
Discrete-Time Polynomials Sampling Period = 4 seconds		
Ship	$B(q^{-1}) = -0.07341q^{-1} - 0.01182q^{-2} + 0.03979q^{-3}$	
	$A(q^{-1}) = 1 - 2.27600q^{-1} + 1.63844q^{-2} - 0.36244q^{-3}$	
Wind	$C_1(q^{-1}) = 0.02$	
	$A_d(q^{-1}) = 1 - q^{-1}$	
Waves	$C_2(q^{-1}) = 0.06q^{-1} - 0.06q^{-2}$	
	$A_n(q^{-1}) = 1 - 0.65270q^{-1} + 0.78662q^{-2}$	

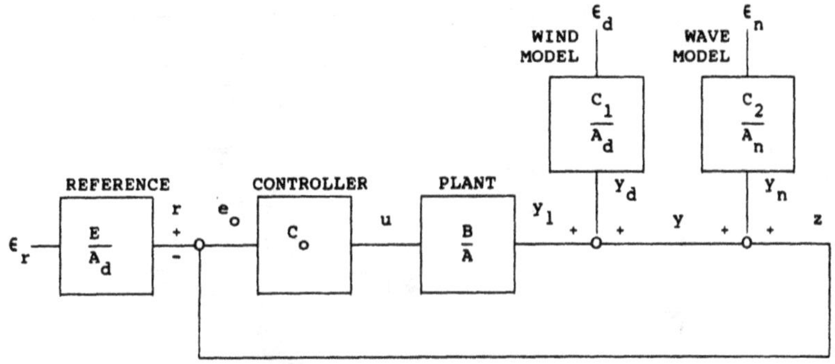

Fig. 6.1 Closed Loop System Description

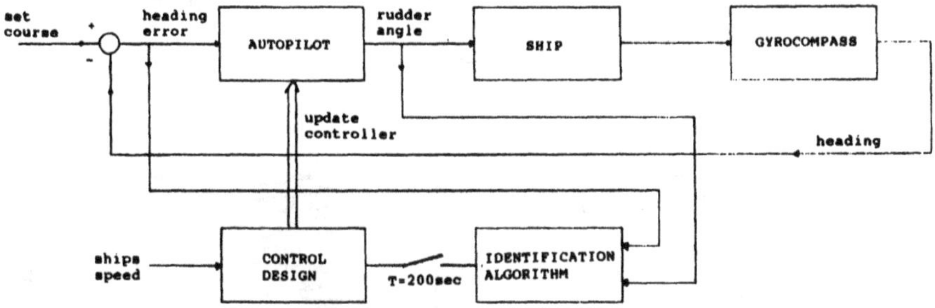

Fig 6.2 Adaptive Autopilot Schematic Diagram

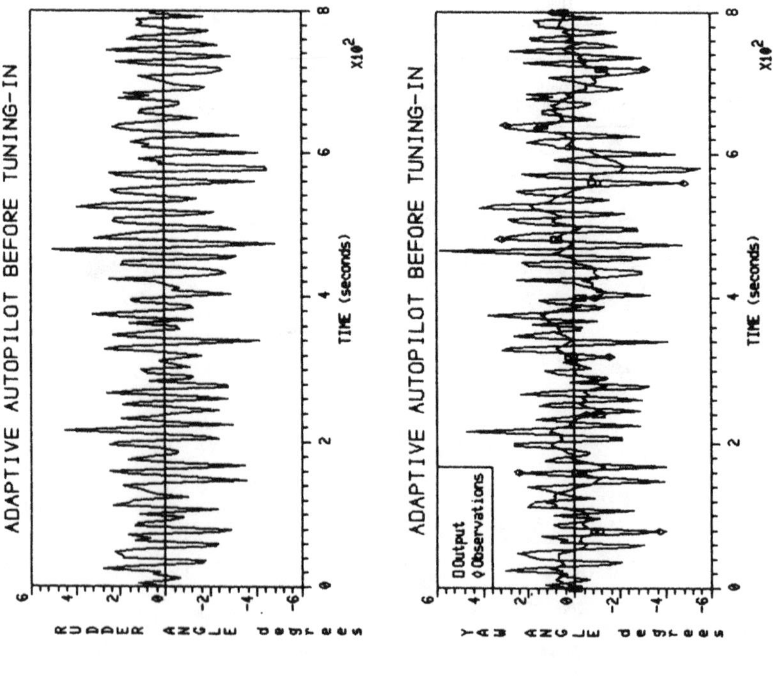

Fig. 6.3 System Responses Before Tuning-In

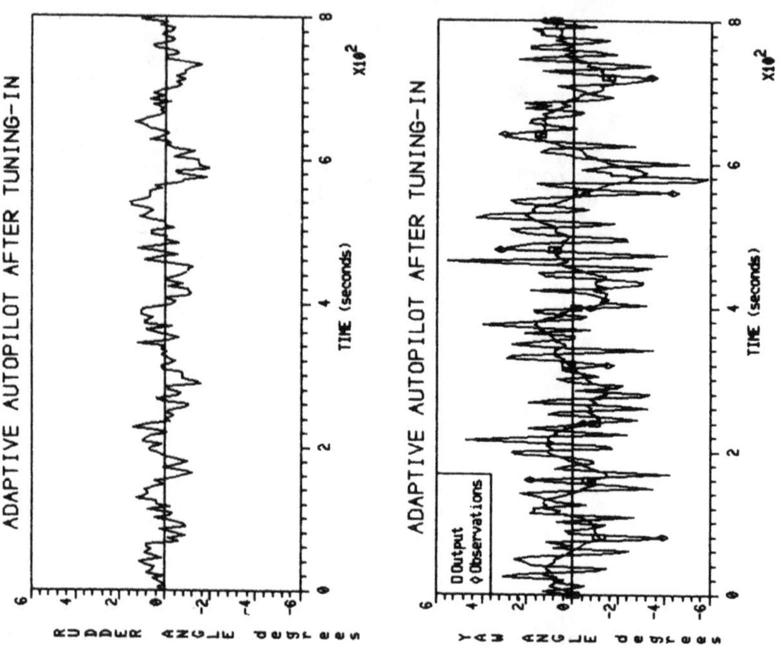

Fig. 6.4 System Responses After Tuning-In

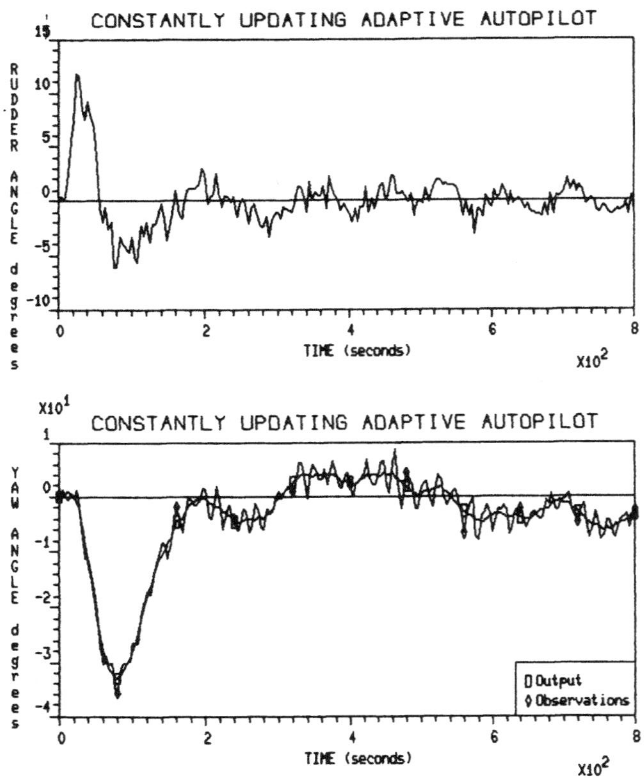

Fig. 6.5 System Responses When Constantly Updating Controller

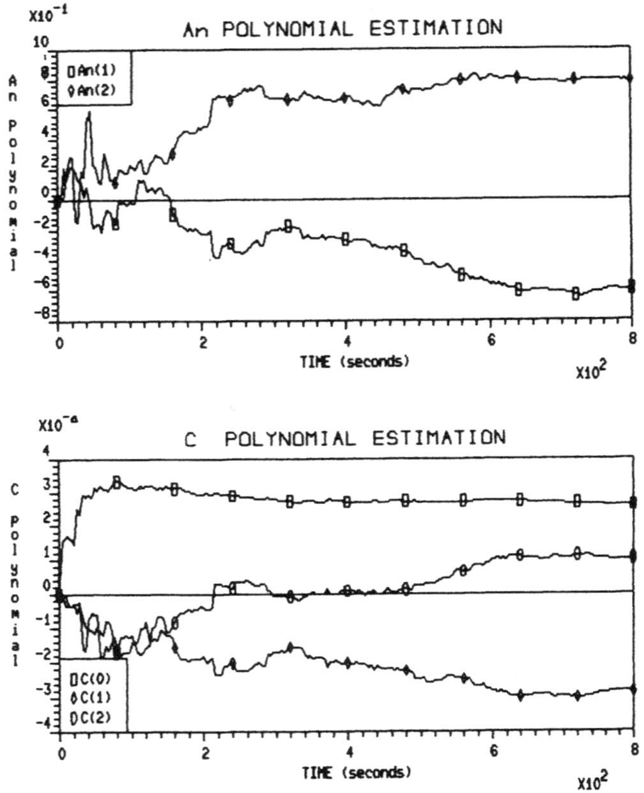

Fig. 6.6 Parameter Identification

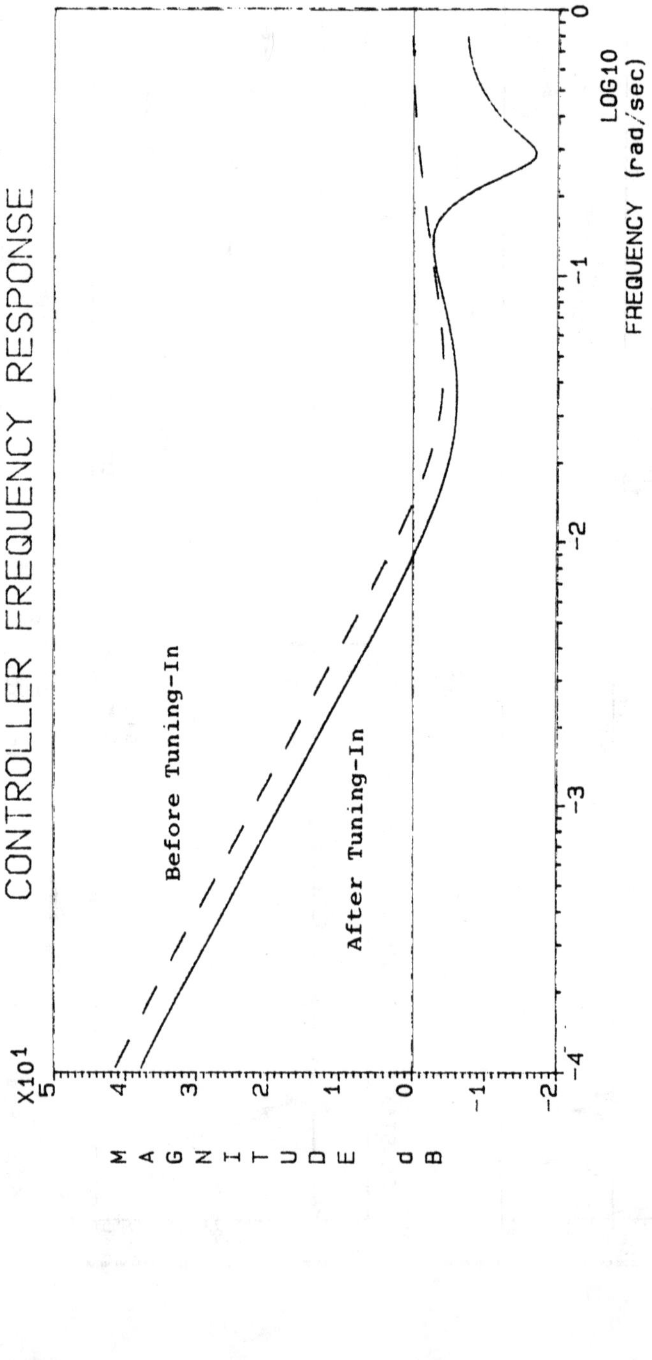

Fig. 6.7 Controller Frequency Response

Chapter 13

Self-tuning control: case studies

D. Peel and M. Tham

13.1 Introduction

Since the re-emergence of self-tuning control in the early 1970's when implementation of such algorithms became a realistic goal, a number of theoretical developments and practical applications have been reported which represent milestones in the field. Useful reviews highlighting both the progress made in theory and practice have been presented by Parks et al (1981), Astrom et al (1983), Seborg et al (1983) and Unbehauen (1985).

Self-tuning controllers were originally developed in the SISO (single-input, single-output) environment where they have been found to lead to noticable improvements in the length of the controller commissioning period. However, improvements in subsequent closed-loop performance may only be possible at the expense of an unacceptable increase in control activity. The control of SISO processes which face problems such as time-varying process dynamics or significant time-delays can be improved by the use of the correct self-tuning techniques as described in the following studies.

Applications of self-tuning control algorithms are becoming more widespread due to the development of a number of commercial self-tuning PID controllers. These controllers offer the advantage that their 'tuned' parameters can be easily interpreted by industrial control engineers familiar with the performance of fixed parameter PID controllers. However their main use has been as commissioning aids for existing PID controllers, i.e. the on-line tuned controller settings are transferred directly to the resident PID controller. Although this is a reasonable use of self-tuning PIDs, it is possible that alternative self-tuning strategies which possess more flexible control laws may enable still further improvements in closed-loop performance. It is the application of these controllers which is considered here, and this chapter is aimed primarily at industrialists who wish to gauge the performance of such self-tuning controllers in practical situations. The applications which are cited have been chosen to highlight a range of practical benefits which can be realised using self-tuning techniques.

13.2 Cement Mill (Al-Assaf, 1987; Al-Assaf et al 1987)

The control problem associated with the cement mill is primarily that of cement temperature regulation. A secondary requirement is to achieve acceptable mill start-up. The cement temperature is measured as the product exits the mill but before it is 'cut' (separation of 'fines' for product and 'coarse' fractions which form the recycle). The manipulative variable is the flowrate of cold water which is atomised and sprayed into the back-end of the mill (counter-current to the flow of the cement), see figure 1.

The regulation of cement temperature has important repercussions on product quality. If the temperature of the cement drops below a given lower limit. (10 deg. C. less than set-point), one of the interlock mechanisms of the control system will automatically stop the pump which normally delivers the cooling water to the mill. This fail-safe is necessary in order to prevent the cement from becoming wet (during normal operation the water is completely evaporated) and setting prematurely either in the mill or in subsequent storage. Alternatively, as the temperature of the cement increases, the action of the mill is such that the cement grinding operation results in the production of a greater amount of dust. Normally this dust is removed by static precipitators but at high loads this dust can escape into the atmosphere. In addition, the gypsum which is added to the clinker feed as a set-regulator can lose its water of crystallisation at high temperatures resulting in severe changes in the properties of the product cement.

The implementation of a SISO structured, fixed parameter GPC (generalized predictive control - Clarke et al (1987a), (1987b)), was expected to perform the cement mill control task and to reduce the variance of the outlet cement temperature during regulation. It was also expected to achieve mill start-up both safely and with a reduced overshoot compared with that of the resident PI controller. In addition, the flexibility of the control law structure would allow changes in feedrate or feed temperature, returns rate or returns temperature to be accommodated as measurable feedforward disturbances within the controller.

The results which are shown in figures 2a and 3a demonstrate the usefulness of a predictive control law in achieving the desired control tasks. However, in comparison with those results obtained using the resident PI controller shown in figures 2b and 3b it is clear that there is little advantage to be gained by using predictive control purely on the basis of performing the regulatory task. However, the comparison of process start-ups using both controllers are more favourable towards the fixed-parameter self- tuner. The result achieved using the PI controller is characterised by a large overshoot (part of which is above the recommended upper limit) and a settling time of 45 minutes. The result achieved with the self-tuning algorithm, on the other hand, demonstrates a considerably reduced settling time of about 10 minutes and no overshoot.

In this application, the GPC was found to successfully achieve both control tasks. Although the regulatory task is carried out with similar degrees of success for both the GPC and the resident PI controller, mill start-up was greatly improved using the GPC algorithm. In addition, it was never found necessary to include any of the extra temperature and flowrate signals as feedforward components of the controller in order to improve performance. This is probably due to the fact that these signals seldom vary other than as a result of measurement noise or process problems which normally lead to complete mill stoppages.

13.3 Turbo-Tray Dryer (Peel, 1986; Peel et al 1987)

13.3.1 The Drying Process

Figure 4 shows the turbo-tray dryer which is considered in this investigation. Typically, 30-40 tonnes/hr. of clay with a moisture content of consistently 24-26 % enters the dryer from a conveyor belt onto the top tray which rotates once every 1 minute 45 secs. At the end of each revolution, the clay is transferred off its present shelf and onto the shelf directly below by the action of fixed scrapers. The same mechanism is repeated 26 times (the number of trays in the dryer) until the clay is scraped from the bottom tray onto the dryer floor. The dryer floor is stationary but is scraped by 16 equally spaced arms which direct the product onto the outlet conveyor belt.

Drying is achieved by the forced circulation of the heating medium (hot air) over the clay on the shelves by means of turbo fans resident in the central shaft of the dryer. These fans 'suck' and 'blow' the hot air over alternate sections of shelves thereby ensuring a continuous flow of hot air over the clay as it passes through the dryer. The hot air is generated by the complete combustion of natural gas in an excess of ambient air.

13.3.2 Motivation

At the outset of this study, the resident control scheme consisted of the PI control of a 'middle' gas temperature by manipulation of the flowrate of natural gas to the air heater. Self-tuning techniques were considered for a number of reasons:

1. A controller which could overcome the time-delays and feedforward disturbances associated with the moisture control loop was required.

2. The dryer was frequently stopped and restarted to dry different clay products to different set-points thus requiring the use of a controller that can adapt to changes in process conditions.

3. During the Winter months, the natural gas fuel is occasionally replaced by fuel oil requiring the controller to adapt to the new process dynamics.

13.3.3 The Control Problem

The control objective was to regulate the outlet moisture content to within its specification limits. Product china-clay is marketted at 10 % moisture content with a specification limit of ± 1 %. The major process disturbances which are encountered during operation are changes in the feedrate of the inlet clay, (the moisture content of the inlet clay was found to be approximately constant). Due to the fixed rotation time of the dryer shelves and the plug flow nature of the solids flow through the dryer, the residence time of the solid in the dryer. and consequently the time-delay, remains constant.

13.3.4 Original Control Scheme

The original control scheme which was in use at the beginning of this collaborative project consisted of the PI control of a gas temperature measured approximately half-way up the height of the dryer. The process operators were observed to manually perform additional feedforward compensation for changes in clay flowrate and feedback trim tasks to ensure the steady-state moisture content is correct.

Figure 5 shows a typical period of PI control logged during normal dryer operation. This run shows an initial settling period of the controller as the temperature approaches set-point between 100 and 150 minutes into the run. At 150 minutes, a reduction in clay flowrate occurred and the operator consequently reduced the temperature set-point. This was intuitively the correct procedure as in the same way as a lower clay flowrate requires less energy to achieve the same desired moisture content due to the reduction in drying duty, a lower temperature also requires less energy to be generated. The result of this preventative action can be seen approximately 45 minutes later when the outlet moisture content drops by about 0.5 % which indicates that the corrective action was insufficient in this case. Further corrective action by means of an outlet moisture feedback trim was not carried out by the operator during the remaining period of this run.

13.3.5 Temperature Control

Middle gas temperature has certain attractions which make it a suitable controlled variable. Firstly, the gas phase possess faster dynamics than those of the solid phase. Secondly, temperature measurements can usually be obtained cheaply and easily. However, it is necessary to determine whether control of the middle gas temperature is sufficient to provide adequate control over the outlet moisture content.

Figure 6 shows the result of controlling the temperature of the non-linear model of the process during periods of changing clay feedrate. Although the temperature is shown to be easily regulated during this period, requiring only a small change in steady-state control output, the outlet moisture content is observed to deviate significantly from its previous steady- state value. Figure 7 demonstrates a similar result obtained from the actual dryer. It is clear from these results that the use of temperature feedback alone is insufficient to achieve the necessary regulation of outlet moisture content. It can now be concluded that operator alteration of temperature set-points is necessary to compensate for changes in feedrate and to align the steady-state outlet outlet moisture content with its desired level, i.e. the set-point trim mechanism.

13.3.6 Direct Moisture Control

The obvious way of attempting to control the outlet moisture content of the dryer is to use a feedback of its measurement as the controlled variable. However, moisture content is notoriously difficult to measure and the dynamics relating outlet moisture content to changes in gas flowrate are characterised by relatively large time constants and significant time-delays. In order to achieve reasonable control performance, it is necessary to consider the use of controllers which can be designed to overcome the detrimental effects of time-delays and disturbances due to changes in clay flowrates.

In simulation, using the k-incremental GMV (generalized minimum variance) controller, Clarke et al (1983), it is possible to achieve a high standard of control. Figure 8 displays a typical result of a period of simulation during which the clay feedrate is shown to change step-wise between two typical steady-states. The structure of the k-incremental controller is based on a SISO process model without reference to the change in clay flowrate information. It is seen in the result that the moisture content is controlled with a settling time of almost 4 hrs (based on the second transient period). This run is repeated for the case where the controller is extended to accommodate clay flowrate as a measurable feedforward signal, figure 9. In this simulation, the settling time of the second transient is reduced to about 2 hrs and the magnitude of the initial deviation from set-point is also reduced by a factor of about 0.7.

13.3.7 Summary of Dryer Application

As a result of this study, it was found that the original control scheme based on feedback of middle gas temperature was inappropriate to achieve the control objective of outlet moisture content regulation. However, it was noted that the additional tasks required to achieve this objective were carried out in practice by the operators. The feedback of outlet moisture content to a self-tuning controller which was designed to accommodate the clay flowrate as a feedforward disturbance, and to make predictions over the time-delay, was found to be the most satisfactory way of regulating product moisture content.

13.4 Multivariable Self-Tuning Controllers

Many methods of multivariable parameter adaptive control have been proposed during the past few years. The initial designs were based on minimum variance design objectives, e.g. Borisson (1979), and hence were only applicable to minimum phase systems. The multivariable versions of many of the SISO self-tuning algorithms were also developed: the GMV self-tuning controller of Clarke and Gawthrop (1975), was extended to the multivariable case by Koivo (1980); Clarke and Gawthrop's (1979) algorithm was extended to the multivariable case by Morris et al (1982). Tham (1985) studied the performance of the multivariable k-incremental GMV controller developed for the SISO case by Hodgson (1982). All the above self-tuning control methods employ an implicit approach to control law implementation i.e. the control law parameters are estimated rather than the parameters of a model which is the case in explicit methods. Explicit methods have also been developed which fulfil pole-placement objectives, Prager and Wellstead (1981), and similarly the state-feedback scheme of Shieh et al (1982). Implicit and explicit STCs based on LQG philosophies have also been derived by Grimble (1984). More recently, new developments in SISO self-tuning control, e.g. GPC, have led to multivariable counterparts being developed independently by Montague et al (1986) and Mohtadi et al (1986).

13.4.1 Spray Drying Tower (Lambert E. 1987)

The application of self-tuning control to a pilot scale spray drying process shown in figure 10 has been attempted. In this process, the product slurry to be dried is atomised and sprayed into the top of a spray drying tower where it passes counter-current through a hot air stream which is introduced at the tower base. The control problem under investigation consists of controlling two of the inner loops of the spray dryer, viz. hot air flowrate and tower pressure. These outputs are controlled by the flowrate of inlet air and outlet air respectively. During normal operation, the hot air flow set-point is specified based on the output of the moisture control loop in the complete dryer control scheme. Tower pressure is regulated to minimise leakage and heat losses by ensuring a slightly negative pressure within the dryer tower. The air flows are generated by fixed-capacity fans and modulated by mechanical dampers in the inlet and outlet ducts. The control loops are coupled in one direction, i.e. between the flowrate of the inlet air and the tower pressure. Changes in the inlet air flow affect both the air temperature and the tower pressure. This interaction is particularly undesirable as hot material may be ejected from the dryer if the pressure rises above atmospheric.

A multivariable control strategy was considered to be applicable to the problem of loop interaction. However, as the loops are only coupled in one direction (i.e. there is no coupling between exhaust air flowrate and the hot air flowrate) it is possible to consider the control of the inlet air flow/air temperature loop using a SISO structured controller, and the exhaust air flow/tower pressure loop using a multi-input, single-output (MISO) controller representation. In the latter case, the inlet air flow is considered as an additional, measurable feedforward signal to an otherwise SISO control loop. The process was thought to be non-linear prior to the application as the resident control scheme made use of a gain-scheduled PID controller which switched between a number of controller settings according to the operating range. The control study was carried out using both a full multivariable representation of GPC Mohtadi et al (1986) and a combination of a SISO and MISO GPC controllers.

Practical trials have demonstrated that loop interaction can be reduced by using either a multivariable GPC, (figure 11), or a combination of MISO and SISO GPC's, (figure 12). In addition the application of two SISO GPC controllers also results in adequate performance, (figure 13). The commissioning of the control scheme using any of the previously described structures is both simplified and less time consuming due to the self-tuning mechanism inherent in the controller.

13.4.2 Multivariable-Multirate Self-Tuning Control

Those multivariable STCs described in Section 13.4.0 which are based on a P-canonical process structure (assuming loops are coupled from inputs to outputs) can be operated at different sampling rates for different process loops. The rationale behind this control strategy is that the sampling time for each loop within a multivariable control system should be chosen to suit the dynamic characteristics of individual loops. As a result, improved control performance can be achieved as demonstrated by practical results, Tham et al (1985). The only restriction is that the sampling intervals must be integer multiples of the smallest (base) sampling interval.

13.4.3 Multivariable Self-Tuning Control of a Binary Distillation Column

13.4.3.1 The Distillation Process

The column separates a methanol-water feed mixture of composition 50 wt% methanol (MeOH) introduced at a rate of 18 g/s. Normal operation requires that the top product has a composition of 95 wt% MeOH while the bottom product is to have a composition of 5 wt% MeOH. Reflux and steam flow rates serve as the manipulative inputs for the top composition (XD) and bottom composition (XB) loops respectively. These two loops exhibit a high degree of interaction. In order to achieve tight dual composition control a multivariable control strategy has to be employed. Top composition is measured by a capacitance probe while the bottom composition is determined by an on-line gas chromatograph (GC). The cycle time of the GC of 150 secs, and the time taken by the control program to assimilate bottom product composition data of approximately 20 secs determines the rate of sampling of the bottom loop. A schematic of the column with the major control loops is shown in figure 14.

For this highly interactive distillation column, the most severe disturbances to normal steady-state operation are step changes in feed flow rate. The on-line evaluation tests thus centre on the regulatory problem. Load disturbances were 'created' by having the feed flow rate follow a series of pre-programmed step changes as shown in figure 15.

13.4.3.2 Setting up the Multivariable Self-tuning Controller (MVSTC)

From the open loop responses (see figure 16) it was clear that the distillation process possesses high order non-linear dynamics. Parameterising the process assuming high order models would result in a large number of parameters to estimate requiring a prolonged period of initial parameter tuning. The time taken by parameter estimation algorithms to initially tune-in a set of estimates or to track changes in process conditions increases with the number of parameters. A trade-off therefore exists between incurring inaccuracies due to model-order mismatch and the speed of parameter adaptation. In self-tuning control, however, processes can often be adequately approximated by first or second order models. Here orders of the controller polynomials, were specified by assuming first order plus time-delay relationships between input and output pairs.

13.4.3.3 Conditioning of Control Signals

Control signals sent to the process must be conditioned to ensure safe process operation. The multivariable GMV type controllers permit a weighting, Q_i, to be placed on the control signal of each loop which penalises excessive control effort. Q_i, can be chosen to have the structure of an inverse discretised PI compensator, Morris and colleagues, (1982). This admits the interpretation of the self-tuning closed loop as being similar to a conventional control loop with dead time compensation, Gawthrop, (1977). More importantly, the parameters of the control weighting for each loop can be specified in terms of proportional bands (PB%) and integral times (Ti) which are familiar to industrial users. The design of Q_i as an inverse PI compensator need not, however, consider time-delays since time-delay compensation is provided as a result of using predicted outputs for feedback. Fine tuning of Q_i may be necessary on-line although our experience has indicated that Q_i weightings tuned (as inverse PI compensators) to minimise the IAE criteria, Lopez and colleagues (1967), are usually adequate. In this study, the Q_is were calculated from the linearised distillation column model, shown in figure 17.

The control weightings used in this study are tabulated below in terms of PB% and Ti:

Loop	PB%	Ti(secs)
Top	15.0	80.0
Bottom	50.0	150.0

Table 1. Control Weightings

To prevent valves from sticking at the 'fully closed' or 'fully open' positions, 'hard' limits were imposed on the calculated control signals. Rate limits, which prevent large changes in control signals were also specified. This is important for certain processes which are sensitive to sudden changes in manipulative inputs. In the case of this distillation column for example, sudden changes in steam flowrate can cause the process to 'cycle' due to pressure surges. Since the tuning transients of adaptive controllers are notoriously poor, the imposition of rate limits is also necessary to prevent unstable process behaviour during this phase of operation. These limits are listed below in terms of percentage valve opening:

Manipulative Input	Reflux flow	Steam flow
Minimum (% valve)	5.0	10.0
Maximum (% valve)	90.0	90.0
Max. change (% valve)	20.0	20.0

Table 2. 'Hard' and Rate Limits Imposed on Manipulative Inputs

The limits listed in Table 2 have been chosen conservatively so as not to overly constrain the already 'detuned' control signals which result from the use of control weightings. On being commissioned, the multivariable self-tuning algorithms thus have the freedom to utilise their calculated control signals.

13.4.3.4 Commissioning the MVSTC

The commissioning phase of any adaptive control algorithm, whether in a single or multi-variable configuration, can be considered to be the most crucial stage of the implementation. This is especially true when starting with zero parameter values, and when a large number of parameters are to be estimated, i.e. processes with large time-delays. If an approximate model of the process is available, the controller parameters (which would normally be estimated) can be calculated and used as initial values. The STC can then be commissioned in the closed loop with considerable confidence in the initial estimates. If the process is unknown, STCs have to be commissioned with parameters initialised to zero and with no confidence in the estimates. The concept of estimate confidence is intrinsic to parameter estimation.

There are alternative methods for commissioning STCs for the control of unknown processes. One technique involves the addition of a PRBS signal on the manipulative input when the system is open loop. The difficulty here lies in the choice of a suitable amplitude and frequency for the PRBS signal such that the process will be sufficiently excited to result in parameter adaptation. Another method is to first control the process using conventional control mechanisms such as PID, with the estimator running in the 'background'. This has the advantage that the process remains in closed-loop throughout. However, if the conventional scheme provides reasonably good control, insufficient dynamic information may be generated to allow valid estimation to take place. To overcome this problem, a deliberate disturbance of acceptable magnitude can be introduced to the process. When it is observed that the estimated parameters have reached some steady values, and a reasonable prediction is being generated, the switch over to full self-tuning mode is made.

13.4.3.5 Parameterisation of the MVSTCs

The MVSTCs considered here were started with controller parameters set to zero and were commissioned with the process under conventional multi-loop PI control. The switch over to self-tuning was made when the estimators gave good predictions and when the estimated parameters were relatively stationary. For the multivariable self-tuning studies, the sample rate chosen for both top and bottom composition loops were chosen as three minutes. The positional and k-incremental algorithms will be referred to as PMVSTC and KMVSTC respectively. For the multivariable-multirate evaluations, however, the sample time for the top loop was chosen as one minute while the sample rate for the bottom loop was limited at 3 minutes. The latter is the fastest sampling interval possible due to the use of a gas chromatograph with fixed cycle times.

13.4.3.6 Results of On-line Applications

Firstly, the performance of the multivariable controllers were evaluated when the feedrate was included as a measured feedforward term in the control law calculation, i.e. with feedforward compensation. Figure 18 shows the regulated top and bottom loop responses using the PMVSTC. An initial offset was observed in the bottom loop response. However, the offset was eliminated as the estimator for the bottom loop adjusted to the new operating conditions. The responses of the top and bottom loops under KMVSTC regulation are shown in figure 19. Apart from the absence of offsets, the performance of the KMVSTC was very similar to that obtained by the PMVSTC. With this controller, however, offset removal is guaranteed without controller parameters having to adapt to eliminate the effect of any disturbances.

Next, the performance of the controllers when regulating against unknown disturbances were investigated. Although feed flow was still monitored, the feedforward compensation terms were removed from the control law calculation to simulate conditions of unmeasurable disturbances. Figure 20 shows the PMVSTC regulated top and bottom responses during which there was a prolonged period of offset in the top loop. At the onset of the worst disturbance condition, i.e. feed step down from its normal operating rate (NOR) of 18g/s where inverse response conditions exist, Tham (1985), regulation was rather poor. Although bottom loop performance was poorer than the case with feedforward load compensation there was, somewhat surprisingly, no observed offset. As expected, when the column was regulated using the KMVSTC, there were no offsets in either loop, (figure 21). The KMVSTC however, showed more sensitive behaviour than the PMVSTC especially when the feedrate was reduced from its NOR and when it was brought back to its NOR. Output deviations in both top and bottom loops were, in general, also larger than those observed during the experiment using the PMVSTC. When the disturbance is unknown, the controller cannot begin regulatory action until a change in process output is detected. For this particular feed step, because of the inverse response characteristics, the initial control signal is therefore in a direction opposite to that which should be calculated when the dominant dynamics come into effect. The controller thus had to 'retune' in order to maintain regulation. This retuning resulted in poorer performances, compared to the corresponding cases when the feed disturbance was known and accounted for in the calculation of control. The higher gain of the KMVSTC, due to its integrating properties, further degraded output responses.

Next, multivariable-multirate self-tuning control was applied to the dual composition control of the distillation column. Only the cases with feedforward compensation will be considered. Figure 22 shows the performance of the PMRSTC (positional multirate self-tuning control). Comparing the responses with those of figure 18 reveals that regulation of the top product composition was significantly better when the multirate strategy was employed. The bottom composition under PMRSTC regulation remains similar to that obtained with the PMVSTC.

The responses obtained with the KMRSTC (k-incremental multi-rate self-tuning controller) are shown in figure 23. Again, the multirate strategy provided for better regulation of the top composition loop whilst the bottom loop response remained similar to that obtained by the KMVSTC (see figure 19). The multirate k-incremental strategy also appeared to give better performances regulating the top composition loop when compared with the responses obtained under PMRSTC control.

13.4.3.7 Concluding Remarks

It is acknowledged that the comparison of control algorithms through applications to real processes is difficult because of changes in conditions beyond the control of the experimenter. The following remarks, based upon the trends observed during the on-line implementation of the various self-tuning policies, can nevertheless be made:

1.Compared to conventional PI controllers, the self-tuning controller can provide for better control of processes which exhibit non-linearities and large time-delays.

2.Both the PMVSTC and the KIMVSTC can regulate against a disturbance which leads to inverse response characteristics in the open loop, regardless of whether the disturbance is known or unknown.

3.The experiments have shown that if disturbances are known and used in the calculation of the control law, the performance of the resulting self-tuning controller is better than when the disturbances are unknown. Whilst the tuning of conventional feedforward control elements is known to be difficult, the advantage here lies in the automatic tuning of these elements. In this case, use of the KMVSTC does not appear to offer significant improvements over the PMVSTC. By the same token, the problem of designing multivariable self-tuning controllers is therefore very much simpler than conventional schemes, since interactions are considered as known feedforward load disturbances.

4.The k-incremental self-tuning approach has very good offset removal capabilities. No offsets in output responses were observed at any time when the algorithm was used. However, adopting a positional self-tuning control law will not necessarily result in offsets, even if unknown disturbances are affecting the system. This arises from the adaptation of the controller parameters to take up values enabling good output prediction and hence good control. With incremental control however, offset removal does not depend on parameter adaptation and is guaranteed.

5.The k-incremental algorithms appeared to be more sensitive to changes in unmodelled dynamics, and possibly to nonlinearities, than their corresponding 'positional' counterparts. This sensitivity can however be reduced by an appropriate choice of control weighting Tham and colleagues, (1985). Recent work also reveals that this approach may degrade the integrity of the self-tuned closed loop system Tham, (1985).

6.It has also been shown that by adopting a multivariable- multirate strategy, better control can be achieved. This is because the sampling intervals for the different loops within the multivariable system have been chosen to suit the dynamics of these individual loops. A faster sample rate also enables the earlier detection of disturbance effects and leads to enhanced disturbance rejection.

13.5 Implementation Aspects

In many of the applications of self-tuning control described above the self-tuner required its own hardware to be specified. The need for some kind of 'safe' and 'versatile' hardware/software configuration led to a final applications structure for the turbo-tray dryer, for instance, of a microcomputer, running the control software. being serially linked to an industrial standard analogue interface (which was typically a TCS 6350 process controller). The presence of a PID loop controller as functional part of the analogue interface provided a backup controller to the computer based advanced control algorithm and thereby introduced an element of safety into the control system structure. The TCS interface also provided a number of signal processing tasks, such as filtering.

A number of the other applications were carried out using the FAUST package, Tuffs et al (1985), which runs on PDP 11/23 hardware and is interfaced to the process using standard A/D and D/A converters. In the distillation column studies, the control computer was a PDP 11/03, interfaced to local analogue PI controllers, via A/D and D/A converters.

13.6 Conclusions

The application studies which have been described in this chapter reflect the authors' experiences and involvement in the implementation of self-tuning controllers carried out at the Universities of Newcastle-upon-Tyne and Oxford. Clearly, many other application studies have been carried out elsewhere and the interested reader is directed towards the summary papers referenced to in the introduction to this chapter and to recent conference publications for further details.

13.7 Acknowledgements

The authors would like to thank the Department of Engineering Science at Oxford University, the Department of Chemical and Process Engineering at the University of Newcastle-upon-Tyne, U.K. and the Department of Chemical Engineering at the University of Alberta, Canada, for the use of their facilities in carrying out this work. Special thanks goes to Professor R.K. Wood from the University of Alberta for making the distillation column available for the self-tuning experiments (under NSERC Grant A- 1944). The authors also wish to thank the many companies who have provided access to process plant so that the results presented in this chapter could be obtained.

13.8 Bibliography

Al-Assaf, Y., (1987), Self-Tuning Control: theory and applications. D.Phil. thesis, In preparation.

Al-Assaf Y., Peel, D., Clarke, D.W., (1987), Generalized Predictive Control of a Cement Mill. In preparation.

Astrom, K.J., (1983), Theory and applications of adaptive control - a survey. Automatica, vol.19, no.5, pp.471-486.

Borisson, U., (1979), Self-tuning regulator for a class of multivariable systems. Automatica, 15, 209-215.

Clarke, D.W., Gawthrop, P.J., (1975), Self-tuning controller. Proc. IEE, vol. 122, no.9, pp.929-934.

Clarke, D.W., and Gawthrop, P.J., (1979), Self-tuning control. Proc. IEE, Vol.126, 6.

Clarke, D.W., Hodgson, A.J.F., and Tuffs, P.S., (1983), The offset problem and k-incremental predictors in self-tuning control. Proc. IEE, vol. 130, pt.D, no. 5, pp. 217-225.

Clarke, D.W., Mohtadi, C., Tuffs, P.S., (1987a), Generalized Predictive Control - part 1: the basic algorithm. Automatica, vol. 23, no. 2., pp.137-148.

Clarke, D.W., Mohtadi, C., Tuffs, P.S., (1987b), Generalized Predictive Control - part 2: extensions and interpretations. Automatica, vol. 23, no. 2., pp.149-160.

Gawthrop,P.J., (1977), Some interpretations of the self-tuning controller. Proc. IEE, Vol.124, No.10.

Grimble, M.J., (1984), Implicit and explicit LQG self-tuning controllers. IFAC 9th World Congress, Budapest, Hungary.

Hodgson,A.J.F.,(1982), Problems of integrity in applications of adaptive controllers. O.U.E.L. Report No. 1436/82.

Koivo, H.N., (1980), A multivariable self-tuning controller. Automatica, Vol.16, 351-366.

Lambert, E.P., (1987), The Industrial Application of Long Range Prediction. D.Phil. thesis in preparation, Oxford University.

Lopez, A.M., Miller, J.A., Smith, C.L. and Murrill, P.W., (1967), Tuning controllers with error integral criteria. Instrumentation Technology, Nov.

Mohtadi, C., Shah, S.L., Clarke, D.W. (1986), Generalized Predictive Control of Multivariable Systems. OUEL report 1640/86.

Montague, G.A., Tham, M.T. and Morris, A.J., (1986), A comparison of multivariable long range predictive control with GMV control in a highly nonlinear environment. American Control Conference. Seattle.

Morris, A.J., Nazer, Y. and Wood, R.K., (1982), Multivariate self-tuning process control. Optimal Control Applications and Methods, vol.3, 363-387.

Parks, P.C., Schaufelberge, W., Schmid, Chr. and Unbehauen, H., (1980), Applications of adaptive control systems. In Methods and Applications in Adaptive Control. Ed. H. Unbehauen, Springer- Verlag.

Peel, D., (1986), Self-tuning process control with application to an industrial turbo-tray dryer. Ph.D. thesis, Dept. Chemical Engineering, University of Newcastle-upon-Tyne.

Peel, D., Morris, A.J., (1987), Control of a turbo-tray dryer. In preparation.

Prager, D.L., Wellstead, P.E., (1980), Multivariable pole assignment self-tuning regulators. Proc. IEE, vol.128, pt.D, no.1, pp.9-18.

Seborg, D.E., Shah, S.L. and Edgar, T.F., (1983), Adaptive control strategies for process control: a survey. AIChE Diamond Jubilee Meeting, Washington DC.

Shieh, L.S., Wang, C.T., Tsay, Y.T., (1982), Multivariable state-feedback self-tuning controllers. IFAC symposium on Identification and system parameter estimation, Washington DC.

Tham, M.T., (1985), Some aspects of multivariable self-tuning control, PhD Thesis, University of Newcastle upon Tyne, U.K.

Tham, M.T.. Morris, A.J., Vagi, F. and Wood, R.K., (1985), An on- line comparison of two multivariable self-tuning controllers. Proc. 7th IFAC Conf. on Digital Applications to Process Control, Vienna, September.

Tuffs, P.S., Clarke, D.W., (1985), FAUST: a software package for self-tuning control. IEE Conference, Control '85, Cambridge, UK.

Unbehauen, H., (1985), Theory and application of adaptive control, Proc. 7th IFAC Conf. on Digital Applications to Process Control, Vienna, September.

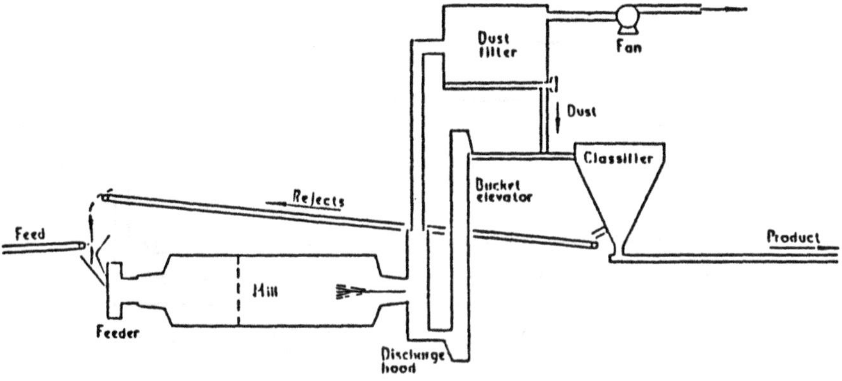

Figure 1. Diagram of Cement Mill.

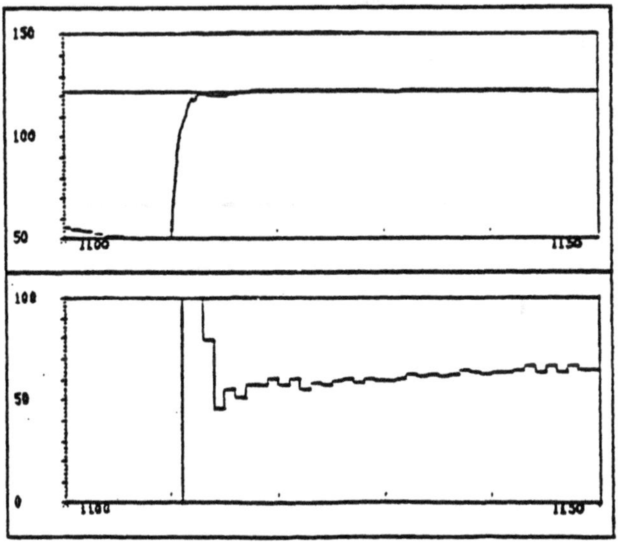

Figure 2a. Mill start-up using GPC.

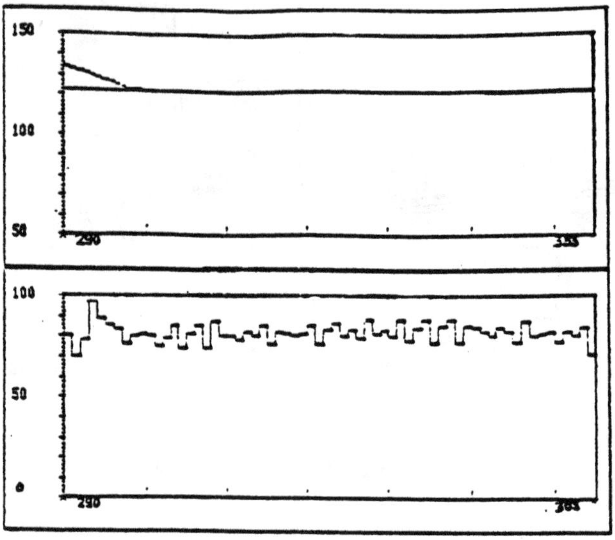

Figure 3a. Regulation of Cement Temperature using GPC.

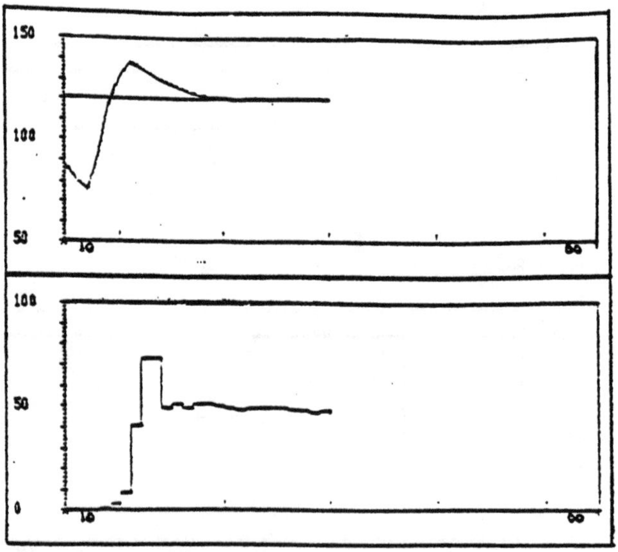

Figure 2b. Mill start-up using PI.

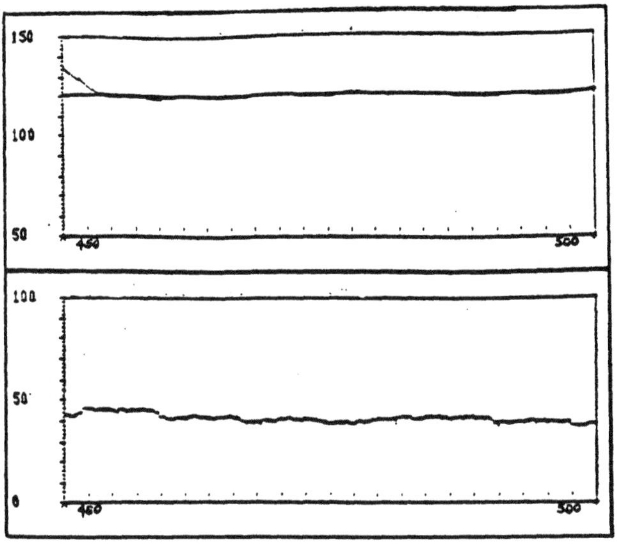

Figure 3b. Regulation of Cement Temperature using PI.

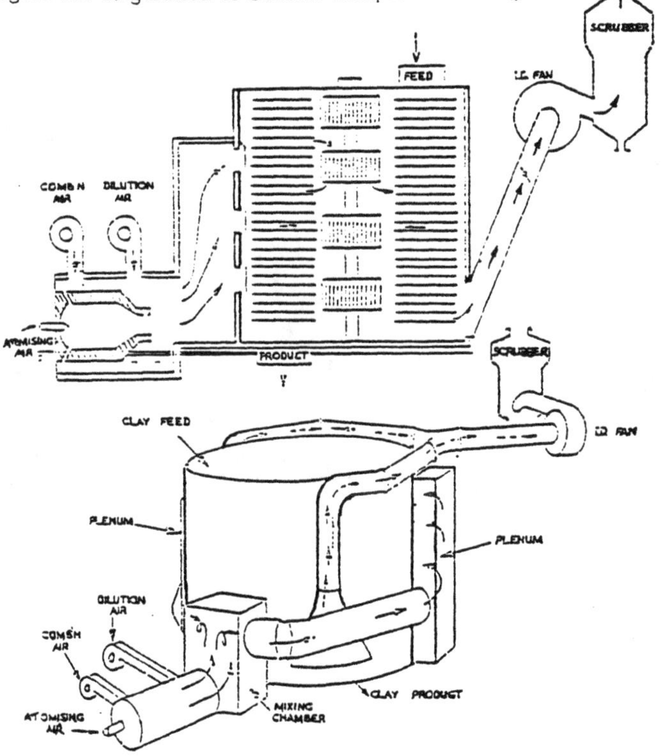

Figure 4. Diagram of Turbo-Tray Dryer.

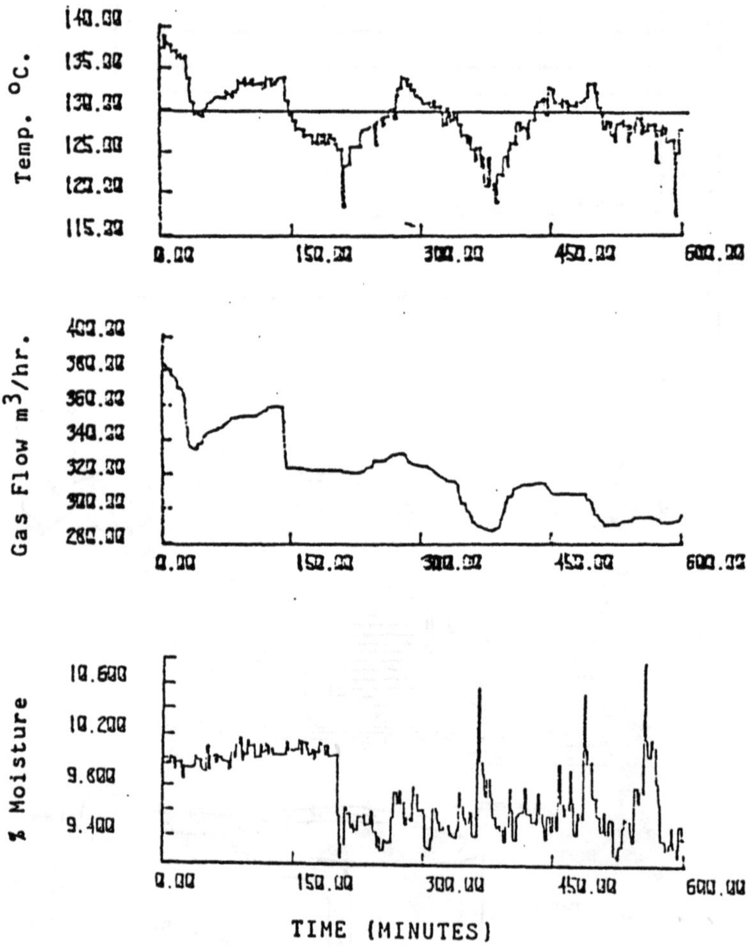

Figure 5. Dryer Control using PI.

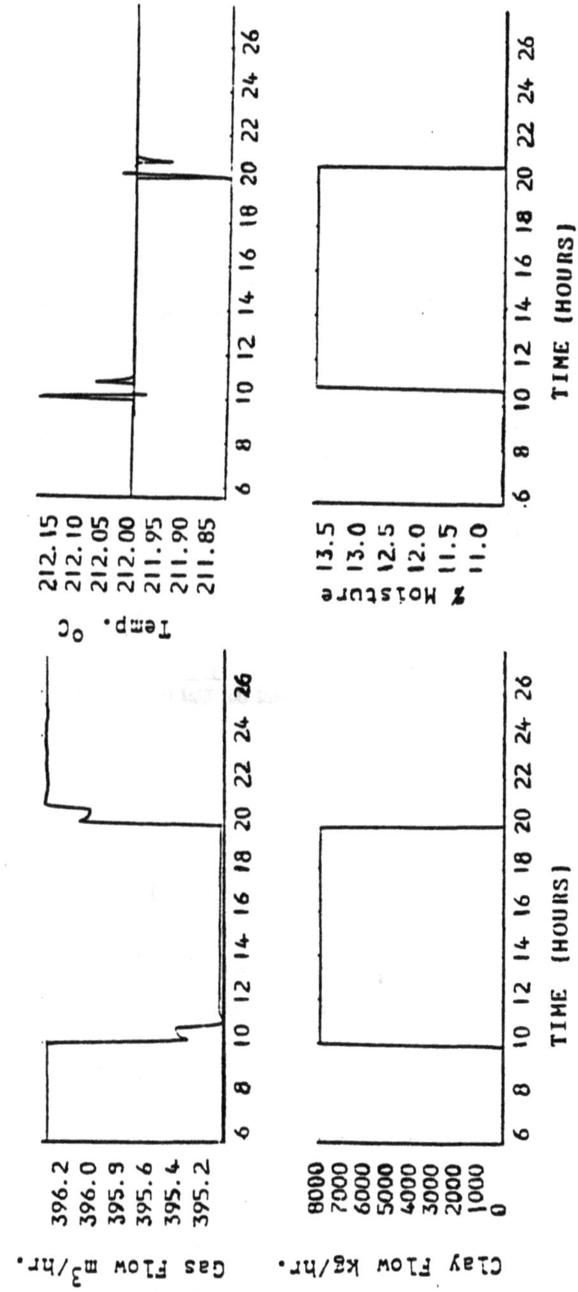

Figure 6. k-incremental GMV Control of Dryer Model (middle gas temp.).

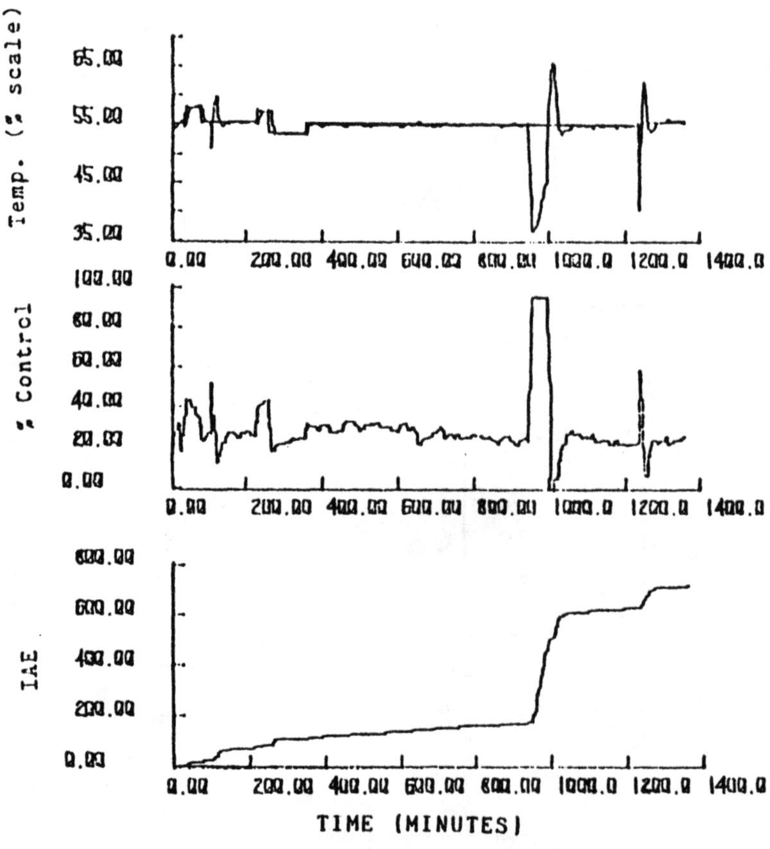

Figure 7. k-incremental GMV Control of Dryer (middle gas temp.).

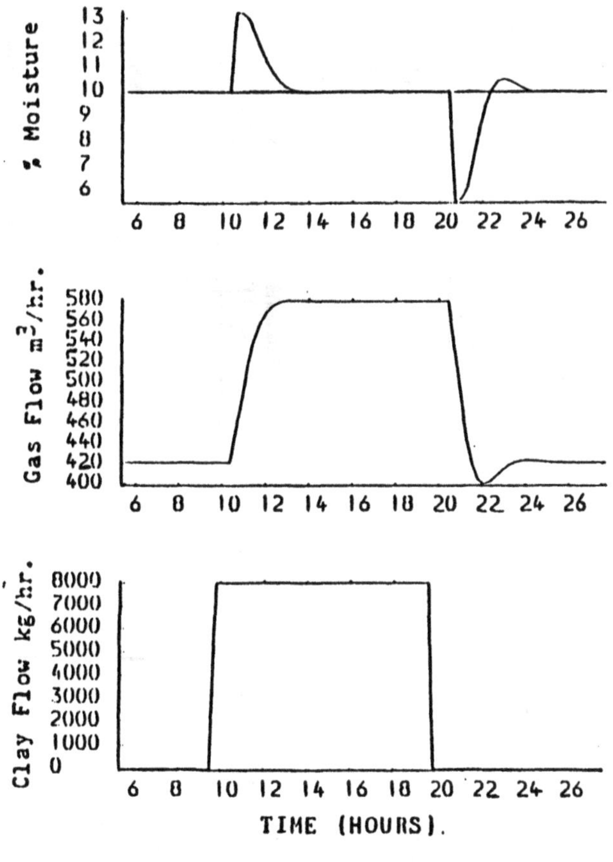

Figure 8. k-incremental GMV Control of Dryer Model (outlet moisture)

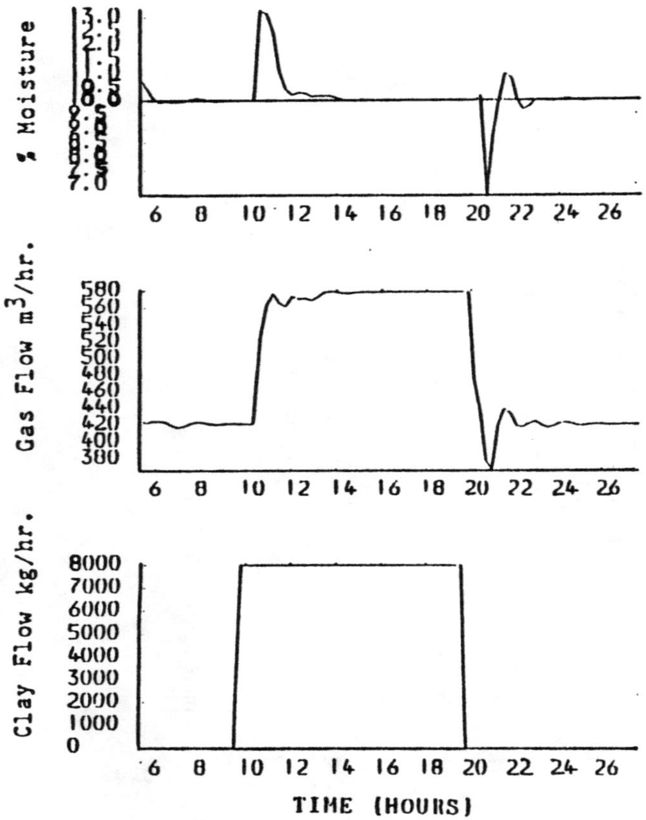

Figure 9. k-incremental GMV (including feedforward) Control of Dryer Model (outlet moisture).

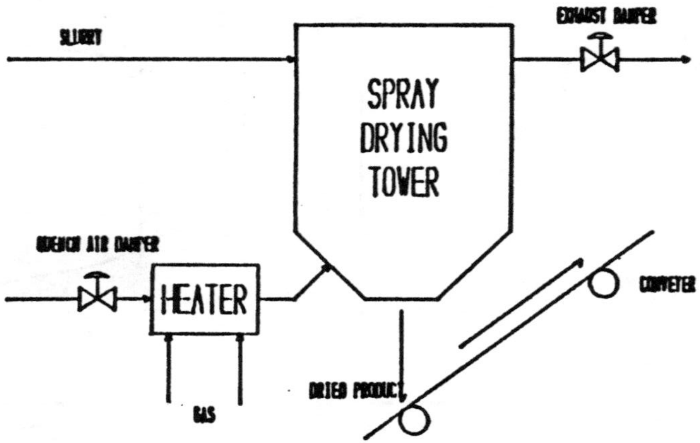

Figure 10. Diagram of Spray Dryer.

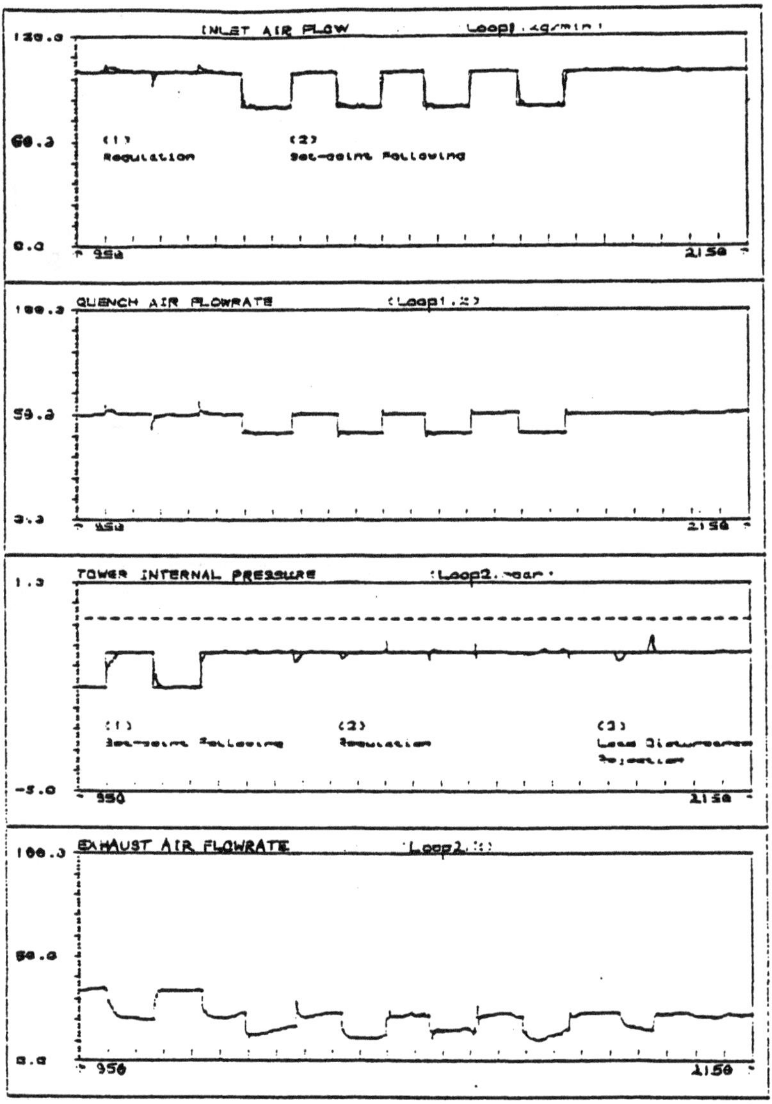

Figure 11. Spray Dryer Control using Multivariable GPC.

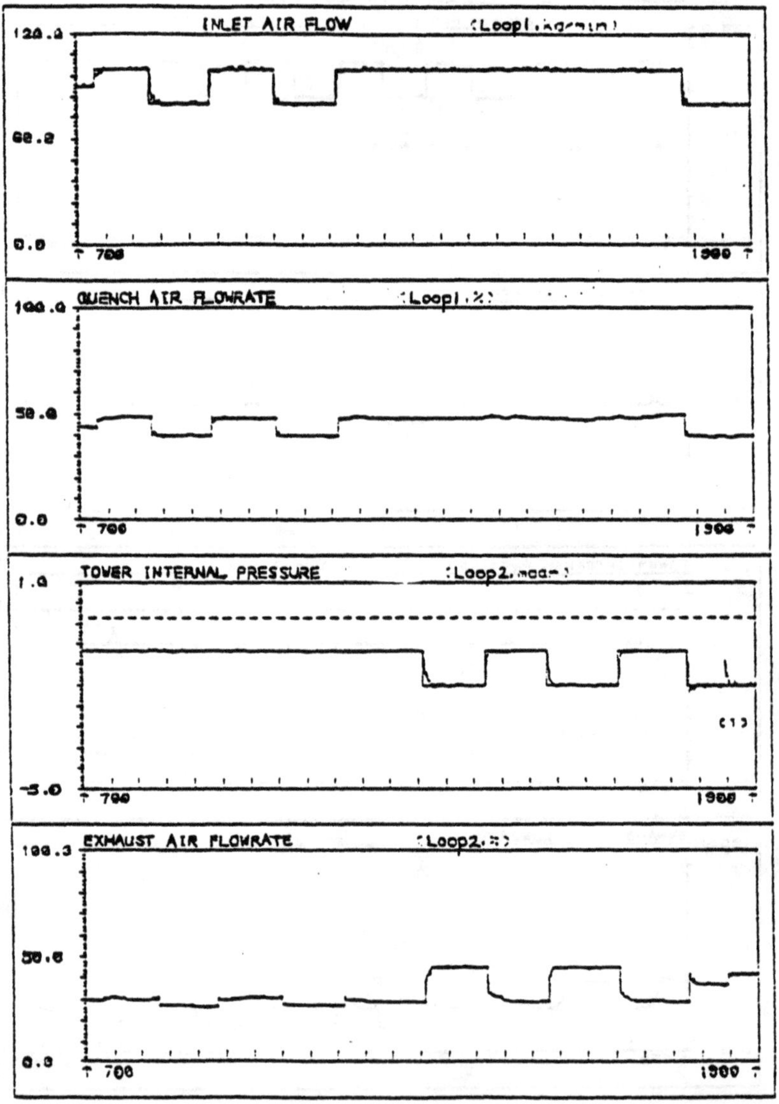

Figure 12. Spray Dryer Control using a combination of a SISO GPC and a SISO GPC including feedforward.

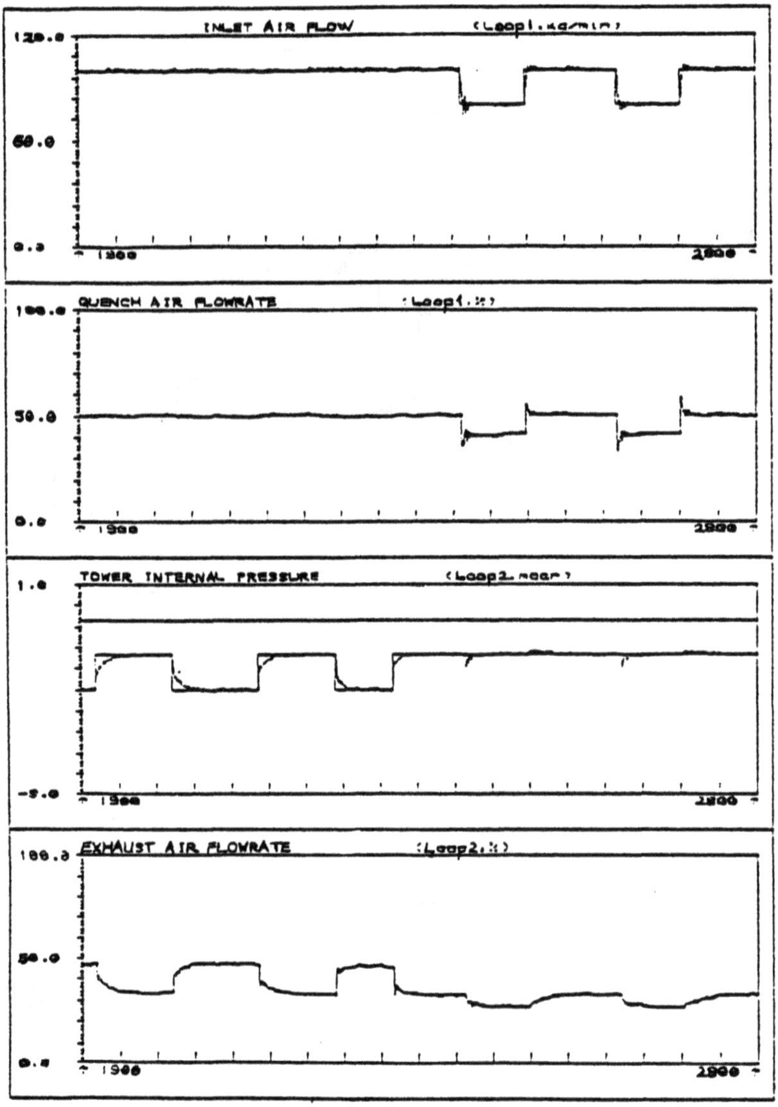

Figure 13. Spray Dryer Control using individual SISO GPCs.

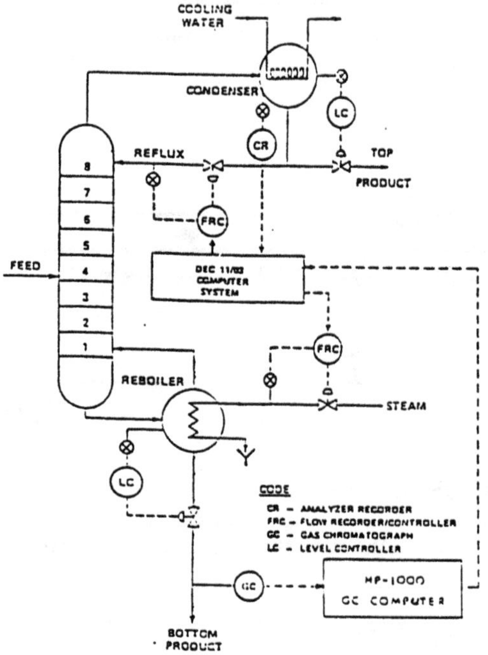

Fig.14. Schematic of Binary Distillation Column

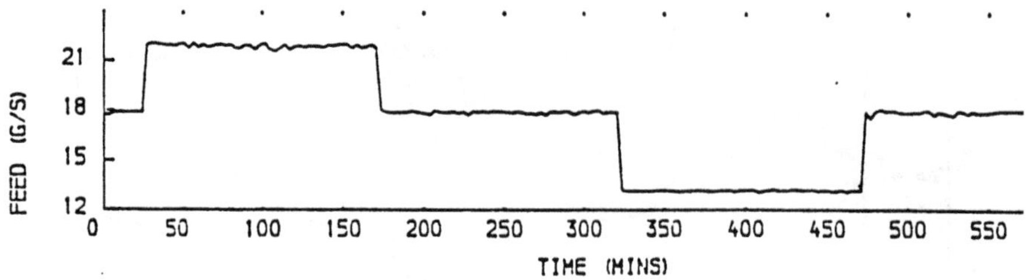

Fig.15. Feed Flowrate Disturbance Sequence

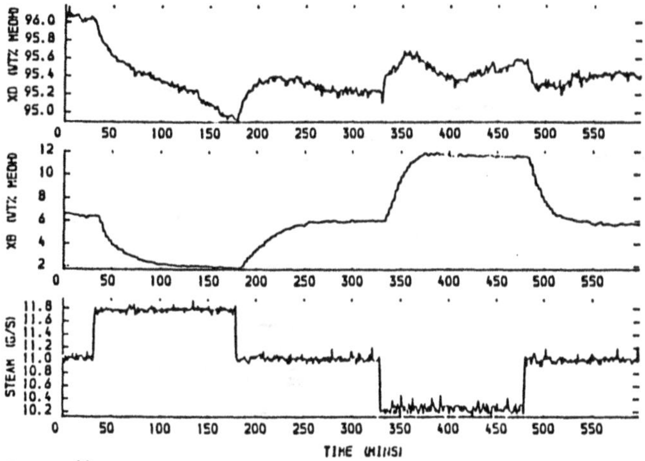

Figure 16a Open Loop Responses to Step Changes in Steam Flow

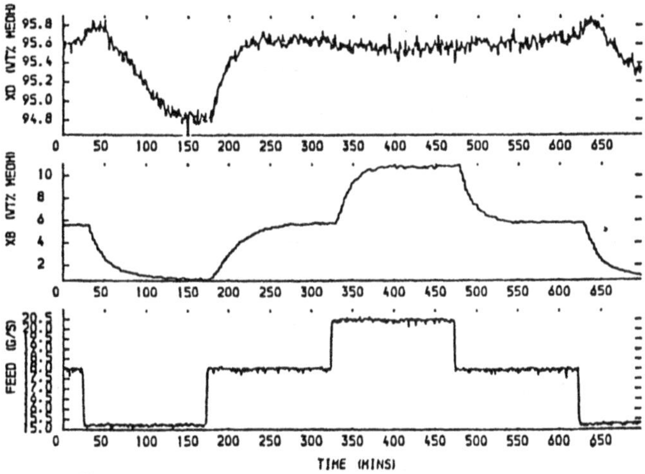

Figure 16b Open Loop Responses to Step Changes in Feed Flow

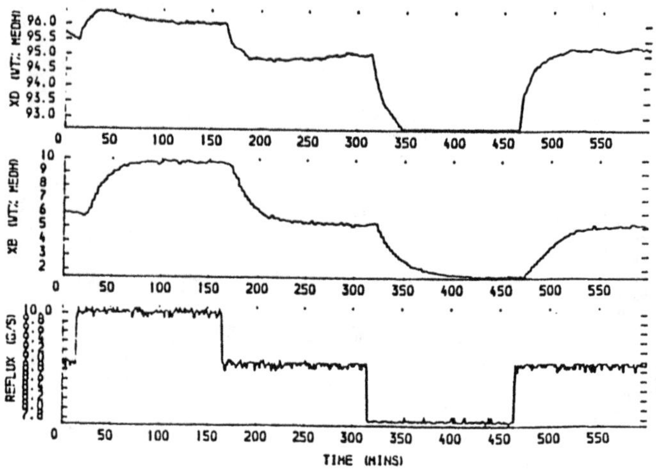

Figure 16c Open Loop Responses to Step Changes in Reflux Flow

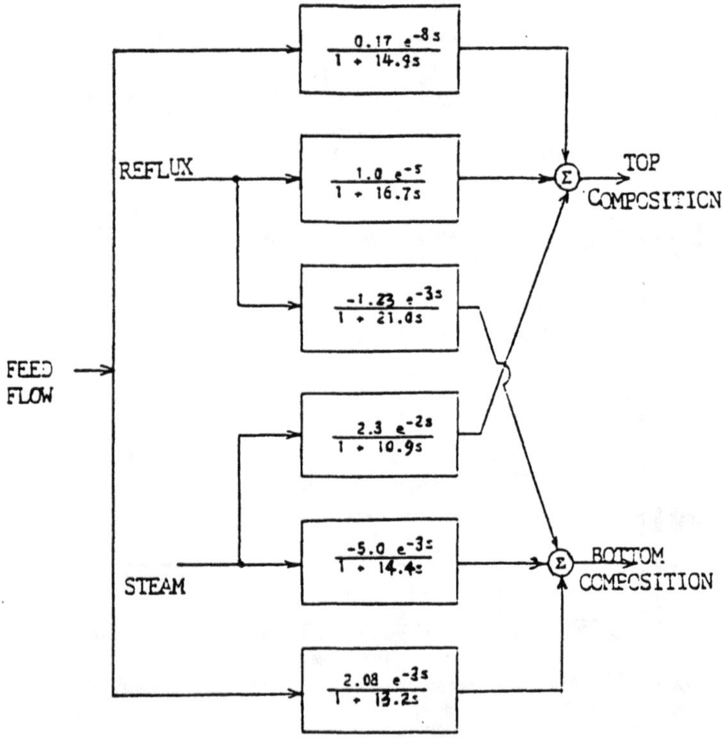

Fig.17 .Linear Transfer-function Model of the Binary Distillation Column

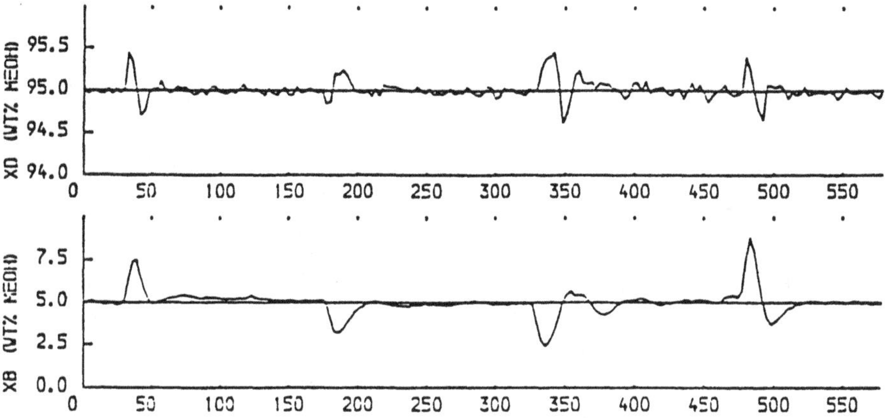

Fig.18. Positional Multivariable Selftuning Control with Feedforward Compensation

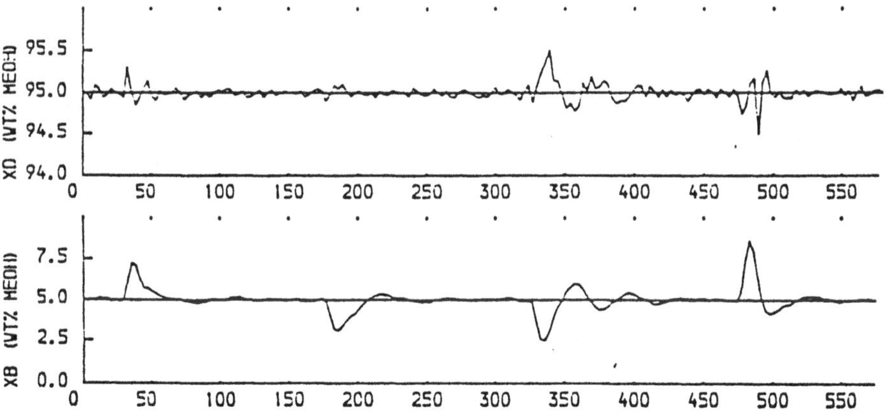

Fig.19. k-incremental Multivariable Selftuning Control with Feedforward Compensation

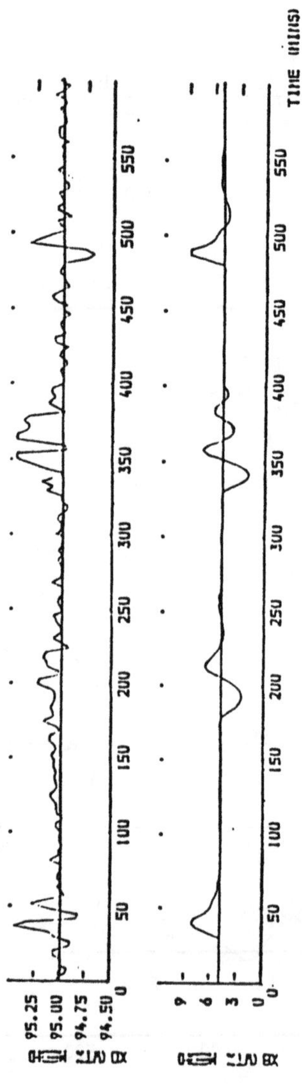

Fig.20. Positional Multivariable Selftuning Control without Feedforward Compensation

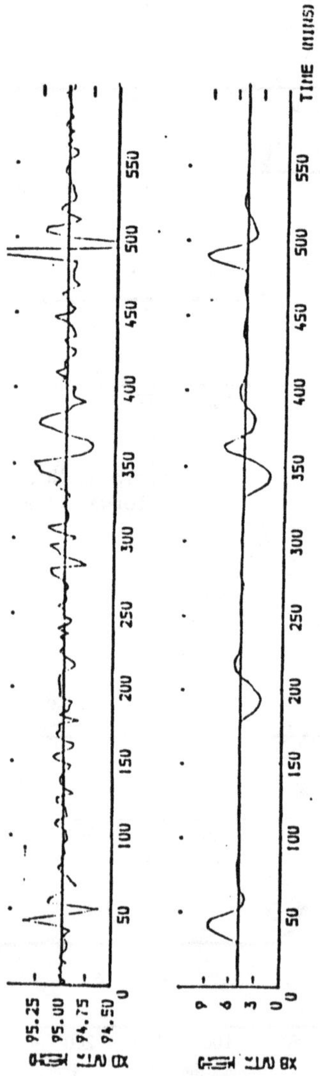

Fig.21. k-incremental Multivariable Selftuning Control without Feedforward Compensation

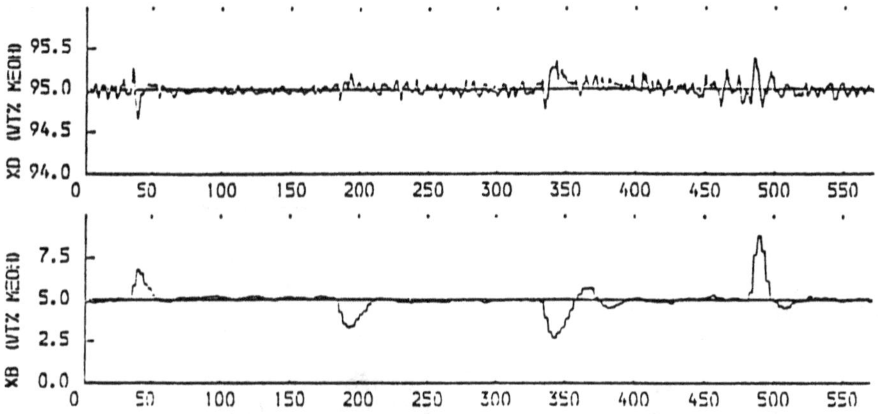

Fig.22. Positional Multivariable-multirate Selftuning Control with Feedforward Compensation

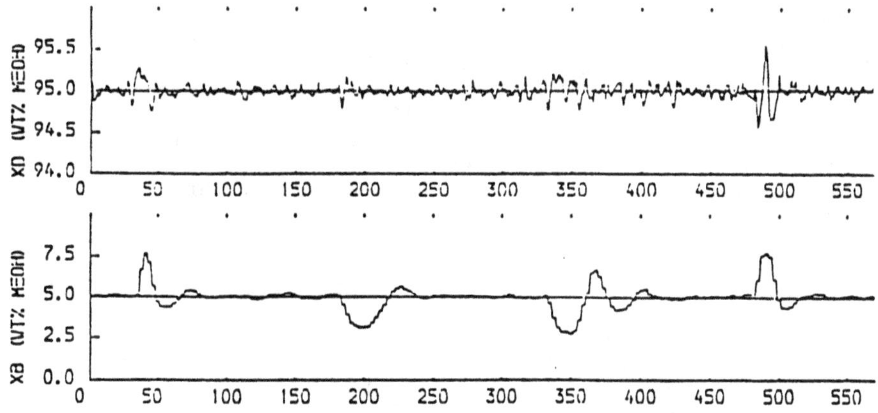

Fig.23. k-incremental Multivariable-multirate Selftuning Control with Feedforward Compensation

Index

www.ingramcontent.com/pod-product-compliance
Lightning Source LLC
Chambersburg PA
CBHW050522190326
41458CB00005B/1635